Remotely Sensed Cities

Remotely Sensed Cities examines how the world's urban areas can be located, measured and analysed using information from the latest airborne and satellite remote sensors, including breakthroughs in the use of LIDAR and IKONOS data for precision mapping, and DMSP OLS night-time imagery for establishing global distributions of population and economic activity. Emphasis is also given to techniques on urban feature extraction using wavelet transforms, and graph-based structural pattern recognition; along with population mapping using zone-based dasymetric models, pixel-based entropy maximization models, and Bayesian classifications. In addition, novel applications include links with geodemographics, crime prevention, and the importance of monitoring urban heat islands of cities in the developing world.

The urban emphasis of this book helps to redress the balance with the many on environmental remote sensing. Only recently has remote sensing assumed equal importance in the monitoring of the world's centres of population. Dynamic changes in global urban growth and economic activity, especially in the developing world, are now routinely measured by night-time images. More localized high resolution data are rapidly attaining prominence in applications requiring precise delineation of urban features, and when coupled with socio-economic information, encourage understanding of urban configurations, changes and behaviour. The book is designed for upper-level undergraduate and graduate students, along with research scientists in geo-mapping disciplines, environmental issues and social sciences.

Victor Mesev is a Lecturer in Geography at the University of Ulster, UK.

Remotely Sensed Cities

Edited by Victor Mesev

CRC Press
Taylor & Francis Group
Boca Raton London New York

CRC Press is an imprint of the
Taylor & Francis Group, an **informa** business

CRC Press
Taylor & Francis Group
6000 Broken Sound Parkway NW, Suite 300
Boca Raton, FL 33487-2742

First issued in paperback 2019

ISBN-13: 978-0-415-26045-9 (hbk)
ISBN-13: 978-0-367-39539-1 (pbk)

British Library Cataloguing in Publication Data
A catalogue record for this book is available
from the British Library

Library of Congress Cataloging in Publication Data
A catalog record has been requested

Visit the Taylor & Francis Web site at
http://www.taylorandfrancis.com

and the CRC Press Web site at
http://www.crcpress.com

Contents

Tables

Figures

Plates (between pages 198–9)

Contributors

Paul Aplin is Lecturer in Geographical Information Science at the School of Geography, University of Nottingham, Nottingham NG7 2RD, UK

Emmanuel P. Baltsavias is Associate Professor at the Institute of Geodesy and Photogrammetry, Swiss Federal Institute of Technology (ETH), 8093 Zurich, Switzerland

Michael J. Barnsley is Research Professor of Remote Sensing and GIS at the Department of Geography, University of Wales Swansea, Swansea SA2 8PP, UK

Stuart L. Barr is Lecturer at the School of Geography, University of Leeds, Leeds LS2 9JT, UK

Kimberly E. Baugh is Research Associate at the Analytical Imaging and Geophysics, Boulder, CO 80303, USA

Budhendra L. Bhaduri is Research Scientist at the Geographic Information Science and Technology, Oak Ridge National Laboratory, Oak Ridge, TN 37831, USA

Edward A. Bright is Research Scientist at the Geographic Information Science and Technology, Oak Ridge National Laboratory, Oak Ridge, TN 37831, USA

Phillip R. Coleman is Senior Research Scientist at the Geographic Information Science and Technology, Oak Ridge National Laboratory, Oak Ridge, TN 37831, USA

John B. Dietz is Research Associate at the Cooperative Institute for Research on the Atmosphere, Colorado State University, Fort Collins, CO 80523, USA

Jerome E. Dobson is Research Professor at the Department of Geography, University of Kansas, Lawrence, KS 66047, USA

Christopher N. H. Doll is Research Student at the Department of Geomatic Engineering, University College London, London WC1E 6BT, UK

Christopher D. Elvidge is Physical Scientist at the NOAA National Geophysical Data Center, Boulder, CO 80305, USA

Armin Gruen is Professor at the Institute of Geodesy and Photogrammetry, Swiss Federal Institute of Technology (ETH), 8093 Zurich, Switzerland

Richard Harris is Lecturer at the School of Geography, Birkbeck College, University of London, London W1P 2LL, UK

Jack T. Harvey is Senior Lecturer at the School of Information Technology and Mathematical Sciences, University of Ballarat, Ballarat 3353, Australia

V. Ruth Hobson is Research Associate at the Cooperative Institute for Research in Environmental Sciences, University of Colorado, Boulder, CO 80303, USA

Mitchel Langford is Lecturer at the Department of Geography, University of Leicester, Leicester LE1 7RH, UK

Chor Pang Lo is Professor at the Department of Geography, University of Georgia, Athens, GA 30602, USA

Victor Mesev is Lecturer in Geography at the School of Biological and Environmental Sciences, University of Ulster, Coleraine BT52 1SA, UK

Soe W. Myint is Assistant Professor at the Department of Geography, University of Oklahoma, Norman, OK 73019, USA

Ingrid L. Nelson is Research Associate at the Cooperative Institute for Research in Environmental Sciences, University of Colorado, Boulder, CO 80303, USA

Janet Nichol is Associate Professor at the Department of Land Surveying and Geo-Informatics, The Hong Kong Polytechnic University, Kowloon, Hong Kong

Jeffrey M. Safran is Research Associate at the Cooperative Institute for Research in Environmental Sciences, University of Colorado, Boulder, CO 80303, USA

Alan M. Steel is Senior Research Officer at the Department of Geography, University of Wales Swansea, Swansea SA2 8PP, UK

Paul C. Sutton is Assistant Professor at the Department of Geography, University of Denver, Denver, CO 80208, USA

Benjamin T. Tuttle is Research Associate at the Cooperative Institute for Research in Environmental Sciences, University of Colorado, Boulder, CO 80303, USA

John R. Weeks is Professor of Geography and Director of the International Population Center at the Department of Geography, San Diego State University, CA 92182, USA

Preface

At first *Remotely Sensed Cities* may seem a contradiction in terms. Surely remote sensing technology is more relevant to parts of our world that are just that, *remote*, the natural environment and the atmosphere. Cities are anything but remote. The great majority of the world's population live, work and relax within large settlements. We have immediate access and can observe our cities from the ground at any time we want. Why therefore would we want to observe them from sensors suspended many kilometres in the sky or space? Observation from such a height would surely mean that individual people and buildings are no longer distinguishable (assuming of course that clouds and atmospheric haze do not completely obliterate the view). And if individual people and buildings cannot be observed then what use is remote sensing to our understanding of the demographic profiles, socio-economic characteristics, mobility patterns, domestic structures and commuting behaviour of our cities? All of these facets of urban areas are indubitably more observable and measurable from detailed population, housing and business surveys conducted from on the ground. If that is the case then all we are left with from remote sensing is the observation of the physical structure of blocks and streets of buildings. However, these may or may not be accurately delineated, and in any case would be less precise than measurements made by land surveyors.

So, why have you just bought/borrowed a book on *Remotely Sensed Cities*, and not one of the many on environmental remote sensing or atmospheric remote sensing. In those books sophisticated imaging sensors and wide-area snapshots at repeated time intervals have elevated remote sensing to the vanguard of global awareness by monitoring major changes in deforestation, ozone depletion and river flooding to name but a few. Indeed, environmental/ atmospheric monitoring has been the cornerstone in the acceptance and credibility of remote sensing technology in scientific circles, and the establishment of many prestigious research centres around the world. That is not to say that these same environmental and atmospheric books have completely ignored the remote sensing of cities but neither have they focused on the potential of remote sensing to reveal specific urban characteristics beyond wide-ranging human/environmental and human/atmospheric interactions.

Again, why a book solely dedicated to the remote sensing of cities? To begin with, size. A quick flick will reveal fifteen main chapters, authored or co-authored by twenty-seven visionaries (some may say pioneers) of urban remote sensing. With so many contributors there must be something important to say about urban remote sensing. Many are at the beginning of their careers, adding refreshing new perspectives to a fast-developing area of research. Most work in the USA and UK but there are also representatives from Europe, Asia and Australia. All have their own ideas but as editor I had to compartmentalize chapters to ensure structure and consistency. Fortunately, all fitted within the following three parts:

- Part 1: High spatial resolution data
- Part 2: Cities by day
- Part 3: Cities by night

Coarse spatial resolutions have traditionally been seen as one of the major obstacles to precise urban mapping from remote sensor data. The first part of the book reviews the potential as well as the limitations of new sources of high spatial resolution data (including LIDAR [Laser-Induced Detection And Ranging] and IKONOS) for measuring the exact physical structure and composition of cities. Why is this important? Well, unlike the natural environment, cities are highly spatially dynamic. The consequences of human actions are evident everywhere but they are no more apparent and accountable than within centres of population. As such, changes in the physical size, shape and density brought about by a juxtaposition in land cover configuration can be meaningfully observed in most cities at an annual, if not monthly, cycle. Remote sensing can help us to monitor these physical changes at the required temporal rate.

At this point, it would be sensible to make sure we know the difference between land cover and land use. Land cover refers to the physical composition of tracts of land (i.e., concrete, asphalt, grass and trees); while land use is the anthropogenic constructs of mixtures of land cover (i.e., residential buildings, commercial buildings, or even agriculture). Using remote sensor data we can measure, with some degree of accuracy, the layout of urban structures (land cover) but it is very difficult to determine how these structures are being used by people (land use) unless we employ additional information specific to socio-economic and housing attributes from other sources. As we are operating in a digital framework, data from other sources are invariably handled by Geographical Information Systems (GISs), and links with remote sensing has recently coined the term RS-GIS integration. Most of the chapters in the second part of this book on "cities by day" refer to this integration partnership where exogenous information from predominantly population censuses is combined with Earth observation image data. The purpose is to establish (usually linear) relationships between socio-economic attributes from censuses with urban physical properties from image data. Because human/environmental interactions are symbiotic any changes in one will affect the other.

The types of remote sensor data used in this second part are captured at the visible and infrared ranges of the electromagnetic spectrum during the day. That is because land use patterns of cities are observable during the day but so are many other uninhabited areas of land. What if we used the same visible and infrared wavelengths but sensed urban areas at night instead? At night, only artificial lights and fires are discernible from space. We could then, as in "cities by day", instigate relationships between light intensity and either population or economic activity, where the discernible applications would be to estimate population censuses or measures of affluence (Gross Domestic Product – GDP) faster and cheaper than conventional methods. Discussions behind these processes, as well as their limitations, form the basis of the last part on "cities by night".

By reaching this point of the preface you obviously haven't been swayed by my earlier negative comments on the appropriateness of the title *Remotely Sensed Cities*. I hope by now that I have touched upon the merits and immense potential of urban remote sensing to convince you to continue reading. If you need more persuasion I can only reiterate the three most important reasons for the book's inception:

- the precise delineation of urban structures from new high spatial resolution image data;
- the critical links with GIS in estimating populations, socio-economic characteristics, geodemographic profiles and urban heat islands;
- the scope of night-time data for emulating censuses, measures of affluence and urban pollution.

Taken together these are reason enough to stir interest in urban planners, governments, social scientists, environmental managers, GIS consultants, computer scientists and, yes, even traditional environmental remote sensing scientists of the theoretical justifications and practical applicabilities of *Remotely Sensed Cities*. The book is also designed with students in mind by ensuring that remotely sensed data are systematically evaluated, methodologies are carefully explained and demonstrated and unnecessary technical jargon is kept to a minimum.

Finally, time devalues research. So I would like to take this opportunity to apologize to all the contributors for imposing a ludicrously short time frame on the completion of their chapters, and at the same time thank them for sticking to the schedule without (many) complaints. My thanks also go to the numerous referees who were given even less time to review chapters, to Ivan Gault for compiling the index and Sarah Kramer at Taylor & Francis for her support and understanding during what for me has been a steep learning curve.

Victor Mesev
June 2002

1 Remotely sensed cities

An introduction

Victor Mesev

1.1 Introduction

Sensing cities remotely is difficult – very difficult! It is tricky enough to interpret the intricate interplay of artificial structures, economic activity, government policies, land tenure, social class, culture and the biosphere, typical of all cities, from on the ground let alone from the sky or space (Brugioni 1983). Cities are complex enough, providing many challenges for planners, civil engineers, environmentalists, government agencies, sociologists, demographers, economists, psychologists and political scientists without adding remote sensing scientists to the list. Complex and at times conflicting decisions by private individuals, commerce and government policies shape not only human behaviour but also the arrangement of physical structures and intervening natural surfaces. Whether the decisions are on where to locate new out-of-town retail outlets, the building of residential estates, the mapping of school catchment zones, the renovation of city parks, the expansion of transport infrastructure, how to reduce air pollution or simply what households grow in their gardens, cities will always be characterized by heterogeneous, convoluted and unpredictable land patterns.

The obvious difference between cities and the "natural" landscape is people, and the scale of human activity. It is people who make cities complex and difficult to measure (not to mention understand) but it is also people, from a pragmatic sense, that make cities important enough to draw the attentions of remote sensing scientists (Davreau *et al.* 1989; Liverman *et al.* 1998; Donnay 1999). However, even though many remote sensing scientists are interested in the urban landscape, it does seem strange (at first) that remotely sensed data and technology[1]*can* add

---

1 Whether remote sensing is a technology/tool or field is a moot point. Some people would argue that because remote sensing has an appeal to interdisciplinary applications it should be referred to as a technology; others claim it has its own identity, its own set of theories, and even its own conferences and journals, and therefore deserves to be regarded as a separate field or discipline. The author takes the view that both are correct but for convenience and from the pragmatic standpoint of most chapters in this book the role of remote sensing is very much one of a technology.

anything new to our knowledge of cities. Frankly, from such an elevated viewpoint, remote sensors (on either airplanes or satellites, and depending on the type of sensor) can only "see" a plan layout of features, mostly rooftops and treetops, which may or may not be concealing lower features such as road surfaces, grass and water bodies (Forster 1985, Ji and Jensen 1999). These are all important components of a city that need to be measured and monitored over time, but are nonetheless only the physical (tangible) consequences of human activity, not the immediate explanations of behaviour. After saying that, even though people are responsible for the arrangement of urban physical patterns, it is only through the analysis of the spatial configuration of physical structures (by remotely sensing) that we can begin to understand the human behaviour that brought about such spatial patterns (Geoghegan *et al.* 1998). In this way, the argument becomes cyclical – knowledge of the physical structure and form of cities contributes to an understanding of socio-economic functioning, and knowledge of socio-economic characteristics dictates urban layout (Martin and Bracken 1993, Masser 2001). In addition, information from outside the realms of remote sensing (if you like "direct" sensing, of which the most prominent sources are population censuses, postal addresses and house and business surveys) is increasingly being used to assist the remote sensing process. Such exogenous information is digital and invariably handled by Geographical Information Systems (GISs). Given that reliable urban remote sensing can be erratic, strong links forged with GIS are vital (Barker 1988; Ehlers *et al.* 1989), and encompass all scenarios where data from both remote sensing and GIS are used in any magnitude, direction or order to improve the measurement and understanding of cities (Mesev 1997).

Our discussion now is getting ahead of us. We need to pause and first review why cities are important enough to sense remotely before reviewing the principles of remote sensing, how remotely sensed data and technology can be applied to the study of cities, what is actually measured, what the limitations are and what insights into urban patterns and processes can be gained from the spatial modelling of remote sensor data and interactions with exogenous data held by a GIS.

1.2   Why study cities?

This question is already answered. Cities contain the majority of the world's people and people make decisions that sustain cities, so cities need to be studied. The world's population was an estimated 750 million in the mid-1700s, is presently (in 2003) well over six billion, and could hit twelve billion before 2100 if current pace in growth continues. Ever-increasing growth in the size and density of cities, especially the "mega-cities" of much of the developing world has major repercussions not only on the quality of human life but also on the environment and atmosphere (Epstein 1998).

Rapid urbanization and exploitation of natural resources raise concerns over such matters as air pollution, deforestation, habitat destruction and climatic imbalances (Yeh and Li 2001). Within cities, growth and competition of resources affect issues relating to housing demand, industrial development, waste disposal and transportation requirements (Ogrosky 1975; Michalak 1993; Kent *et al.* 1993; Polle 1996; Ridd and Liu 1998; Couloigner and Ranchin 2000), among many others. The monitoring of such concerns is critical for measuring sustainable levels of urban expansion that are tolerable to both society and the environment. In response to these global needs, remote sensing (despite its inherent limitations) is an important technology for rapid, up-to-date and frequent snapshots of the world's distribution of cities. Furthermore, since the tragedies of 11 September 2001 the ability of remote sensing to locate populations at risk is now seen as essential in the rapid deployment of personnel, medical equipment and emergency supplies from blankets to bandages (Chapter 12, this volume; Dobson *et al.* 2000; Bjorgo 2000).

Before examining precisely how remote sensing can contribute to the study of cities it is important first to establish a couple of definitions. Cities themselves are generally defined as concentrated centres of buildings with minimum populations, or minimum geographical areas, or minimum levels of economic activity, or minimum levels of administrative functions, or some combinations of all. Interpretations of the thresholds for these minimum criteria vary internationally although schemes, such as the Nomenclature of Units for Territorial Statistics (NUTS) (Doll, Chapter 15 in this book), have been introduced to facilitate comparisons. In terms of defining where cities start and end, the criteria are even less precise. Cartographically, cities can be represented across a range of scales from points to areas, dependent on the geographical size of the settlement. High-density urban cores along with established residential suburbs are less ambiguous, but delineation at the urban fringe is open to interpretation, especially during periods of sprawl when adjacent settlements are assimilated (Tobler 1969; Jensen and Cowen 1999). In Europe, the Morphological Urban Area (MUA) is an attempt to establish consistent measurements of the compactness of the urban space and "holes" in the urban fabric as benchmarks for urban growth comparisons (see Weber 2001). This is appropriate in Europe where, on the whole, growth is focused more on density, but in North America cities are less constrained (e.g., a 1 per cent increase in population represents a 6–12 per cent increase in developed area: Elvidge *et al.*, Chapter 13 in this book; Garreau 1991), and in the developing world cities are experiencing rapid growth in both space and density (Chen *et al.* 2000; Fazal 2001; Ji *et al.* 2001; Lo 2002).

The study of cities is therefore dependent on the scale of the remote sensing application. Regional scales of approximately 1 : 250,000 and 1 : 100,000 from Landsat MSS (Multispectral Scanning Systems) and TM (Thematic Mapper), respectively, may be adequate for general urban delineation but scales of 1 : 20,000 and finer may be necessary for more detailed intra-urban land use patterns (see Section 1.3.2 for discussion on spatial resolutions) (Cowen and Jensen 1998). These variations are echoed by classification schemes, such as the one instigated by Anderson *et al.* (1976), which established specific hierarchical criteria for detecting levels of land use/land cover at various scales. In Europe, CLUSTER (Classification for Land Use Statistics, Eurostat Remote Sensing Project) developed urban land use criteria from the CORINE (Coordination et Recherche d'INformations sur l'Environnement) land cover scheme (Weber 2001). Although designed for consistency and application transferability these schemes also suffer from inflexibility, ambiguity and implementation difficulties.

1.3   Inferring urban land cover

The focus of this book is on cities but remotely sensed images do not automatically discriminate between different types of surface or phenomena. As far as the electronic remote sensor is concerned urban surfaces are no different from the rest of the landscape or even the atmosphere (see Gurney *et al.* 1993 for many environmental and atmospheric examples). It does mean of course that principles that are applicable to the remote sensing of the environment or the atmosphere are just as relevant to the remote sensing of cities.

The starting point in all remote sensing applications is to define the object/ surface/phenomena of interest. In the urban case, this is twice as difficult. As noted above, cities are made up of physical materials, such as brick, slate, asphalt, vegetation, etc. (the land cover definition), and how these materials are combined to serve the city's population (the land use definition). Because remote sensing is a process of physical detection the inference of land cover is a lot more straightforward than land use (Dobson 1993; Webster 1996; Bibby and Shepherd 1999).

1.3.1   *Some basic principles of remote sensing*

Sensing cities remotely is not new. Aerial photography has been prominent in military reconnaissance as well as in civil use for most of the twentieth century (Carls 1947; Avery 1965; Ranjan and Rastogi 1990), and is still relevant today, primarily for routine digital photogrammetric large-scale map production (Baltsavias *et al.* 2001; Baltsavias and Gruen, Chapter 3 in this book). Traditional photo-interpretational skills (shape, size, pattern, hue, texture, shadows, site, association and resolution) are able to identify and measure the outline and height (stereoscopic views) of urban features, as well as any changes over time. This is because aerial photos possess the quality to identify features clearly but at a highly localized scale.

Alternatively, digital multispectral sensors on board satellites at higher altitudes are able to record the reflectance properties of an entire city at various spectral levels. Remember, the amount of reflected solar irradiance (and emitted energy) is detected and digitally represented at the pixel level – the smallest component of a satellite sensor image, which, in turn, is based on the instantaneous field-of-view capabilities of the sensor. Different sensors can detect reflected energy at different wavelengths of the electromagnetic spectrum, allowing ease of differentiation between similar objects, surfaces or phenomena; the ability to tailor to specific applications. However, for urban applications, the immense variety in the type and composition of objects and surfaces inevitably requires a much broader spectral range (from visible to infrared to microwave) than most other applications (see Henderson and Xia 1997; and Grey and Luckman 2001, for radar interpretation). Instead of wavelengths, the most important technical concern, over the years in urban applications, has been the pursuit of finer spatial resolutions of image pixels (Welch 1982; Curran and Williamson 1986; Atkinson and Curran 1997). The argument goes like this – because cities are made up of a complex mixture of artificial and natural surfaces of variable sizes, shapes and arrangements, the only way to map cities with any degree of precision and accuracy is by increasing the spatial resolution of images. This single-minded pursuit of photographic quality in digital satellite imagery was, as we shall see, oblivious to any difficulties that may arise when pixel dimensions begin to approach or even undershoot the scale of urban objects. In addition to multispectral variability, and reductions in spatial resolutions, satellite images also offer rapid availability, comprehensive coverage, high temporal frequency and lower relative costs (entire cities) compared with aerial photography. These have all elevated satellite remote sensing to the forefront of research and practical applications (as exemplified by the majority of the chapters in this book).

1.3.2 *Ever-shrinking spatial resolution*

Consecutive generations of satellite sensor technology have successfully satisfied the demand for higher spatial resolution of image data: from 80 m (Landsat MSS), to 30 m (Landsat TM), to 20 m and 10 m (SPOT HRV – High Resolution Visible, multispectral and panchromatic, respectively), and finally to 4 m and 1 m (EOSAT Space Imaging IKONOS-2, multispectral and panchromatic, respectively). IKONOS-2 was launched on 24 September 1999 and marked the advent of commercial ventures that also include: Orbimage's OrbView-3 at 4-m multispectral and 1-m panchromatic; OrbView-4's 200-band hyperspectral scanning system; and EarthWatch's QuickBird at 2.5-m multispectral and 0.61-m panchromatic in 2002 (Barnsley and Hobson 1996; Corbley 1996). In terms of ownership, the USA's monopoly of Earth observation satellite sensor data capable of identifying urban features was challenged by France's SPOT (Système Pour l'Observation de la Terre) programme in 1986, and later joined by India's IRS programme in 1988, culminating in IRS-1C in 1995 and IRS-1D in 1997 carrying the LISS-III (Linear Imaging Self-scanning Sensor) at 23-m mid-infrared and 6-m panchromatic; Japan's JERS-1 radar and optical sensors at 18 m × 25 m in 1992; declassified Russian SPIN-2 (Space Information-2 metre) digitized KUR-1000 panchromatic photographs in 1998; and by the European Space Agency (ESA) ERS (European Remote Sensing) radar series in 1991 and 1997, and ENVISAT-1 in 2001 (Corbley 1996; Aplin *et al.* 1997; Lillesand and Kiefer 2000; Jensen 2000; Aplin, Chapter 2 in this book). The "skies" are literally filling up with a whole host of countries, including Canada, Brazil/China, Israel and Australia among many others, launching, or queuing up to launch, their own land observation and meteorological satellites.

1.3.3   *Properties of urban surfaces*

Despite commendable reductions in the spatial resolution of image data over the years, urban surfaces remain one of the more challenging surfaces to sense remotely. The expected proliferation of highly detailed and "realistic" maps of cities from high spatial resolution satellite imagery has not as yet materialized. The reasons may be more methodological, alluding to conceptual, rather than technical factors. Simple spatial refinements in the models of representation (pixel dimensions) do not necessarily lead to a better fit of reality (actual urban properties) (Forster 1993; Fisher 1997). In other words, city structures do not fit neatly within the idealized representation of image pixels: bluntly, if using, say SPOT-P, most buildings are not 10-m by 10-m squares! As noted already, the typically complex mixture and irregular size, density and arrangement of artificial and natural urban land cover types will invariably produce reflectance levels that are the result of the interaction of more than one surface material, regardless of the spatial resolution. Of course, images with a 30-m spatial resolution will contain more heterogeneous pixels (representing radiometric variance) than images with, say, a 4-m spatial resolution, but at the same time many 4-m pixels will be smaller in size than many urban features, introducing spectral "noise" and thus downgrading the classification (Atkinson and Curran 1997). Aplin (Chapter 2 in this book) reviews this paradox in relation to comparisons between simulated IKONOS and SPOT data advocating a technique based on per-parcel, rather than per-pixel, classifications to reduce noise and improve thematic accuracy. Indeed, studies by Baltsavias and Gruen (Chapter 3 in this book) have shown that satellite images, even in high spatial resolution mode, perform less than expected when compared with digital aerial photography, certainly in terms of detail and precision when used in creating digital surface models and orthoimages (Meaille and Wald 1994). They also warn of organizational friction to the widespread adoption of new high spatial resolution data, citing low availability, slow delivery, incomplete global coverage, lack of stereo images and lack of choice of the date of acquisition as major uptake obstacles.

Instead, Baltsavias and Gruen point to the future of high precision data from LIDAR (Laser-Induced Detection and Ranging), an active airborne system used for detailed and accurate identification and mapping of buildings in the vertical as well as in the horizontal plane. They applaud the ability of LIDAR data to produce three-dimensional models of the urban landscape but also highlight their inability to determine land cover types not to mention land use. Both shortcomings are addressed by Barnsley *et al.* (Chapter 4 in this book) who separate urban land cover from vegetation by linking LIDAR data with multispectral IKONOS image data, and then develop a graph-based pattern recognition system to infer land use by height and structural configuration. Understanding inherent textural patterns within satellite images is also the target of wavelet research. As texture becomes even more intricate with finer spatial resolution data, Myint (Chapter 5 in this book) explores the potential of wavelet transforms as techniques for unravelling building complexity at multiscale resolutions.

In all, research thus far is demonstrating that high spatial resolution image data are a welcome addition to the pursuit of building identification, but to unlock their full potential we need to link them with data from other sources and/or expose them to spatial data analysis. Only in this way can high spatial resolution remotely sensed data be part of how we visualize, monitor and manage our cities in the future.

1.4 Inferring urban land use

Our discussions, so far, have concentrated on how remote sensing, particularly high spatial resolution data, is able to identify the physical properties of urban objects and surfaces – a city's land cover. Mapping land cover is a venerable achievement in itself but as noted already cities are more than simply a tangled mass of artificial and natural surfaces. They are the containment of human occupancy and the consequences of human decision-making. Such socio-economic characteristics are near impossible to infer directly from remote sensing. Instead, a more complete picture of city structure *and* city habitation is only possible by land use deduction from land cover (Webster 1995). The process of deduction is assisted by reference to additional, non-remote sensing, sources of information on the socio-economic characteristics of cities. Typically, population censuses are employed to "populate" land cover distributions inherent or classified from remotely sensed images.

1.4.1   To "populate" or even "socialize" the pixel?

Knowing the population distribution of cities at frequent time intervals is important in unravelling changes in social and economic behaviour and understanding mobility patterns. So how can remote sensing infer population numbers from raw digital numbers of fixed geometric representation? Surely people cannot be seen in their homes from the sky or space, even if the spatial resolution was 1 cm! Of course not. The only reliable, consistent, accurate and comprehensive data set, at the national level, that supplies population details is a population census. However, a population census, for reasons of confidentiality and expediency, is represented by zonal units, representing groups of houses rather than individual households. In order to combine population numbers from the census with satellite image data an areal interpolation process is needed that redistributes the numbers from each census zonal unit within the land cover limits of the image at corresponding locations (Harvey 2002). The process, known as dasymetric mapping, is demonstrated by Langford and Dobson *et al.* (Chapters 6 and 12 in this book, respectively), the former in generating urban population density models by both eliminating non-urban land cover (through image classification) and establishing a density relationship between census counts and built land cover. Further relationships between a population census and residential land use are evaluated by Lo (Chapter 7 in this book). Using linear regression models and an allometric growth model Lo develops a methodology for the accurate monitoring of urban growth by calibrating population and housing statistics with spatially aggregated image data. The approaches of both Langford and Lo (as well as Yuan *et al.* 1997; Chen 2002) are simple, cost-effective and provide a seamless integration between remote sensing and socio-economic data.

However, zone-based models are limited by the heterogeneity of both population density and land cover within each zone. In response to the first of these, Harvey (Chapter 8 in this book) explores the possibility of estimating populations at the individual pixel level using an expectation maximization technique to fit linear models between zonal population counts and untransformed Landsat TM bands iteratively. Pixel-based models use simpler and more robust mathematical formulations, provide greater spatial flexibility in application and are more amenable to spatially targeted refinement than zone-based models. However, as with zone-based models, pixel-based models suffer from an overestimation of populations at low density and underestimation at high density.

Both zone- and pixel-based models are critically dependent on pixels being accurately classified as residential land use as opposed to commercial, industrial, institutional or recreational land use. The whole debate on the classification of urban land use from satellite imagery would fill a book in itself (Hutchinson 1982; Baraldi and Parmiggiani 1990; Heikkonen and Varfis 1998), but for now it is just important to appreciate the variety of statistical techniques, methodologies and heuristics used, ranging from contextual (Gurney and Townshend 1983) and textural measures (using wavelet transforms, Myint Chapter 5 in this book), cover-type frequencies (Eyton 1993), artificial neural networks (Dreyer 1993; Zhang and Foody 2001), fuzzy sets (Fisher and Pathirana 1990), mixture models (Kressler and Steinnocher 1996; Mitchell *et al.* 1998) to the calculation of fractal dimensions. Some image classification techniques are directly inferred from sensor data (Iisaka and Hegedus 1982), while others rely on assistance from ancillary data sources. One such ancillary-based technique, outlined by Mesev (Chapter 9 in this book), calls on census data to establish land use class frequencies as prior probabilities of a Bayesian-modified discriminant function (Strahler 1980; Berthod *et al.* 1996; Mesev 1998, 2001). Another methodology introduces other, non-census, sources of socio-economic information into image classification. Harris (Chapter 10 in this book) reviews the potential of using data on postal deliveries as a means to identify not only residential land use but also to couple the geodemographic profiles of residents with the structural patterns of their occupancy. In all, the goal is to improve classification accuracy: how well the distribution of thematic urban classes represented by a classified image corresponds to the actual distribution of land use within cities. For many this improvement is marginal at best and, for the amount of extra processing and strict statistical assumptions, may not be practical (reflected by the lack of uptake by major image processing proprietary software).

In "populating" the pixel (see Geoghegan *et al.* 1998 for original expression), the pixel does not necessarily have to represent the built structure of people's homes. It can also represent people's workplaces. All the techniques and applications on residential land use mentioned so far have relied on the interpretation of pixels that represent built land cover from reflected energy at the visible and infrared portions of the electromagnetic spectrum, or active microwave signals; both types taken during the day. However, these approaches are frequently limited by similar spectral response patterns between built structures (stone, brick, etc.) within cities and exposed rocky outcrops and soil outside cities. What if satellite sensors are capable of recording cities better at night rather than during the day? For the past 30 years the US Air Force Defense Meteorological Space Program (DMSP) has operated satellites with Operational Linescan System (OLS) sensors that are capable of detecting visible and near-infrared emissions from towns and cities. These faint emissions are amplified using a photo-multiplier tube (PMT) to generate 1-km spatial resolution pixels that represent night-time lights, mostly from cities but also from gas flares and ephemeral lights (mostly fires). Technical details of DMSP OLS night-time image data and scope of applications are covered by Elvidge *et al.* (Chapter 13 in this book). The world at night is a striking site (just look at the cover of this book) (Croft 1978; Southwell 1997). Cities are represented by specks of bright illumination set against an unambiguous dark background.

Nocturnal lighting is not simply a representation of residential land use but also an indicator of economic activity, such as street lighting, commercial and recreational buildings remaining open in the evenings, security lights and lights from vehicles. The coarse spatial resolution of 1 km is necessary to detect these faint light emissions but also allows global coverage within manageable storage requirements. Models of world population are therefore readily constructed and Sutton (Chapter 14 in this book) demonstrates how such models can be calibrated with past censuses to forecast population growth, especially in developing countries where censuses are costly, difficult to administer and unreliable. Another world model, LandScan, is one constructed by Dobson *et al.* (Chapter 12 in this book and Dobson *et al.* 2000) from not only night-time imagery and population censuses but also other digital cartographic features and digital elevation models. Based on dasymetric principles, LandScan has been employed extensively by the USA, foreign governments and the UN for estimating populations at risk in actual or anticipated natural and man-made disasters.

1.4.2   Consequences of human activity

As noted above, data from the DMSP OLS not simply record residential land use but all nocturnal indicators of economic activity. The density of lights generated at night is an excellent marker for both the scale of human interaction and the level of economic performance. Studies by Doll (Chapter 15 in this book) have calibrated regression models between the intensity of lit areas from night-time sensor data with levels of Gross Domestic Product (GDP) at global and subnational scales. Results of dependency are encouraging and represent essential information on the prospects for using night-time imagery to forecast levels of power consumption and, by inference, comparisons of global relative prosperity. Similar reliable and consistent relationships using night-time data can be extended to the monitoring of fossil fuel trace gas emissions (Doll *et al.* 2000). The mapping of CO_2 emissions is an important step toward understanding not only the pollution levels over major cities but also global changes in temperatures and climates (Gallo *et al.* 1995). This concern is echoed in the work by Nichol (Chapter 11 of this book) who employs thermal remote sensing to evaluate urban heat islands for Singapore and Kano in Nigeria (for other examples see Quattrochi and Ridd 1994). The examples are used to demonstrate the important differences between ground-based and remotely sensed heat islands, including relationships between surface and air temperatures, the satellite-"seen" surface and the complete radiating surface, and the remotely sensed temperature and the actual surface temperature of objects.

Other studies using remote sensing to assess human consequences on the environment include land use changes resulting from deforestation in the Amazon (Moran and Brondizio 1998), famine early-warning models

(Hutchinson 1982) and linking climate models to the monitoring of disease outbreaks (Epstein 1998). Finally, one other consequence of human activity is the novel application of using night-time data to resolve the question of whether street lighting deters crime (Painter 1996). The relationship between intra-urban lit areas and crime incidents is developed by Weeks (Chapter 16 in this book) but the results are a little surprising, indicating that more crime is located in brighter areas. The most apparent reason for this seems to be that local authorities had already allocated more lighting within "problem" neighbourhoods but, for many other social reasons, crime persists. Many more applications can be found in the growing body of literature and in specialized edited volumes such as Liverman *et al.* (1998) and Donnay *et al.* (2001).

1.5   Modelling remotely sensed cities

General remote sensing research is dominated by the pursuit of radiometric interpretation at various spectral resolutions. Urban remote sensing is no different except that the dynamic and multifaceted nature of cities demands more than simply the measurement of the layout of urban physical structures *in situ*. Of course, high temporal frequencies facilitate change detection, but to understand some of the human processes behind the arrangement of built form there is also the demand to analyse the spatial configuration of interpreted radiometric interpretations at various spectral resolutions.

Many of the studies in this book allude to urban modelling: the spatial analysis of classified remotely sensed data. Other work not covered in this book include: the segmentation of urban boundaries by Delaunay triangulation (Bähr 2001); the implementation of geostatistical techniques using the semi-variogram to explore urban spatial patterns (Brivio and Zilioli 2001; Atkinson 2001); and the exploitation of mathematical morphology as indicators of image texture and landscape metrics (Pesaresi and Bianchin 2001). Finally, the spatial configuration of various urban land uses has been used to develop and empirically support theories of fractal geometry by relating urban form to function (Batty and Longley 1994; Mesev *et al.* 1995; Longley and Mesev 2002). In all, the goal of modelling is to develop indicators and descriptors of population distributions and urban land use patterns and densities, which are then, in turn, related to explanations of urban processes.

1.6   Remote sensing/GIS integration

If there is one theme running through this book that connects the various methodological strands within the chapters then it is surely the link between remote sensing and GIS. There is no one agreed and unambiguous definition of remote sensing/GIS integration (RS-GIS) (Star *et al.* 1991; Davis *et al.* 1991; Star *et al.* 1997; Mesev 1997). Instead it has been used to refer to any type of linkage between remote sensing and GIS, ranging from seamless and hybrid RS-GIS databases (Zhou 1989), through implicit Boolean links, to cursory discussions on data sharing. Integration between remote sensing and GIS is not new. Admittedly, the inclusion of non-image spatial data was not of paramount importance in the days of the Earth Orbital Program (Marble 1981), but many digital systems and applications since then have investigated how these data could be linked with data from the more established technology of remote sensing. A succession of individual pieces of work in the 1970s and 1980s examined the more obvious mutual benefits from linking a data collection technology (remote sensing) with a predominantly data handling and decision support technology (GIS). Benefits such as extended data inventories and land use updating quickly established remote sensing, with its high temporal frequency and immediate access, as a vital source of input into GIS. A more bidirectional flow was strengthened when GIS data were used to aid image segmentation, improve image classification and assist the spectral and spatial analysis of radiometric and classified image data (Wang *et al.* 1992; Harris and Ventura 1995; Williamson 1996; Thomson and Hardin 2000; Zhan 2000). Sustained research into RS-GIS was rewarded by the devotion of an entire National Center for Geographic Information and Analysis (NCGIA) initiative to looking at the organizational and technical challenges for an improved and mutually beneficial integration (Star *et al.* 1991). The initiative was comprehensive, non-application specific and attempted to broaden awareness of the critical pitfalls of merging complementary, yet conceptually and organizationally different, technologies. More recently, exponential growth in the volume and range of spatial data now requires systems that can handle disparate data in variable formats from dissimilar sources. Today, GIS and RS data are more readily available (but not always fully accessible) by Internet browsing facilities, as well as from data-sharing contracts, decommissioned remotely sensed data (McDonald 1995) and operational links with GPSs (Global Positional Systems). Despite this data explosion and the solid foundations of the NCGIA initiative, progress in RS-GIS integration for specifically urban problems has been cautious and arduous.

Technical incompatibilities between remote sensing and GIS are numerous but not insurmountable, and represent challenging areas of research for both geocomputational scientists and social scientists. However, because integration issues are numerous and incongruent, closer links between the two technologies/disciplines require formalized conceptual frameworks that detail structured integration considerations, such as the land cover/land use dichotomy and field/object-based representations (Mesev 1997). Beyond technical assimilation, remote sensing and GIS also need to contribute to theory and models within interdisciplinary research (Rindfuss and Stern 1998; Longley and Mesev 2002; Longley 2002).

1.7 Structure of the book

The reasons for compiling this three-part volume at this time are twofold. First, the need to assess the progress of high spatial resolution remote sensor instrumentation for precise measurement of dynamic, yet highly scale-sensitive city structures; and, second, to investigate enhancements to remote sensor data interpretation of population distributions and urban land use patterns. The potential of high spatial resolution sensor imagery is explored in Part I, together with conceptual and technical limitations. Parts II and III are subdivided conveniently by the presence (Part II) or absence (Part III) of the Sun. The division determines whether remote sensing methodologies and applications are based on predominantly reflected energy (during the day) or predominantly emitted energy (during the night) – incidentally, the LandScan model (Dobson *et al.*, Chapter 12 in this book) uses both day-time and night-time information. In both parts the goals are to generate rapid, up-to-date and frequent maps of population, urban land use structures or urban environmental considerations using spatial modelling techniques and/or ancillary data from GIS. In doing so, a number of pressing questions on population growth, urban sprawl, wealth distribution, socio-economic profiling, land use structure and urban pollution are addressed. The list is extensive but not complete. The reader is urged to consult specialized volumes edited by Liverman *et al.* (1998), Baltsavias *et al.* (2001), Donnay *et al.* (2001) and Davreau *et al.* (1989) as well as dedicated issues in *Applied Geography*, Volume 13, 1993 and *Computers, Environment and Urban Systems*, Volume 21, 1997. To the credit of the urban remote sensing community there is, apart from a couple of stalwarts, minimal overlapping of authors and material in these sources. This volume is no exception, but unlike the others contains a good number of early career researchers, no doubt attracted to the field of urban remote sensing by its potential and technical challenge. Over recent years a series of international forums have been established to cater for the rise in interest. In 1995, the European Science Foundation sponsored a seminar of Remote Sensing and Urban Analysis as part of its GISDATA programme, coupled with the Remote Sensing and Urban Statistics initiative organized by the European Statistical Agency (Eurostat). More recently, "Urban 2001" was an IEEE/ISPRS joint workshop on Remote Sensing and Data Fusion over Urban Areas. The ISPRS (International Society for Photogrammetry and Remote Sensing) has also supported a new working group VII-4 on Human Settlements and Impact Analysis as well as sponsoring the latest symposium on Remote Sensing of Urban Areas in 2002. Both initiatives are set to become established annual events. In the USA, the ASPRS (American Society for Photogrammetry and Remote Sensing) is currently compiling a volume on Remote Sensing and Human Settlements as part of its "Manual of Remote Sensing" series.

1.8   References

Brivio and Zilioli, 2001, Urban pattern characterization through geostatistical analysis of satellite images. In J.-P. Donnay, M. J. Barnsley and P. A. Longley (eds) *Remote Sensing and Urban Analysis* (London: Taylor and Francis) 39–53.

Anderson, J. R., Hardy, E. E., Roach, J. T. and Witmer, R. E., 1976, *A land use and land cover classification system for use with remote sensor data*, US Geological Survey Professional Paper 964 (Washington, DC: US Government Printing Office).

Aplin, P., Atkinson, P. M. and Curran, P. J., 1997, Fine spatial resolution satellite sensors for the next decade. *International Journal of Remote Sensing*, 18, 3873–81.

Atkinson, 2001, Geostatistical regularization in remote sensing. In N. J. Tate and P. M. Atkinson (eds) *Modelling Scale in Geographical Science* (Chichester: Wiley) 237–60.

Atkinson, P. M. and Curran, P. J., 1997, Choosing an appropriate spatial resolution for remote sensing investigations. *Photogrammetric Engineering and Remote Sensing*, 63, 1345–51.

Avery, G., 1965, Measuring land use changes on USDA photographs. *Photogrammetric Engineering*, 31, 620–4.

Bähr, H-P., 2001, Image segmentation for change detection in urban environments. In J.-P. Donnay, M. J. Barnsley and P. A. Longley (eds) *Remote Sensing and Urban Analysis* (London: Taylor and Francis), pp. 95–113.

Baltsavias, E. P., Gruen, A. and Van Gool, L. (eds), 2001, *Automatic Extraction of Man-Made Objects from Aerial and Space Images III* (Lisse, The Netherlands: Balkema).

Baraldi, A. and Parmiggiani, F., 1990, Urban area classification by multispectral SPOT images. *IEEE Transactions, Geosciences & Remote Sensing*, 28, 674–80.

Barker, G. R., 1988, Remote sensing: The unheralded component of geographic information systems. *Photogrammetric Engineering and Remote Sensing*, 54, 195–9.

Barnsley, M. J. and Hobson, P., 1996, Making sense of sensors. *GIS Europe*, 5(5), 34–6.

Batty, M. and Longley, P. A., 1994, *Fractal Cities: A Geometry of Form and Function* (London: Academic Press).

Berthod, M., Kato, Z., Yu, S. and Zerrubia, L., 1996, Bayesian image classification using Markov random fields. *Image and Vision Computing*, 14, 285–95.

Bibby, P. and Shepherd, J., 1999, Monitoring land cover and land use for urban and regional planning. In P. A. Longley, M. F. Goodchild, D. J. Maguire and D. W. Rhind (eds) *Geographical Information Systems*, 2nd edn (New York: John Wiley & Sons), pp. 953–65.

Bjorgo, E., 2000, Using very high spatial resolution multispectral satellite sensor imagery to monitor refugee camps. *International Journal of Remote Sensing*, 21, 611–16.

Brugioni, D. A., 1983, The census: It can be done more accurately with space-age technology. *Photogrammetric Engineering and Remote Sensing*, 49, 1337–9.

Carls, N., 1947, *How to Read Aerial Photographs for Census Work* (Washington, DC: US Government Printing Office).

Chen, K., 2002, An approach to linking remotely sensed data and areal census data. *International Journal of Remote Sensing*, 23, 37–48.

Chen, S., Zheng, S. and Xie, C., 2000, Remote sensing and GIS for urban growth analysis in China. *Photogrammetric Engineering and Remote Sensing*, 66, 593.

Corbley, K. P., 1996, One-meter satellites. *Geo Info Systems*, July, 28–42.

Couloigner, I. and Ranchin, T., 2000, Mapping of urban areas: A multiresolution modelling approach for semi-automatic extraction of streets. *Photogrammetric Engineering and Remote Sensing*, 66, 867–74.

Cowen, D. J. and Jensen, J. R., 1998, Extraction and modeling of urban attributes using remote sensing technology. In D. Liverman, E. F. Moran, R. R. Rindfuss and P. C. Stern (eds) *People and Pixels: Linking Remote Sensing and Social Science* (Washington, DC: National Academy Press), pp. 164–88.

Croft, T. A., 1978, Night-time images of the Earth from space. *Scientific American*, 239, 68–79.

Curran, P. J. and Williamson, H. D., 1986. Sample size for ground and remotely sensed data. *Remote Sensing of Environment*, 20, 31–41.

Davis, F. W., Quattrochi, D. A., Ridd, M. K., Lam, N. S-M., Walsh, S. J., Michaelsen, J. C., Franklin, J., Stow, D. A., Johannsen, C. J. and Johnston, C. A., 1991, Environmental analysis using integrated GIS and remotely sensed data: Some research needs and priorities. *Photogrammetric Engineering and Remote Sensing*, 57, 689–97.

Davreau, F., Barbary, O., Michel, A. and Lortic, B., 1989, *Area Sampling from Satellite Image for Socio-Demographic Surveys in Urban Environments* (Paris: Editions de l'ORSTOM).

Dreyer, P., 1993, Classification of land cover using optimized neural nets on SPOT data. *Photogrammetric Engineering and Remote Sensing*, 59, 617–21.

Dobson, J. E., 1993, Land cover, land use differences distinct. *GIS World*, 6(2), 20–2.

Dobson, J. E., Bright, E. A., Coleman, P. R., Durfee, R. C. and Worley, B. A., 2000, LandScan: A global population database for estimating populations at risk. *Photogrammetric Engineering and Remote Sensing*, 66, 849–57.

Doll, C. N. H., Muller, J-P. and Elvidge, C. D., 2000, Night-time imagery as a tool for mapping socio-economic parameters and greenhouse gas emissions. *Ambio*, 29, 159–64.

Donnay, J-P., 1999, Use of remote sensing information in planning. In J. Stillwell, S. Geertman and S. Openshaw (eds) *Geographical Information and Planning* (Berlin: Springer), pp. 242–60.

Donnay, J-P., Barnsley, M. J. and Longley, P. A., 2001, *Remote Sensing and Urban Analysis* (London: Taylor & Francis).

Ehlers, M., Edwards, G. and Bédard, Y., 1989, Integration of remote sensing with geographic information systems: A necessary evolution. *Photogrammetric Engineering and Remote Sensing*, 55, 1619–27.

Epstein, P. R., 1998, Health applications of remote sensing and climate modelling. In D. Liverman, E. F. Moran, R. R. Rindfuss and P. C. Stern (eds) *People and Pixels: Linking Remote Sensing and Social Science* (Washington, DC: National Academy Press), pp. 197–207.

Eyton, R. J., 1993, Urban land use classification and modelling using cover-type frequencies. *Applied Geography*, 13, 111–21.

Fazal, S., 2001, Application of remote sensing and GIS techniques in urban sprawl and land use change mapping: A case study of a growing urban centre in India. *Asian Profile*, **29**, 45–62.

Fisher, P., 1997, The pixel: A snare and a delusion. *International Journal of Remote Sensing*, **18**, 679–85.

Fisher, P. F. and Pathirana, C., 1990, The evaluation of fuzzy membership of land cover classes in the suburban zone. *Remote Sensing of Environment*, **34**, 121–32.

Forster, B. C., 1980, Urban residential ground cover using Landsat digital data. *Photogrammetric Engineering and Remote Sensing*, **46**, 547–58.

Forster, B. C., 1985, An examination of some problems and solutions in monitoring urban areas from satellite platforms. *International Journal of Remote Sensing*, **6**, 139–51.

Forster, B. C., 1993, Coefficient of variation as a measure of urban spatial attributes, using SPOT HRV and Landsat TM data. *International Journal of Remote Sensing*, **14**, 2403–9.

Gallo, K. P., Tarpley, J. D., McNab, A. L. and Karl, T. R., 1995, Assessment of urban heat islands – a satellite perspective. *Atmospheric Research*, **37**, 37–43.

Garreau, J., 1991, *Edge City: Life on the New Frontier* (New York: Doubleday).

Geoghegan, J., Pritchard, Jr L., Ogneva-Himmelberger, Y., Chowdhury, R. R., Sanderson, S. and Turner, B. L., 1998, "Socializing the pixel" and "pixelizing the social" in land use and land cover-change. In D. Liverman, E. F. Moran, R. R. Rindfuss and P. C. Stern (eds) *People and Pixels: Linking Remote Sensing and Social Science* (Washington, DC: National Academy Press), pp. 51–69.

Grey, W. and Luckman, A., 2001, Monitoring urban developments using radar remotely sensed data. *Swansea Geographer*, **36**, 53–62.

Gurney, C. M. and Townshend, J. R. G., 1983, The use of contextual information in the classification of remotely sensed data. *Photogrammetric Engineering and Remote Sensing*, **49**, 55–64.

Gurney, R. J., Foster, J. L. and Parkinson, C. L., 1993, *Atlas of Satellite Observations Related to Global Change* (Cambridge: Cambridge University Press).

Harris, P. M. and Ventura, S. J., 1995, The integration of geographic data with remotely sensed imagery to improve classification in an urban area. *Photogrammetric Engineering and Remote Sensing*, **61**, 993–8.

Harvey, J. T., 2002, Estimating census district populations from satellite imagery: Some approaches and limitations. *International Journal of Remote Sensing*, **23**, 2071–95.

Heikkonen, J. and Varfis, A., 1998, A land cover/land use classification of urban areas: A remote sensing approach. *International Journal of Pattern Recognition and Artificial Intelligence*, **12**, 475.

Henderson, F. M. and Xia, Z. G., 1997, SAR applications in human settlement detection, population estimation and urban land use pattern analysis: A status report. *IEEE Transactions on Geosciences and Remote Sensing*, **35**, 79–85.

Hutchinson, C. F., 1982, Techniques for combining Landsat and ancillary data for digital classification improvement. *Photogrammetric Engineering and Remote Sensing*, **48**, 123–30.

Iisaka, J. and Hegedus, E., 1982, Population estimation from Landsat imagery. *Remote Sensing of Environment*, **12**, 259–72.

Jensen, J. R., 2000, *Remote Sensing of the Environment: An Earth Resource Perspective* (Englewood Cliffs, NJ: Prentice-Hall).

Jensen, J. R. and Cowen, D. C., 1999, Remote sensing of urban/suburban infrastructure and socio-economic attributes. *Photogrammetric Engineering and Remote Sensing*, 65, 611–22.

Ji, C. Y., Liu, Q., Sun, D., Wang, S., Lin, P. and Li, X., 2001, Monitoring urban expansion with remote sensing in China. *International Journal of Remote Sensing*, 22, 1441–55.

Ji, M. and Jensen, J. R., 1999, Effectiveness of subpixel analysis in detecting and quantifying urban imperviousness from Landsat Thematic Mapper imagery. *Geocarto International*, 14(4), 33–41.

Kent, M., Jones, A. and Weaver, R., 1993, Geographical information systems and remote sensing in land use planning: An introduction. *Applied Geography*, 13, 5–8.

Kressler, F. and Steinnocher, K., 1996, Change detection in urban areas using satellite images and spectral mixture analysis. *International Archives of Photogrammetry and Remote Sensing*, 31, 379–83.

Lillesand, T. M. and Kiefer, R. W., 2000, *Remote Sensing and Image Interpretation*, 4th edn (New York: John Wiley & Sons).

Liverman, D., Moran, E. F., Rindfuss, R. R. and Stern, P. C. (eds), 1998, *People and Pixels: Linking Remote Sensing and Social Science* (Washington, DC: National Academy Press).

Lo, C. P., 2002, Urban indicators of China from radiance-calibrated digital DMSP-OLS night-time images. *Annals of the Association of American Geographers*, 92, 225–40.

Longley, P. A. 2002, Geographical information systems: Will developments in urban remote sensing and GIS lead to 'better' urban geography? *Progress in Human Geography*, 26, 231–9.

Longley, P. A., and Mesev, V., 2002, Measurement of density gradients and space-filling in urban systems. *Papers in Regional Science*, 81, 1–28.

Masser, I., 2001, Managing our urban future: The role of remote sensing and geographic information systems. *Habitat International*, 25, 503–12.

McDonald, R. A., 1995, Opening the cold war sky to the public: Declassifying satellite reconnaissance imagery. *Photogrammetric Engineering and Remote Sensing*, 61, 385–90.

Marble, D. F., 1981, Some problems in the integration of remote sensing and geographic information systems. Paper given at Conference of the 2nd Australasian Remote Sensing, Brisbane, Australia.

Martin, D. J. and Bracken, I., 1993, The integration of socioeconomic and physical resource data for applied land management information systems. *Applied Geography*, 13, 45–53.

Meaille, R. and Wald, L., 1994, Using geographical information system and satellite imagery within a numerical simulation of regional urban growth. In MacLean (ed.) *Remote Sensing and Geographic Information Systems* (Bethesda, MD: ASPRS), pp. 210–21.

Mesev, V., 1997, Remote sensing of urban systems: Hierarchical integration with GIS. *Computers, Environment and Urban Systems*, 21, 175–187.

Mesev, V., 1998, The use of census data in urban image classification. *Photogrammetric Engineering and Remote Sensing*, 64, 431–8.

Mesev, V., 2001, Modified maximum likelihood classifications of urban land use: Spatial segmentation of prior probabilities. *Geocarto International*, 16(4), 39–46.

Mesev, T. V., Batty, M., Longley, P. A. and Xie, Y., 1995, Morphology from imagery: Detecting and measuring the density of urban land use. *Environment and Planning A*, 27, 759–80.

Michalak, W. Z., 1993, GIS in land use change analysis: Integration of remotely sensed data into GIS. *Applied Geography*, 13, 28–44.

Mitchell, R., Martin, D. and Foody, G., 1998, Unmixing aggregate data: Estimating the social composition of enumeration districts. *Environment and Planning A*, 30, 1929–41.

Moran, E. F. and Brondizio, E., 1998, Land-use change after deforestation in Amazônia. In D. Liverman, E. F. Moran, R. R. Rindfuss and P. C. Stern (eds) *People and Pixels: Linking Remote Sensing and Social Science* (Washington, DC: National Academy Press), pp. 94–120.

Ogrosky, C. E., 1975, Population estimates from satellite imagery. *Photogrammetric Engineering and Remote Sensing*, 41, 707–12.

Painter, K., 1996, The influence of street lighting improvements on crime, fear and pedestrian street use, after dark. *Landscape and Urban Planning*, 35, 193–201.

Pesaresi, M. and Bianchin, A., 2001, Recognizing settlement structure using mathematical morphology and image texture. In J.-P. Donnay, M. J. Barnsley and P. A. Longley (eds) *Remote Sensing and Urban Analysis* (London: Taylor and Francis), pp. 55–67.

Polle, V. F. L., 1996, Planning urban services in developing countries: Quantification of community service needs using remote sensing indicators. *The ITC Journal*, 1, 64.

Quattrochi, D. A. and Ridd, M. K., 1994, Measurement and analysis of thermal energy responses from discrete urban surfaces using remotely sensed data. *International Journal of Remote Sensing*, 15, 1991–2022.

Ranjan, A. and Rastogi, N., 1990, Application of remote sensing/aerial photography for urban planners. *Civic Affairs*, 38(5), 57.

Ridd, M. K. and Liu, J., 1998, A comparison of four algorithms for change detection in an urban environment. *Remote Sensing of Environment*, 63, 95–100.

Rindfuss, R. R. and Stern, P. C., 1998, Linking remote sensing and social science: The needs and challenges. In D. Liverman, E. F. Moran, R. R. Rindfuss and P. C. Stern (eds) *People and Pixels: Linking Remote Sensing and Social Science* (Washington, DC: National Academy Press), pp. 1–27.

Southwell, K., 1997, Remote sensing; night lights. *Nature*, 390, 21.

Star, J. L., Estes, J. E. and Davis, F., 1991. Improved integration of remote sensing and geographic information systems: A background to NCGIA initiative 12. *Photogrammetric Engineering and Remote Sensing*, 57, 643–5.

Star, J. L., Estes, J. E. and McGwire, K. C., 1997, *Integration of Geographic Information Systems and Remote Sensing* (Cambridge: Cambridge University Press).

Strahler, A. H., 1980, The use of prior probabilities in maximum likelihood classification of remotely sensed data. *Remote Sensing of Environment*, 10, 135–63.

Thomson, C. N. and Hardin, P., 2000, Remote sensing/GIS integration to identify potential low-income housing sites. *Cities*, 17, 97–109.

Tobler, W., 1969, Satellite confirmation of settlement size coefficients. *Area*, 1, 30–4.

Wang, F., Treitz, P. M. and Howarth, P. J., 1992, Road network detection from SPOT imagery for updating geographical information systems in the rural-urban fringe. *International Journal of Geographical Information Systems*, 6, 141–57.

Weber, C., 2001, Urban agglomeration delimitation using remotely sensed data. In Donnay, Barnsley and Longley (eds) *Remote Sensing and Urban Analysis* (London: Taylor and Francis), pp. 145–59.

Webster, C. J., 1995, Urban morphological fingerprints. *Environment and Planning B*, 22, 279–97.

Webster, C. J., 1996, Population and dwelling estimates from space. *Third World Planning Review*, 18, 155–76.

Welch, R., 1982, Spatial resolution requirements for urban studies. *International Journal of Remote Sensing*, 3, 139–46.

Williamson, G. G., 1996, A review of current issues in the integration of GIS and remote sensing. *International Journal of Geographical Information Systems*, 10, 85–101.

Yeh, A. G.-O. and Li, X., 2001, Measurement and monitoring of urban sprawl in a rapidly growing region using entropy. *Photogrammetric Engineering and Remote Sensing*, 67, 83–90.

Yuan, Y., Smith, R. M. and Limp, W. F. 1997, Remodelling census population with spatial information from Landsat TM imagery. *Computers Environment and Urban Systems*, 21, 245–58.

Zhan, Q., 2000, Urban land use classes with fuzzy membership and classification based on integration of remote sensing and GIS. *International Archives of Photogrammetry and Remote Sensing*, 33, 751–9.

Zhang, J. and Foody, G. M., 2001, Fully-fuzzy supervised classification of sub-urban land cover from remotely sensed imagery: Statistical and artificial neural network approaches. *International Journal of Remote Sensing*, 22, 615–28.

Zhou, Q., 1989, A method for integrating remote sensing and geographic information systems. *Photogrammetric Engineering and Remote Sensing*, 55, 591–6.

Part I
High spatial resolution data

2 Comparison of simulated IKONOS and SPOT HRV imagery for classifying urban areas

Paul Aplin

2.1 Introduction

Back in the early 1980s, in a seminal article, Welch (1982) outlined the spatial resolution requirements of remotely sensed imagery for classifying urban areas. In terms of spaceborne instruments, the contribution of remote sensing to urban studies was deemed limited at the very least. Simply put, satellite sensors were not capable of generating imagery with a fine enough spatial resolution to identify the majority of urban features. In fact, not only were instruments "at the time" insufficient for accurate urban classification, but Welch (1982) predicted that other finer spatial resolution instruments planned for launch later in the 1980s would not significantly increase classification accuracy. It is only relatively recently, at the start of the third millennium, that the spatial resolution of satellite sensors has increased to the extent that accurate urban classification can be performed, in line with Welch's (1982) recommendations.

Donnay *et al.* (2001) refer to three generations of satellite sensor. First generation instruments, such as the Landsat Multispectral Scanning Subsystem (MSS), were operating in the 1970s and early 1980s. Landsat MSS had a spatial resolution of 79 m, the limitations of which should be clear for detailed urban analysis. Second generation instruments, launched in the 1980s, included the Landsat Thematic Mapper (TM), which had a spatial resolution of 30 m, and Systeme Pour l'Observation de la Terre (SPOT) High Resolution Visible (HRV) instrument, which comprised a 20-m spatial resolution multispectral sensor and a 10-m spatial resolution panchromatic sensor. Somewhat surprisingly, these second generation instruments, although providing significantly finer spatial resolution imagery than Landsat MSS, were not (as Welch 1982 had predicted), significantly more accurate for urban classification. In fact, some studies demonstrated that urban classification was more accurate using Landsat MSS imagery than Landsat TM imagery (Toll 1985) or even SPOT HRV imagery (Martin *et al.* 1988) because although spatial resolutions were still coarser than most urban features there now would be more pixels erroneously classified, adding to the overall increase in "noise". The third generation of satellite sensors emerged in the late 1990s, having benefited from the end of the Cold War and the resulting relaxation of legislation governing commercial involvement in remote sensing. This had several consequences, including the declassification of military technology (e.g., Corona "spy" satellite sensor images from the 1960s to 1980s were made publicly available: McDonald 1995) and, importantly, the award of licences to several commercial organizations to produce very fine spatial resolution satellite sensors (Fritz 1996). It led to the development of instruments with spatial resolution capabilities an order of magnitude higher than second generation satellite sensors. In particular, IKONOS, launched in 1999, comprises a 4-m spatial resolution multispectral sensor and a 1-m spatial resolution panchromatic sensor (Aplin *et al.* 1997). Such imagery finally meets Welch's (1982) general recommendation for a spatial resolution of around 5 m or finer for urban studies, and should, therefore, enable relatively accurate urban classification.

The development and subsequent operation of third generation satellite sensors has attracted considerable interest from the remote sensing community (Zhou and Li 2000; Fortier *et al.* 2001; Fowler 2001; Franklin *et al.* 2001; Guindon 2001; Key *et al.* 2001). In particular, much attention has been focused on the potential of instruments such as IKONOS for classifying urban areas (Bjorgo 2000; Dare and Fraser 2001; Kontoes *et al.* 2001; Tanaka and Sugimura 2001). However, as yet, few studies have been documented that demonstrate the benefits of fine spatial resolution satellite sensors in this respect.

This chapter provides a starting point for quantifying the capabilities of third generation satellite sensors for urban classification. Specifically, IKONOS and SPOT HRV (the most detailed second generation instrument in terms of spatial resolution) images are simulated, and comparative classification analysis is performed to demonstrate the relative merits of each for classifying urban areas. This involves per-pixel and per-parcel land cover classification using four multispectral image data sets: (i) 4-m spatial resolution 4-band imagery, (ii) 4-m spatial resolution 3-band imagery, (iii) 20-m spatial resolution 4-band imagery and (iv) 20-m spatial resolution 3-band imagery. While the first and fourth data sets represent IKONOS and SPOT HRV imagery, respectively, the second and third data sets were included to enable investigation of changes in spatial properties independent of changes in spectral properties, and vice versa.

2.2 Classifying urban areas

Notwithstanding Welch's (1982) "general" recommendations regarding spatial resolution requirements for urban studies, it should be recognized that urban areas are far from uniform and it is, therefore, impractical to advocate a general classification system. It is, of course, the diversity within and between urban landscapes that makes them so difficult to classify. Not only do urban areas vary internally (i.e., "within" cities) according to the different land uses present (e.g., residence, commerce, industry), but they also vary externally (between cities) according to location. For instance, the expansive urban landscapes characteristic of North America (Masek *et al.* 2000) contrast markedly with densely packed European cities (Forster 1983) and sprawling Asian "mega-cities" (Chen *et al.* 2000; Ji *et al.* 2001). Therefore, spatial resolution requirements for urban areas will vary considerably according to land use and location and, as such, any example of urban classification analysis should be considered in the context of its study area.

At this point it is important to distinguish between land cover and land use, a distinction often overlooked in classification analysis (Barnsley *et al.* 2001), especially in the context of urban studies. Land cover refers to "the description of the physical nature of the land surface", while land use "describes the same features in terms of their socio-economic significance" (Wyatt *et al.* 1993). For instance, land cover classes of vegetation, buildings and water may refer to land use classes of gardens, houses and canals, respectively. It is important to note the difference because, while land cover can be classified directly from remotely sensed imagery, land use cannot. Instead, it may be possible to infer land use from land cover observations, although this will generally depend on additional data (e.g., socio-economic data: Jensen and Cowen 1999; Mesev *et al.* 2001) or processing (e.g., contextual analysis: Caetano *et al.* 1997; Cortijo and De La Blanca 1998). This issue is significant because, of the two classification schemes, land use is a more valuable source of information and is usually the desired end product of analysis, but it is also harder to obtain. As a general rule, land cover and/ or land use classification analysis should be performed with due regard to the limitations of remotely sensed imagery for these purposes. In particular, attempts should not be made to infer land use classes directly from spectral classification.

Many different forms of classification analysis have been tested on urban environments. Traditional statistical classifiers such as maximum likelihood classification have been used widely (Paola and Showengerdt 1995; Chan *et al.* 2001; Stefanov *et al.* 2001), while other algorithms such as neural networks are increasingly being adopted (Gamba and Houshmand 2001; Zhang and Foody 2001). Given the high spatial frequency of urban land cover, many studies have incorporated texture measures to aid classification (Dawson and Parsons 1994; Karathanassi *et al.* 2000; Shaban and Dikshit 2001). Similarly, various forms of ancillary data (e.g., map or census data) have been combined with remotely sensed imagery to increase urban classification accuracy (Anys *et al.* 1998; Mesev 1998; Yu *et al.* 1999; Imhoff *et al.* 2000). Finally, multitemporal imagery has been used widely in urban classification to monitor development (Kwarteng and Chavez 1998;

Ridd and Liu 1998; Weng 2001; Yeh and Li 2001), particularly at the urban–rural fringe (Wang 1993; Gao and Skillcorn 1998).

2.3   Comparison of IKONOS and SPOT

Prior to investigating the relative benefits of IKONOS and SPOT HRV imagery for classifying urban areas, it is useful to consider the differing characteristics of these instruments. From a historical perspective, the SPOT satellite series has an established track record, operating successfully since the 1980s. Currently, the fourth SPOT satellite is in orbit. In contrast, the IKONOS satellite, launched in September 1999, is less well established, although it has performed successfully thus far and is intended to operate for a period of at least seven years in the first instance (Table 2.1a).

In terms of urban classification, the most significant features of IKONOS and SPOT HRV are their spatial and spectral properties. As mentioned above, IKONOS has significantly finer spatial resolutions than SPOT HRV for both panchromatic and multispectral image production (Table 2.1b). A further difference, however, is that multispectral IKONOS imagery has four visible and Near-InfraRed (NIR) spectral wavebands, while multispectral SPOT HRV imagery has only three. It should be noted that the HRV sensor was operated on the first three SPOT satellites (SPOT-1, SPOT-2, SPOT-3). The latest SPOT satellite (SPOT-4) carries the High Resolution Visible and Infrared (HRVIR) sensor which has the same three visible and NIR spectral wavebands as the HRV sensor, but also has a Mid-InfraRed (MIR) spectral waveband (Rees 1999). Commonly, however, classification analysis undertaken using SPOT HRVIR imagery only involves the first three (visible and NIR) spectral wavebands. For instance, much analysis involves comparing SPOT-1, -2 or -3 HRV imagery with SPOT-4 HRVIR imagery.

Further IKONOS and SPOT missions are planned, with SPOT-5 scheduled for launch in 2002 and a subsequent IKONOS satellite scheduled for launch around 2005/2006. Each of these satellites carry instruments of increased spatial resolution capabilities compared with their respective predecessors (Table 2.1c). SPOT-5 will carry the High Resolution Geometry (HRG) sensor, capable of generating 2.5-m spatial resolution panchromatic imagery and 10-m spatial resolution multispectral imagery (Couloigner *et al.* 1998). The subsequent IKONOS sensor will provide panchromatic and multispectral imagery with spatial resolutions of 0.5 m and 2 m, respectively.

2.3.1   *Spatial and spectral considerations*

Clearly, the benefit of increasing the spatial resolution of satellite sensor imagery for land cover classification is that finer spatial resolution imagery provides greater detail than coarser spatial resolution imagery. The implica-

Table 2.1a Specifications of the IKONOS and SPOT satellite platforms

Satellite series	IKONOS	SPOT			
Organization	Space Imaging EOSAT	Spot Image			
Country	USA	France			
Website	www.spaceimaging.com	www.spotimage.fr			
Orbit altitude (km)	681	822			
Orbit time (min)	98	101			
Orbit inclination (°)	98.1	98.7			
Orbit type	Sun synchronous	Sun synchronous			
Nodal crossing time	10:30	10:30			
Repeat cycle (days)	140	26			
Satellite name	IKONOS	SPOT-1	SPOT-2	SPOT-3	SPOT-4
Launch date	24 September 1999	22 February 1986	22 January 1990	26 September 1993	24 March 1998
Lifetime (years)	7 (projected)	5	8	3	5 (projected)
Weight (kg)	720	1907	1907	1907	2700
Instrument name	IKONOS	HRV	HRV	HRV	HRVIR

Table 2.1b Specifications of remote sensor instruments on board the IKONOS and SPOT satellites

Instrument name		IKONOS	HRV	HRVIR
Number of instruments		Single	Dual	Dual
Panchromatic mode	Spatial resolution (m)	1	10	10
	Number of channels	1	1	1
	Spectral waveband (ηm)	450–900	510–730	610–680
Multispectral mode	Spatial resolution (m)	4	20	20
	Number of channels	4	3	4
	Spectral wavebands (ηm)	450–520, 520–600, 630–690, 760–900	500–590, 610–680, 790–890	500–590, 610–680, 790–890, 1580–1750
Pan-sharpened data	Spatial resolution (m)	1	NA	10
	Number of channels	4	NA	4
	Spectral wavebands (ηm)	450–520, 520–600, 630–690, 760–900	NA	500–590, 610–680, 790–890, 1580–1750
Positional accuracy (m)		<2	<30	<30
Swathe width (km)		11	60	60
Angle of tilt (°)		±45 across/along track	±27 across track	±27 across track
Stereo acquisition		Yes	Yes	Yes
Revisit frequency (days)		<4	<4	<4
Format of data (bit)		11	8	8
On-board storage		64 Gb	44 min (2 × 22)	83 min (2 × 40 + 3)
Downloading speed (Mbps)		320	50 (2 × 25)	50 (2 × 25)

Table 2.1c Specifications of forthcoming satellites in the IKONOS and SPOT programmes

Satellite series		IKONOS	SPOT
Satellite name		?	SPOT-5
Launch date		2005/2006	2001/2002
Orbital characteristics		Lower than earlier IKONOS satellite	Identical to earlier SPOT satellites
Instrument name		?	HRG
Panchromatic mode	Spatial resolution (m)	0.5	2.5 or 5
	Number of channels	1	1
	Spectral waveband (ηm)	450–900	510–730
Multispectral mode	Spatial resolution (m)	2	10 (20 for 1.58–1.75-μm channel)
	Number of channels	4	4
	Spectral wavebands (ηm)	450–520, 520–600, 630–690, 760–900	500–590, 610–680, 790–890, 1580–1750
Positional accuracy (m)		?	<10
Swathe width (km)		?	60
Tilting/stereo capabilities		Identical to earlier IKONOS instruments	Identical to HRVIR
Revisit frequency (days)		?	<4
Format of data (bit)		11	8
On-board storage		Greater than earlier IKONOS instruments	Greater than HRVIR
Downloading speed (Mbps)		Faster than earlier IKONOS instruments	Faster than HRVIR

tions for classifying urban areas seem fairly obvious. Since urban landscapes tend to comprise numerous small features, increasing the spatial resolution should increase the ability to identify features. Therefore, IKONOS should enable more accurate urban classification than SPOT HRV. However, this assertion raises certain general issues. First, there are limits to the degree to which spatial resolution can be increased. For instance, given the complexity of urban landscapes, even 1-m spatial resolution IKONOS imagery is likely to result in mixed pixels. In response to this problem, various methods of fuzzy classification (Small 2001; Zhang and Foody 2001) and subpixel mapping (Verhoeye and De Wulf 2000; Aplin and Atkinson 2001; Ranchin *et al.* 2001; Tatem *et al.* 2001) have been developed to provide more detailed representations of complex urban scenes.

Paradoxically, under certain circumstances, increasing spatial resolution can lead to a reduction in classification accuracy by increasing within-feature variation (Irons *et al.* 1985; Cushnie 1987). For instance, where the feature of interest is a road, increasing the spatial resolution to a size considerably finer than the road may result in the misclassification of minor features (e.g., road markings) within the road rather than correct classification of the road itself. This problem is associated with traditional per-pixel classification whereby each individual pixel is assigned to a land cover class or classes. Misclassification occurs where individual pixels within features are incorrectly assigned (Pax-Lenney and Woodcock 1997).

A potential solution to both these problems (mixed pixels and within-feature variation) is to use per-parcel classification, whereby land cover classes are assigned to individual land cover parcels (e.g., buildings) rather than pixels. By operating on entire parcels, spuriously misclassified pixels (arising as a result of mixed pixels or within-feature variation) are removed from the classification process (Harris and Ventura 1995; Ryherd and Woodcock 1996). This has significant potential in the context of classifying urban areas since it is common for relatively large proportions of pixels to be misclassified in urban scenes. Commonly, per-parcel classification is performed by integrating remotely sensed imagery and digital cartographic vector data (Ortiz *et al.* 1997; Priestnall and Glover 1998). This method has the added benefit that analysis can be performed within a Geographical Information System (GIS), the preferred means of storing and analysing spatial data by most urban planners. Where required, therefore, GIS-based per-parcel classification can be integrated relatively easily with other urban planning tasks and data. Generally, multispectral imagery provides more land cover information than panchromatic imagery, since each spectral waveband provides specific information about land cover features (Ben-Dor *et al.* 2001; Roessner *et al.* 2001). However, the benefit of increasing the number of spectral wavebands of remotely sensed imagery for urban land cover classification is less straightforward than that of increasing the spatial resolution. Notably, increasing the number of wavebands does not necessarily lead to a uniform increase in land cover information. Commonly, considerable redundancy is present in multispectral imagery, whereby different spectral wavebands contain similar information about the land cover (Mather 1999). This is particularly the case for wavebands from the same sections of the electromagnetic spectrum (e.g., two or more wavebands from either the visible or the NIR section of the spectrum may share land cover information). Multispectral imagery may be considered in terms of spectral dimensionality where each dimension is a different section of the spectrum. Generally, each spectral dimension holds fundamentally different information about the features of interest (Jensen 1996). Consequently, although multispectral IKONOS and SPOT HRV imagery have four and three spectral wavebands respectively, each have two spectral dimensions, visible and NIR. Therefore, it is unlikely that the difference in spectral properties between these two sources of data would lead to significant differences in land cover classification accuracy between the two. In contrast, adding a new dimension to visible/NIR multispectral imagery (such as the MIR waveband of SPOT HRVIR) may lead to significant increases in classification accuracy, although this would depend on other factors such as the particular features of interest.

2.4   Study area and data

A small section of the village of Arundel, West Sussex, UK was selected for urban land cover classification analysis. This study area is predominantly residential and measures 0.3 km by 0.3 km. The major land cover classes present are grassland, woodland, asphalt and rooftiles. These classes were used for subsequent classification analysis.

Three types of data were acquired: (i) Compact Airborne Spectrographic Imager (CASI) imagery (Babey and Anger 1993), (ii) Land-Line vector data and (iii) ground reference data. Multispectral CASI imagery with a spatial resolution of 2 m and nine (visible and NIR) spectral wavebands was acquired in late July 1997 (Figure 2.1a). This was used to simulate IKONOS and SPOT HRV imagery.

Land-Line digital vector data (Figure 2.1b), supplied by the Ordnance Survey, were used for per-parcel classification. These data comprised feature-coded point and line entities (land cover parcel boundaries) and were registered to the British National Grid (BNG). Prior to classification analysis, the Land-Line data were polygonized such that each land cover parcel was identified as a single feature. Ground reference data were acquired in early September 1997 through a combination of land cover surveys, interviews with residents and airborne videography. These three sources of data were combined to generate a comprehensive reference land cover map of the study area for selecting classes and assessing classification accuracy.

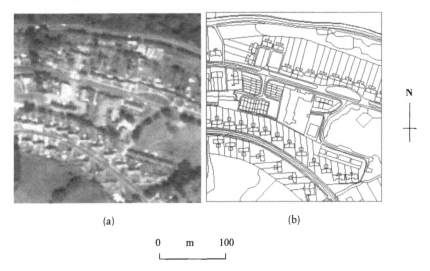

(a)   (b)

0   m   100

Figure 2.1 Arundel study area (a) original 2-m spatial resolution 9-band CASI image and (b) Land-Line vector data.

2.5 Preprocessing

Prior to classification analysis, three preprocessing measures were performed: (i) geometric registration, (ii) spectral merging and (iii) spatial degradation.

2.5.1 *Geometric registration*

To enable integrated analysis between the remotely sensed imagery and the vector data, it was necessary to geometrically register the two data sets. Since the Land-Line data were already registered to the BNG, the imagery was co-registered with the Land-Line data directly rather than registering each data set independently to a common map coordinate system. For this purpose, four well-distributed ground control points (located approximately at the northern, southern, eastern and western extents of the study area) were identified on both data sets. Then, a first-order mathematical transformation function was generated and nearest-neighbour resampling was employed (Lillesand and Kiefer 2000). The Root Mean Square (RMS) error associated with geometric registration was slightly less than 1.2 m.

2.5.2 *Spectral merging*

Nine visible and NIR spectral wavebands of CASI imagery were supplied by the Environment Agency. To simulate multispectral IKONOS and SPOT HRV imagery, it was necessary to merge these nine wavebands to four and

Table 2.2 Spectral merging to simulate IKONOS and SPOT HRV spectral wavebands

Section of electromagnetic spectrum	Original CASI waveband		Simulated IKONOS waveband		Simulated SPOT HRV waveband	
	Number	Spectral λ (ηm)	Number	Spectral λ (ηm)	Number	Spectral λ (ηm)
Blue	1	480–500	1	480–520		
Green	2	500–520			1	500–565*
	3	545–565	2	545–603*		
	4	593–603				
Red	5	660–665	3	660–685	2	660–680
	6	665–680				
	7	680–685				
Near-infrared	8	845–870	4	845–890	3	845–890
	9	870-890				

*Incomplete spectral range.

three wavebands, respectively. First, the spectral characteristics of the desired (output) wavebands were specified. Then, an association was created between the input and output wavebands to determine which of the nine input wavebands should contribute to the output wavebands (Table 2.2). Finally, for each output waveband, a simple model was constructed to average the spectral values of the required input wavebands on a per-pixel basis.

An exact match between the desired and the derived wavebands was not possible because the original CASI image wavebands did not cover all the required parts of the electromagnetic spectrum. Instead, the closest match between the original and the desired wavebands was used to determine the spectral characteristics of the derived wavebands. Although this procedure did not enable an exact simulation of the spectral properties of the IKONOS and SPOT HRV sensors, the resulting wavebands were representative of the desired sections of the electromagnetic spectrum.

2.5.3 Spatial degradation

Originally, the CASI imagery had a spatial resolution of 2 m. To simulate multispectral IKONOS and SPOT HRV imagery, it was necessary to degrade the original imagery to 4 m and 20 m, respectively. A relatively simple method of image degradation was employed whereby the spectral values of the pixels in the original CASI imagery were averaged to derive the spectral values of the pixels in the degraded imagery. This simple technique of image degradation has been practised widely (Heric *et al.* 1996; Cutler

1998), although other investigators have stressed the need to account for additional factors, such as the Signal-to-Noise ratio (SNR) and the Point Spread Function (PSF), in the process (Townshend 1981; Townshend and Justice 1988).

Generally, the SNR of remotely sensed imagery increases as spatial resolution coarsens. Averaging, as a simple means of image degradation, however, increases the SNR by an artificially high value. That is, assuming that image noise is random and normally distributed, averaging will tend to average "out" such noise, by an amount proportional to the level of degradation. For instance, by degrading by a factor of two (e.g., 2-m spatial resolution to 4-m spatial resolution), the level of noise in the degraded image will be four times lower than that of the original image. This means that degraded imagery has an artificially high SNR when compared with imagery with an identical spatial resolution acquired directly by a sensor.

The averaging process also retains the PSF of the original imagery in the degraded imagery, rather than increasing the PSF in proportion to the degree of degradation. This means that the pixels of the degraded imagery approximate more closely a square wave response than those of the original imagery.

As the objective of this study was to identify any benefits of using fine rather than coarse spatial resolution imagery for land cover classification, potential minor sources of error from these two factors (SNR, PSF) merely rendered the results more conservative. In other words, it is likely that the accuracy with which the simulated SPOT HRV imagery (with an artificially high SNR and an artificially small PSF) was used to classify land cover was higher than it would have been using imagery with a lower SNR and a larger PSF (equivalent to real SPOT HRV imagery). Therefore, it is likely that any increase in classification accuracy identified as a result of using fine rather than coarse spatial resolution imagery was an underestimate. Overall, it was believed that the influence of these two factors (SNR, PSF) on the degradation process was relatively minor and the omission of any measures to account for them here did not affect the results significantly. Degradation to 4 m and 20 m was performed on both the simulated IKONOS spectral wavebands and the simulated SPOT HRV spectral wavebands. Consequently, four spatial/spectral combinations of imagery were generated: (i) 4-m spatial resolution 4-band imagery (IKONOS), (ii) 4-m spatial resolution 3-band imagery, (iii) 20-m spatial resolution 4-band imagery and (iv) 20-m spatial resolution 3-band imagery (SPOT HRV).

2.6   Classification

Classification involved three main stages: (i) per-pixel classification, (ii) per-parcel classification and (iii) accuracy assessment. Each one of the four spatial/spectral combinations of imagery was classified.

2.6.1 Per-pixel classification

For each spatial/spectral combination of imagery, class training was performed by selecting a representative sample of each land cover class. The selection of training samples was made with reference to the land cover map. Supervised maximum likelihood per-pixel classification (Thomas *et al.* 1987) was then performed. This was an appropriate classifier since the training classes were normally distributed and discrete. The output of this analysis was four per-pixel classified images, each of which corresponded to one of the spatial/spectral combinations of imagery (e.g., Figures 2.2a and 2.2b, see colour plates).

2.6.2 Per-parcel classification

Following per-pixel classification, the classified images were each integrated with the polygonized Land-Line data for per-parcel classification. Initially, the Land-Line polygons were rasterized. Then, for each classified image, the rasterized Land-Line data and the image were combined to form a single data set. The combined data set was analysed to identify the modal land cover of each parcel. Finally, these modal land cover values were reassigned to entire parcels in the original (non-rasterized) polygon coverage to generate a per-parcel classification. The output at this stage was four per-parcel classifications, one for each of the spatial/spectral combinations of imagery (e.g., Figures 2.2c and 2.2d, see colour plates).

2.6.3 Accuracy assessment

An accuracy assessment was performed on each of the eight classifications by comparing them with the reference land cover map. To enable a straightforward comparison between different classifications, several measures were taken. First, to allow a simple comparison between per-pixel and per-parcel classifications, all accuracy assessments were conducted on a per-pixel basis. That is, each of the per-parcel classifications was rasterized to enable per-pixel accuracy assessment. Although this method had the drawback that two or more points could be selected from the same parcel, it has the advantage that parcels were weighted according to their size (Aplin *et al.* 1999). Further, to maintain a high degree of geometric accuracy in the procedure of assessing the accuracy of the per-parcel classifications, these were rasterized to a relatively small pixel size (0.4 m).

Second, to enable a direct comparison between per-pixel and per-parcel classification accuracy, it was necessary to use a common pixel size for (per-pixel) accuracy assessment. Therefore, the pixel sizes of the per-pixel classified images (4 m, 20 m) were altered to match the 0.4-m pixel size of the rasterized per-parcel classifications. Again, this meant that more than a single point could be selected from the same (original) pixel. However, this

Table 2.3a Confusion matrix of the simulated IKONOS image (4-m resolution 4-band) per-pixel classification

Class (reference) / Class predicted	Grassland	Woodland	Asphalt	Rooftiles	Total
Grassland	81	7	0	12	100
Woodland	48	42	5	5	100
Asphalt	27	8	43	22	100
Rooftiles	47	3	9	41	100
Total	203	60	57	80	400
Producer's accuracy (%)	39.90	70.00	75.44	51.25	
User's accuracy (%)	81.00	42.00	43.00	41.00	

Overall classification accuracy (%) = 51.75

Table 2.3b Confusion matrix of the simulated IKONOS image (4-m resolution 4-band) per-parcel classification

Class (reference) / Class predicted	Grassland	Woodland	Asphalt	Rooftiles	Total
Grassland	94	1	2	3	100
Woodland	20	77	0	3	100
Asphalt	4	2	79	15	100
Rooftiles	19	0	1	80	100
Total	137	80	82	101	400
Producer's accuracy (%)	68.61	96.25	79.21	79.21	
User's accuracy (%)	94.00	77.00	80.00	80.00	

Overall classification accuracy (%) = 82.50

approach, equivalent to the procedure of rasterizing per-parcel classifications, enabled original pixels to be weighted according to their size. This approach also enabled a direct comparison between per-pixel classifications originally at different spatial resolutions.

Third, a straightforward comparison between classes was enabled by selecting a constant number of sample points (100) per class. To summarize the accuracies of the eight classifications, the confusion matrices are presented (Tables 2.3a–h).

Each of the classifications based on 4-m spatial resolution imagery was more accurate than the corresponding classification based on 20-m spatial resolution imagery. The main reason for this was the greater proportion of mixed pixels in the 20-m spatial resolution per-pixel classified images than the 4-m spatial resolution per-pixel classified images. Generally, the features in the study area (e.g., buildings, gardens and roads) were relatively small

Table 2.3c Confusion matrix of the 4-m resolution 3-band image per-pixel classification

Class (reference) Class predicted	Grassland	Woodland	Asphalt	Rooftiles	Total
Grassland	80	8	8	4	100
Woodland	29	66	5	0	100
Asphalt	42	14	33	11	100
Rooftiles	38	4	14	44	100
Total	189	92	60	59	400
Producer's accuracy (%)	42.33	71.74	55.00	74.58	
User's accuracy (%)	80.00	66.00	33.00	44.00	

Overall classification accuracy (%) = 55.75

Table 2.3d Confusion matrix of the 4-m resolution 3-band per-parcel classification

Class (reference) Class predicted	Grassland	Woodland	Asphalt	Rooftiles	Total
Grassland	86	11	0	3	100
Woodland	14	84	1	1	100
Asphalt	39	3	47	11	100
Rooftiles	37	0	0	63	100
Total	176	98	48	78	400
Producer's accuracy (%)	48.86	85.71	97.92	80.77	
User's accuracy (%)	86	84	47	63	

Overall classification accuracy (%) = 70

Table 2.3e Confusion matrix of the 20-m resolution 4-band image per-pixel classification

Class (reference) Class predicted	Grassland	Woodland	Asphalt	Rooftiles	Total
Grassland	85	8	2	5	100
Woodland	25	62	10	3	100
Asphalt	50	2	26	22	100
Rooftiles	63	8	19	10	100
Total	223	80	57	40	400
Producer's accuracy (%)	38.12	77.50	45.61	25.00	
User's accuracy (%)	85.00	62.00	26.00	10.00	

Overall classification accuracy (%) = 45.75

Table 2.3f Confusion matrix of the 20-m resolution 4-band image per-parcel classification

Class (reference) / Class predicted	Grassland	Woodland	Asphalt	Rooftiles	Total
Grassland	90	0	0	10	2
Woodland	9	91	0	0	0
Asphalt	50	2	22	26	1
Rooftiles	66	8	10	16	79
Total	215	101	32	52	82
Producer's accuracy (%)	41.86	90.10	68.75	30.77	
User's accuracy (%)	90.00	91.00	22.00	16.00	

Overall classification accuracy (%) = 54.75

Table 2.3g Confusion matrix of the simulated SPOT HRV image (20-m resolution 3-band) per-pixel classification

Class (reference) / Class predicted	Grassland	Woodland	Asphalt	Rooftiles	Total
Grassland	87	5	3	5	100
Woodland	42	48	9	1	100
Asphalt	46	23	20	11	100
Rooftiles	44	3	22	31	100
Total	219	79	54	48	400
Producer's accuracy (%)	39.73	60.76	37.04	64.58	
User's accuracy (%)	87	48	20	31	

Overall classification accuracy (%) = 46.5

Table 2.3h Confusion matrix of the simulated SPOT HRV image (20-m resolution 3-band) per-parcel classification

Class (reference) / Class predicted	Grassland	Woodland	Asphalt	Rooftiles	Total
Grassland	95	2	0	3	100
Woodland	39	57	3	1	100
Asphalt	43	20	27	10	100
Rooftiles	56	0	17	27	100
Total	233	79	47	41	400
Producer's accuracy (%)	40.77	72.15	57.45	65.85	
User's accuracy (%)	95	57	27	27	

Overall classification accuracy (%) = 51.5

and, importantly, considerably smaller than 20 m. Therefore, the finer spatial resolution imagery resulted in fewer mixed pixels and, consequently, less misclassification than the coarser spatial resolution imagery.

The effect on classification accuracy of varying the spectral properties of the imagery was less straightforward than that of varying the spatial properties. In fact, for per-pixel classification, the 3-band imagery (representing the spectral properties of SPOT HRV) was more accurate than the 4-band imagery (representing the spectral properties of IKONOS), while the opposite was true for per-parcel classification. Generally, the differences in classification accuracy arising from the two spectral waveband combinations were relatively insignificant (<4 per cent). The only exception to this was per-parcel classification based on the 4-m spatial resolution imagery. In this case the 4-band imagery generated a classification accuracy 12.5 per cent greater than that of the 3-band imagery. These two classifications were also more accurate than all other classifications by a considerable margin (>15 per cent). One possible explanation for this is that the effect of varying spectral properties on classification accuracy is only significant where other variables are conducive to classification. That is, the accuracy of classifying 20-m spatial resolution imagery and/or classifying on a per-pixel basis was relatively low (around 50 per cent) such that altering the spectral properties had little impact. In contrast, the accuracy of classifying 4-m spatial resolution imagery on a per-parcel basis was relatively high. In this case, the additional information provided by a fourth spectral band resulted in a significant increase in classification accuracy.

Each of the per-pixel classifications was less accurate than the corresponding per-parcel classification. This increase in accuracy at the per-parcel stage was due to the removal of spuriously misclassified pixels present at the per-pixel stage.

The simulated IKONOS imagery (4-m spatial resolution 4-band imagery) was more accurate than the simulated SPOT HRV imagery (20-m spatial resolution 3-band imagery, respectively) for both per-pixel and per-parcel classification. Overall, per-parcel classification of the simulated IKONOS imagery was the most accurate (82.5 per cent). Per-pixel classification of the simulated SPOT HRV imagery was the second least accurate (46.5 per cent), marginally more accurate than per-pixel classification of the 20-m spatial resolution 4-band imagery (45.75 per cent).

2.7 Conclusions

In the context of the current investigation, four clear conclusions can be drawn. For land cover classification of grassland, woodland, asphalt and rooftiles in a complex residential urban study area:

- 4-m spatial resolution imagery was more accurate than 20-m spatial resolution imagery;

- 4-band imagery was more accurate than 3-band imagery where other factors (spatial resolution, classification method) were conducive to classification;
- per-parcel classification was more accurate than per-pixel classification; and
- simulated IKONOS imagery was more accurate than simulated SPOT HRV imagery.

As mentioned above, caution should be exercised when attempting to draw broad conclusions from specific urban studies, given the variability of urban landscapes. However, the general implications of this study are clear. The greater detail provided by IKONOS imagery over SPOT HRV imagery (in terms of the number of spectral wavebands, but in particular the spatial resolution) enables considerably smaller urban features to be classified accurately. Welch's (1982) recommendation for fine spatial resolution imagery to be used in urban studies has finally been met by third generation satellite sensors. This chapter has provided a starting point for quantifying the capabilities of third generation instruments for urban classification. Further analysis is now required to verify these findings using real satellite sensor data and a variety of urban landscapes.

2.8   Acknowledgements

The preliminary stages of this research were conducted at the Department of Geography, University of Southampton. The support of Peter Atkinson and Paul Curran of the University of Southampton was greatly appreciated. Thanks, too, to the Ordnance Survey and the Environment Agency for supplying the Land-Line data and the CASI imagery, respectively. An earlier version of this chapter was presented at the Remote Sensing Society's Annual Conference in 2000 (Aplin 2000).

2.9   References

Anys, H., Bannari, A., He, D.-C. and Morin, D., 1998, Cartographie des zones urbaines à l'aide des images aéroportées MEIS-II. *International Journal of Remote Sensing*, 19, 883–94.

Aplin, P., 2000, Comparison of simulated IKONOS and SPOT HRV imagery for land cover classification. Paper given at Conference of the Remote Sensing Society, The Remote Sensing Society, Nottingham, CD ROM.

Aplin, P. and Atkinson, P. M., 2001, Sub-pixel land cover mapping for per-field classification. *International Journal of Remote Sensing*, 22, 2853–8.

Aplin, P., Atkinson, P. M. and Curran, P. J., 1997, Fine spatial resolution satellite sensors for the next decade. *International Journal of Remote Sensing*, **18**, 3873–81.

Aplin, P., Atkinson, P. and Curran, P., 1999, Per-field classification of land use using the forthcoming very fine spatial resolution satellite sensors: Problems and potential solutions. In P. M. Atkinson and N. J. Tate (eds) *Advances in Remote Sensing and GIS Analysis* (Chichester, UK: John Wiley & Sons), pp. 219–39.

Babey, S. K. and Anger, C. D., 1993, Compact airborne spectrographic imager (CASI): A progress review. *SPIE*, **1937**, 152–63.

Barnsley, M. J., Moller-Jensen, L. and Barr, S. L., 2001, Inferring urban land use by spatial and structural pattern recognition. In J-P. Donnay, M. J. Barnsley and P. A. Longley (eds) *Remote Sensing and Urban Analysis* (London: Taylor and Francis), pp. 115–44.

Ben-Dor, E., Levin, N. and Saaroni, H., 2001, A spectral based recognition of the urban environment using the visible and near-infrared spectral region (0.4–1.1 µm). A case study over Tel-Aviv, Israel. *International Journal of Remote Sensing*, **22**, 2193–218.

Bjorgo, E., 2000, Using very high spatial resolution multispectral satellite sensor imagery to monitor refugee camps. *International Journal of Remote Sensing*, **21**, 611–16.

Caetano, M., Navarro, A. and Santos, J. P., 1997, Improving urban areas mapping with satellite imagery by contextual analyses and integration of a road network map. Paper given at Conference of the Remote Sensing Society, Nottingham, pp. 106–11.

Chan, J. C.-W., Chan, K.-P. and Yeh, A. G-O., 2001, Detecting the nature of change in an urban environment: A comparison of machine learning algorithms. *Photogrammetric Engineering and Remote Sensing*, **67**, 213–25.

Chen, S., Zeng, S. and Xie, C., 2000, Remote sensing and GIS for urban growth analysis in China. *Photogrammetric Engineering and Remote Sensing*, **66**, 593–8.

Cortijo, F. J. and De La Blanca, N. P., 1998, Improving classical contextual classifications. *International Journal of Remote Sensing*, **19**, 1591–613.

Couloigner, I., Ranchin, T., Valtonen, V. P. and Wald, L., 1998, Benefits of the future SPOT-5 and of data fusion to urban roads mapping. *International Journal of Remote Sensing*, **19**, 1519–32.

Cushnie, J. L., 1987, The interactive effect of spatial resolution and degree of internal variability within land cover types on classification accuracies. *International Journal of Remote Sensing*, **8**, 15–29.

Cutler, M. E. J., 1998, Assessing variation in the relationships between remotely sensed data and canopy chlorophyll composition. Unpublished Ph.D. thesis, University of Southampton, UK.

Dare, P. M. and Fraser, C. S., 2001, Mapping informal settlements using high resolution satellite imagery. *International Journal of Remote Sensing*, **22**, 1399–401.

Dawson, B. R. P. and Parsons, A. J., 1994, Texture measures for the identification and monitoring of urban derelict land. *International Journal of Remote Sensing*, **15**, 1259.

Donnay, J-P., Barnsley, M. J. and Longley, P. A. (eds), 2001, Remote sensing and urban analysis. *Remote Sensing and Urban Analysis* (London: Taylor and Francis), pp. 3–18.

Forster, B. C., 1983, Some urban measurements from Landsat data. *Photogrammetric Engineering and Remote Sensing*, **49**, 1693–707.

Fortier, M-F. A., Ziou, D., Armenakis, C. and Wang, S., 2001, Automated correction and updating of road databases from high-resolution imagery. *Canadian Journal of Remote Sensing*, **27**, 76–89.

Fowler, M. J. F., 2001, Cover. A high-resolution satellite image of archaeological features to the south of Stonehenge. *International Journal of Remote Sensing*, **22**, 1167–71.

Franklin, S. E., Wulder, M. A. and Gerylo, G. R., 2001, Texture analysis of IKONOS panchromatic data for Douglas-fir forest age class separability in British Columbia. *International Journal of Remote Sensing*, **22**, 2627–32.

Fritz, L. W., 1996, The era of commercial earth observation satellites. *Photogrammetric Engineering and Remote Sensing*, **62**, 39–45.

Gamba, P. and Houshmand, B., 2001, An efficient neural classification chain of SAR and optical urban images. *International Journal of Remote Sensing*, **22**, 1535–53.

Gao, J. and Skillcorn, D., 1998, Capability of SPOT XS data in producing detailed land cover maps at the urban-rural periphery. *International Journal of Remote Sensing*, **19**, 2877–91.

Guindon, B., 2001, Application of perceptual grouping concepts to the recognition of residential buildings in high resolution satellite images. *Canadian Journal of Remote Sensing*, **27**, 264–75.

Harris, P. M. and Ventura, S. J., 1995, The integration of geographic data with remotely sensed imagery to improve classification in an urban area. *Photogrammetric Engineering and Remote Sensing*, **61**, 993–8.

Heric, M., Lucas, C. and Devine, C., 1996, The Open Skies Treaty: Qualitative utility evaluations of aircraft reconnaissance and commercial satellite imagery. *Photogrammetric Engineering and Remote Sensing*, **62**, 279–84.

Imhoff, M. L., Tucker, C. J., Lawrence, W. T. and Stutzer, D. C., 2000, The use of multisource satellite and geospatial data to study the effect of urbanization on primary productivity in the United States. *IEEE Transactions on Geoscience and Remote Sensing*, **38**, 2549–56.

Irons, J. R., Markham, B. L., Nelson, R. F., Toll, D. L., Williams, D. L., Latty, R. S. and Stauffer, M. L., 1985, The effects of spatial resolution on the classification of Thematic Mapper data. *International Journal of Remote Sensing*, **6**, 1385–403.

Jensen, J. R., 1996, *Introductory Digital Image Processing* (2nd edn) (Upper Saddle River, NJ: Prentice Hall).

Jensen, J. R. and Cowen, D. C., 1999, Remote sensing of urban/suburban infrastructure and socio-economic attributes. *Photogrammetric Engineering and Remote Sensing*, **65**, 611–22.

Ji, C. Y., Liu, Q., Sun, D., Wang, S., Lin, P. and Li, X., 2001, Monitoring urban expansion with remote sensing in China. *International Journal of Remote Sensing*, **22**, 1441–55.

Karathanassi, V., Iosifidis, C. H. and Rokos, D., 2000, A texture-based classification method for classifying built areas according to their density. *International Journal of Remote Sensing*, **21**, 1807–23.

Key, T., Warner, T. A., McGraw, J. B. and Fajvan, M. A., 2001, A comparison of multispectral and multitemporal information in high spatial resolution imagery for classification of individual tree species in a temperate hardwood forest. *Remote Sensing of Environment*, **75**, 100–12.

Kontoes, C. C., Raptis, V., Lautner, M. and Oberstadler, R., 2001, The potential of kernel classification techniques for land use mapping in urban areas using 5m-spatial resolution IRS-1C imagery. *International Journal of Remote Sensing*, 21, 3145–51.

Kwarteng, A. Y. and Chavez, P. S., 1998, Change detection study of Kuwait City and environs using multi-temporal Landsat Thematic Mapper data. *International Journal of Remote Sensing*, 19, 1651–62.

Lillesand, T. M. and Kiefer, R. W., 2000, *Remote Sensing and Image Interpretation* (4th edn) (New York: John Wiley & Sons).

McDonald, R. A., 1995, Opening the cold war sky to the public: Declassifying satellite reconnaissance imagery. *Photogrammetric Engineering and Remote Sensing*, 61, 385–90.

Martin, L. R. G., Howarth, P. J. and Holder, G., 1988, Multispectral classification of land use at the urban-rural fringe using SPOT satellite data. *Canadian Journal of Remote Sensing*, 14, 72–9.

Masek, J. G., Lindsay, F. E. and Goward, S. N., 2000, Dynamics of urban growth in the Washington DC metropolitan area, 1973–1996, from Landsat observations. *International Journal of Remote Sensing*, 21, 3473–86.

Mather, P. M., 1999, *Computer Processing of Remotely Sensed Images* (2nd edn) (Chichester, UK: John Wiley & Sons).

Mesev, V., 1998, The use of census data in urban image classification. *Photogrammetric Engineering and Remote Sensing*, 64, 431–8.

Mesev, V., Gorte, B. and Longley, P. A., 2001, Modified maximum likelihood classification algorithms and their application to urban remote sensing. In J-P. Donnay, M. J. Barnsley and P. A. Longley (eds) *Remote Sensing and Urban Analysis* (London: Taylor and Francis), pp. 71–94.

Ortiz, M. J., Formaggio, A. R. and Epiphanio, J. C. N., 1997, Classification of croplands through integration of remote sensing, GIS and historical database. *International Journal of Remote Sensing*, 18, 95–105.

Paola, J. D. and Showengerdt, R. A., 1995, A detailed comparison of backpropagation neural network and maximum-likelihood classifiers for urban land use classification. *IEEE Transactions on Geoscience and Remote Sensing*, 33, 981.

Pax-Lenney, M. and Woodcock, C. E., 1997, The effect of spatial resolution on the ability to monitor the status of agricultural lands. *Remote Sensing of Environment*, 61, 210–20.

Priestnall, G. and Glover, R., 1998, A control strategy for automated land use change detection: An integration of vector-based GIS, remote sensing and pattern recognition. In S. Carver (ed.) *Innovations in GIS 5* (Bristol: Taylor and Francis), pp. 162–75.

Ranchin, T., Wald, L. and Mangolini, M., 2001, Improving the spatial resolution of remotely sensed images by means of sensor fusion: A general solution using the ARSIS method. In J-P. Donnay, M. J. Barnsley and P. A. Longley (eds) *Remote Sensing and Urban Analysis* (London: Taylor and Francis), pp. 21–37.

Ridd, M. K. and Liu, J., 1998, A comparison of four algorithms for change detection in an urban environment. *Remote Sensing of Environment*, 63, 95–100.

Roessner, S., Segl, K., Heiden, U. and Kaufmann, H., 2001, Automated differentiation of urban surfaces based on airborne hyperspectral imagery. *IEEE Transactions on Geoscience and Remote Sensing*, 39, 1525–32.

Ryherd, S. and Woodcock, C., 1996, Combining spectral and texture data in the segmentation of remotely sensed images. *Photogrammetric Engineering and Remote Sensing*, 62, 181–94.

Shaban, M. A. and Dikshit, O., 2001, Improvement of classification in urban areas by the use of textural features: The case study of Lucknow city, Uttar Pradesh. *International Journal of Remote Sensing*, 22, 565–93.

Small, C., 2001, Estimation of urban vegetation abundance by spectral mixture analysis. *International Journal of Remote Sensing*, 22, 1305–34.

Stefanov, W. L., Ramsey, M. S. and Christensen, P. R., 2001, Monitoring urban land cover changes: An expert system approach to land cover classification of semiarid to arid urban centers. *Remote Sensing of Environment*, 77, 173–85.

Tanaka, S. and Sugimura, T., 2001, Cover. A new frontier of remote sensing from IKONOS images. *International Journal of Remote Sensing*, 22, 1–5.

Tatem, A. J., Lewis, H. G., Atkinson, P. M. and Nixon, M. S., 2001, Super-resolution target identification from remotely sensed images using a Hopfield neural network. *IEEE Transactions on Geoscience and Remote Sensing*, 39, 781–96.

Thomas, I. L., Benning, V. M. and Ching, N. P., 1987, *Classification of Remotely Sensed Images* (Bristol: Adam Hilger).

Toll, D. L., 1985, Effect of Landsat Thematic Mapper sensor parameters on land cover classification. *Remote Sensing of Environment*, 17, 129–40.

Townshend, J. R. G., 1981, The spatial resolving power of Earth resources satellites. *Progress in Physical Geography*, 5, 32–55.

Townshend, J. R. G. and Justice, C. O., 1988, Selecting the spatial resolution of satellite sensors required for global monitoring of land transformations. *International Journal of Remote Sensing*, 9, 187–236.

Verhoeye, J. and De Wulf, R., 2000, Sub-pixel mapping of Sahelian wetlands using multi-temporal SPOT VEGETATION images. Paper given at Conference of the 28th International Symposium on Remote Sensing of Environment, Information for Sustainable Development. (Cape Town: CSIR Satellite Applications Centre), CD-ROM, Category 4, pp. 14–19.

Wang, F., 1993, A knowledge-based vision system for detecting land changes at urban fringes. *IEEE Transactions on Geoscience and Remote Sensing*, 31, 136.

Welch, R., 1982, Spatial resolution requirements for urban studies. *International Journal of Remote Sensing*, 3, 139–46.

Weng, Q., 2001, A remote sensing-GIS evaluation of urban expansion and its impact on surface temperature in the Zhujiang Delta, China. *International Journal of Remote Sensing*, 22, 1999–2014.

Wyatt, B. K., Greatorex-Davies, J. N., Hill, M. O., Parr, T. W., Bunce, R. G. H. and Fuller, R. M., 1993, *Comparison of land cover definitions* (Institute of Terrestrial Ecology report to the Department of the Environment) (Cambridge: Institute of Terrestrial Ecology).

Yeh, A. G.-O. and Li, X., 2001, Measurement and monitoring of urban sprawl in a rapidly growing region using entropy. *Photogrammetric Engineering and Remote Sensing*, 67, 83–90.

Yu, S., Berthod, M. and Giraudon, G., 1999, Towards robust analysis of satellite images using map information – application to urban area detection. *IEEE Transactions on Geoscience and Remote Sensing*, 37, 1925–39.

Zhang, J. and Foody, G. M., 2001, Fully-fuzzy supervised classification of sub-urban land cover from remotely sensed imagery: Statistical and artificial neural network approaches. *International Journal of Remote Sensing*, 22, 615–28.

Zhou, G. and Li, R., 2000, Accuracy evaluation of ground points from IKONOS high-resolution satellite imagery. *Photogrammetric Engineering and Remote Sensing*, 66, 1103–12.

3 Resolution convergence

A comparison of aerial photos, LIDAR and IKONOS for monitoring cities

Emmanuel P. Baltsavias and Armin Gruen

3.1 Introduction

Remotely sensed data are data taken from above the Earth's surface. They offer a global coverage, with variable spatial, radiometric, spectral and temporal resolutions, and are the major source of geo-spatial information. The importance of cities and their structural complexity and continuous change make the use of such data even more necessary. In this chapter, we will limit ourselves to remotely sensed sensors that can facilitate the extraction of basic geo-spatial information, such as Digital Terrain Models (DTMs), Digital Surface Models (DSMs) and orthoimages, In addition, we will examine the identification of urban objects, such as buildings, roads, vegetation, etc., as well as the mapping of entire cities and the rudiments of three-dimensional city modelling. We will restrict the discussions mainly to imaging sensors with spatial resolutions of up to around 1 m, as well as active systems such as LIDAR (Laser-Induced Detection And Ranging); while other sensors, such as SAR (Synthetic Aperture Radar), thermal and hyperspectral will be mentioned only briefly. We also acknowledge that lower spatial resolution satellite imagery have been often used and are still valuable for various urban applications.

3.2 Sensors

In the last decade we have seen a continuous development of new sensors, which is only expected to increase in the future. These sensors offer a variety of spatial, radiometric, spectral and temporal resolutions, as well as variable spatial and spectral coverage. In addition, there are differences in their data processing, available commercial processing systems, production through-put and costs. We will consider sensors within three categories: (i) airborne cameras, (ii) airborne LIDAR and (iii) high spatial resolution spaceborne optical sensors.

3.2.1   Airborne cameras

Airborne photogrammetric film cameras have not seen any significant new developments other than their increased use in combination with GPS (Global Positioning Systems) for: (i) the reduction of ground control points in aerial triangulation and (ii) in aircraft navigation. Although integrated GPS/INS (Inertial Navigation System) systems are considered too expensive by most users, they are still seen as the workhorses for routine photogrammetric map production and will continue to be so in the near future, especially for large-scale mapping. However, since 1999 the major photogrammetric system manufacturers took a step toward a fully digital production chain by announcing new digital photogrammetric aerial cameras. Examples include the ADS40 by LH Systems (Sandau *et al.* 2000; Fricker 2001) and the digital modular camera (DMC) by Z/I Imaging (Hinz *et al.* 2001). Lesser known is the Three-Line-Scanner (TLS) by Starlabo, Japan (Starlabo 2002), and the high resolution stereo camera (HRSC) (Types A, AX and AXW), a high spatial resolution digital photogrammetric camera based on linear CCDs (Charge-Coupled Devices) with a variable number of CCD lines, FOV (Field Of View) and radiometric resolution (Neukum 1999; Neukum *et al.* 2001; Lehmann 2001). The latter, developed by German Aerospace Centre (DLR) and used extensively by ISTAR in France, was the first non-commercial, operational digital photo-grammetric camera. A major difference between these cameras is that while DMC uses area CCDs the other systems employ linear CCD technology.

All the systems offer multispectral capabilities as default or as an option. The question of which technology (linear or area CCDs) is better has been the topic of heated debates. Clearly, both have advantages and disadvan-tages, but in the opinion of the authors, under current conditions, the use of linear CCDs is preferable. ADS40 and DMC target applications requiring resolutions from 15–20 cm to 1 m, while TLS and HRSC in some applica-tions fulfil requirements down to 3 cm. The most important technical specifications of the digital airborne photogrammetric cameras are listed in Table 3.1. Other digital aerial cameras with smaller formats (e.g., 4K by 4K pixels) have been developed but are still either experimental prototypes or systems with reduced use within one organization (Thom and Souchon 1999; Toth 1999). Lower resolution CCDs, video or even small-format film imagery have occasionally been used for mostly thematic applications requiring low-cost and tolerating low accuracy (e.g., in forestry).

The new digital photogrammetric cameras need further development, testing, fine-tuning and appropriate processing software. The authors estimate that the first mature systems (hardware and software) will be in place in 2003–2004. Although their price is quite high (two to three times more than comparative film cameras) it is expected that such digital cameras, after initial market hesitation, will begin to overtake sales of film-based alternatives, and eventually replace them. The major advantages

Table 3.1 Main technical specifications of digital airborne photogrammetric cameras. Pixel footprint and swathewidth are calculated for 2,000 ft flying height over ground. For the ADS40, the values in parentheses refer to the single CCD lines (two of them comprise a staggered one). The position of the multispectral (MS) lines on the focal plane with respect to the nadir differs for each system

Model	CCD type	Focal length (mm)	Number of pixels	Pixel size (μm)	Ground pixel size (m)	FOV across/along (deg)	Swathe width (km)	Stereo angles (deg)	PAN channels	MS channels	Bits int/ext
DMC	Area	120 PAN 25 MS	13,500 × 8,000 PAN 3,000 × 2,000 MS	12	0.20	74 × 44	2.7	NA, variable	NA	RGB, NIR	12
ADS40	Line	62.5	24,000 (12,000)/ 12,000	6.5	0.21	64	2.4	28.3/4.1	3(6)	RGB, NIR, +2 optional	14
HRSC-A	Line	175	5,184	7	0.08	11.8	0.4	±18.9	5	RGB, NIR	8
HRSC-AX	Line	150	12,000	6.5	0.09	29.1	1.1	±20.5	5	RGB, NIR	12
HRSC-AXW	Line	47	12,000	6.5	0.28	79.4	3.3	±14.4	3	RGB	12
TLS	Line	60	10,200	7	0.23	61.5	2.4	±21	3	RGB	12/8+

of such systems, apart from digital processing, are: the simultaneous image acquisition of both panchromatic and multispectral images (with up to six channels for the ADS40) by one sensor; higher radiometric quality, especially in dark areas; a full coverage of each flight strip with three panchromatic images and possibly additional multispectral ones (especially for linear CCD systems); and that data redundancy can be favourably exploited in developing more accurate and robust processing methods (e.g., for automatic DSM and DTM generation).

3.2.2   Airborne LIDAR

LIDAR data have exploded on to the remote sensing scene since the mid-1990s. After a period of scientific investigations and technological developments in the late 1980s and early/mid-1990s, new system producers and service providers have emerged at the end of the millennium, with currently around seventy firms being active in the field (Airborne Laser Mapping 2002). New system manufacturers have emerged, while older systems are continuously being improved (e.g., at higher altitude, greater repetition, greater sampling intensity in generating images, multiple signal echoes, etc.). Flood (2001) gives a short development overview, describing the current situation and establishing a prognosis for the future. Table 3.2 shows typical values of the most important parameters of current LIDAR systems. Details of each system can be found at manufacturers' websites through links at Airborne Laser Mapping (2002). LIDAR data are employed in a variety of different applications, ranging from classical DTM generation to specific "killer" applications, such as the mapping of power lines. LIDAR is not only competition for conventional aerial cameras and photogrammetry but is also a complement in opening up new applications, for instance, in combining with other sensors (aerial film cameras, multispectral linear CCDs, high spatial resolution area-based CCDs, etc.). LIDAR firms are mainly divided among a handful of system manufacturers with many service providers, who own or lease commercial systems or build customary systems, offering data processing (at least preprocessing) with a combination of commercial and in-house software. Customers vary from large mapping agencies and private firms to smaller public and private organizations, but although they examine quality control they conduct little or no processing. Commercial packages for laser data processing are scarce and the main photogrammetric and remote sensing software systems offer little or nothing (see Chapter 4 in this book by Barnsley *et al.* on LIDAR applications).

3.2.3   High spatial resolution spaceborne sensors

High spatial resolution satellites, after many delays, abandoned plans and failed launches, finally started in September 1999 with IKONOS-2 (www.spaceimaging.com) (see Chapter 2 in this book by Aplin). Other

Table 3.2 Overview of major technical parameters of airborne LIDAR systems, excluding profilers and bathymetric lasers (h = flying height over ground)

	Typical values
Scan angle (deg)	20–40
Pulse rate (kHz)	5–35
Scan rate (Hz)	25–40
Flying height (h) (m)	200–300 (H), 500–2,000 (A)[1]
GPS frequency (Hz)	1–2
INS frequency (Hz)	50
Beam divergence (mrad)	0.3–2
Laser footprint (m)	0.3–2 ($h = 1,000$ m)
Number of echoes per pulse	2–5
Swathe width (m)	0.35–0.7
Across-track spacing (m)	0.5–2
Along-track spacing (m)	0.3–1
Angle precision (roll, pitch/heading) (deg)	0.02–0.04/0.03–0.05
Range accuracy (cm)	5–15
Height accuracy (cm)	15–20
Planimetric accuracy (m)	0.3–1

[1] H = helicopter, A = airplane.

such systems successfully launched include EROS-A1 in December 2000 (www.imagesatintl.com) and QuickBird-2 in October 2002 (www.digital globe.com), while many more are planned for the near future. The older Russian SPIN-2 KVR-1000 imagery (panchromatic, 1–2-m spatial resolution) are not widely used and do not allow stereo processing. Recently, two Russian systems termed DK-1 and DK-2 provide 1-m and 1.5-m spatial resolution imagery but there is very little known about the platform and sensor, including whether it is film-based or electronic (Petrie 2002). IKONOS, EROS and QuickBird have a panchromatic channel with 1-m, 1.8-m (and 1-m with interpolation) and 0.61-m spatial resolutions, respectively. All three have a single camera head and acquire imagery in different directions by rapidly rotating the whole satellite body. IKONOS and QuickBird have very similar cameras, with red, blue, green and infrared multispectral channels in the range 450–900 nm and spatial resolution four times less than their panchromatic channel. In this chapter, EROS satellites will not be considered: as experience with data are limited, there is nothing reported on stereo processing as far as the authors are aware, radiometric quality and spatial resolution are lower that those of IKONOS; and images are only panchromatic. QuickBird currently provides the highest spatial resolution but is too new to be treated here. The US government has already issued licences for commercial systems with 0.5-m spatial resolutions, and these are expected in the near future. In spite of initial high expectations, the use of high spatial resolution satellite imagery has been, at least up to now, limited. Main buyers remain governments and the military. Availability of data and global coverage are still low, prices are extremely high and in most cases not competitive with aerial imagery, delivery times are long, while users have practically no control over important imaging parameters like acquisition date, sensor elevation and azimuth, weather conditions, etc., which significantly influence image quality. Stereo images were only sold to governments and national mapping agencies up until recently, while Space Imaging does not disclose the sensor model of IKONOS. Some positive steps include:

- the provision of Rational Polynomial Coefficients (RPCs) for object-to-image transformation for products other than stereo images (OrthoKit);
- agreement between Space Imaging and manufacturers of commercial systems (ERDAS Imagine, LH Systems Socet Set, Z/I Imaging ImageStation, PCI Geomatics OrthoEngine) to support IKONOS imagery for import, stereo viewing and processing, and orthoimage generation, etc. by using RPCs;
- selling of stereo images to all;
- vigorous competition between an increasing number of commercial systems may lead to better products and lower prices.

3.2.4 GPS/INS systems

Although not a primary sensor, the development of GPS/INS systems has had a tremendous impact. Use of airborne sensors without frame geometry, including LIDAR, would not be possible, practically, without the use of GPS/INS systems, or for lower accuracy applications, arrays of GPS antennas. Even satellites use such systems for orientation in addition to stellar trackers. Among the available commercial systems, Applanix of Canada holds the lion's share in the market and its position and orientation systems (POS) are used for most airborne sensors (both cameras and LIDAR) with high-accuracy requirements. The use of GPS/INS has led to reduced control information requirements. However, for maximum geometric accuracy some control points are still necessary (e.g., for determination of systematic errors in system calibration and camera geometry during aerial triangulation). Finally, GPS/INS has further facilitated sensor integration, in particular the integration of LIDAR with imaging cameras.

3.3 Requirements for urban mapping and applications

3.3.1 General considerations

Urban objects and applications are extremely variable and have very specific data requirements. Even when considering a single object (e.g., a building), the requirements can vary greatly from, for example, a 2-m accurate geometric modelling for telecommunication applications to a 1-dm accurate model with detailed roofs and texture mapping for architectural applications. This situation becomes even more complex as user requirements become stricter. The main sensor parameters relating to these requirements are spatial, spectral and temporal resolution. Radiometric resolution plays by far the minor role, while area coverage relates mostly to costs and data acquisition duration. Spatial resolution for digital imaging sensors is not identical to the pixel footprint. It has the ability to discern an object of a given size, which in turn is influenced by a number of factors from atmospheric effects, the total modulation transfer system (MTF) of the imaging system, the noise of the electronics, the object's reflectance characteristics and those of its neighbours, to the length and orientation of a linear feature or the quality and calibration of the used monitor. However, for reasons of simplicity we will assume these two terms are identical. Spectral resolution relates to the number of multispectral channels, the centre and range of each band and the curve of sensor sensitivity over the spectrum for each channel. Temporal resolution relates to the maximum possible data acquisition frequency. Taken together, these three parameters can rarely be optimized simultaneously, especially for digital sensors, in order to avoid excessive data volume. Usually, one or two of them are sacrificed in favour of the other. So, with spaceborne sensors high temporal or spectral resolution invariably means a lower spatial resolution. This "rule" does not apply to film cameras which can have both high spatial and temporal resolution but are limited to three spectral channels.

3.3.2 Temporal resolution

The mapping of urban objects generally requires a temporal resolution in the order of 1–10 years. This requirement is more than adequately fulfilled by high spatial resolution spaceborne sensors. For example, IKONOS has a revisit capability of around three days, while smaller sensor elevations permit revisiting down to one day (although at the expense of lower spatial resolution and much more pronounced occlusions). Emergency cases and disasters may need an immediate response of up to a few days, calling for the deployment of an airborne system, mainly camera-based. Traffic and parking studies may need continuous observations of at least in the range of minutes. They are mostly dealt with by stationary systems, although in some cases airborne cameras, mainly on board helicopters, have been used.

3.3.3 Spectral resolution

Most urban applications involving object measurement and geometric modelling may be facilitated by panchromatic images, certainly for manual processing of imagery. Colour improves interpretability (and therefore measurement in manual processing) but more importantly contributes to accurate and reliable automated processing from DSM generation to extraction of roads and buildings. Colour is also of course essential in thematic land use/land cover mapping; where TIR (Thermal InfraRed) is useful for heat island detection, pipe leaks, environmental assessment, energy demand and conservation applications, some disaster and emergency cases, etc. However, neither high spatial resolution satellite imagery nor aerial film or digital cameras provide MIR (Middle InfraRed) and TIR channels with the exception of ADS40. Airborne hyperspectral systems may provide a wide and finely resolved spectral coverage but they are not widespread and are only used for specific applications, such as water quality, detection of roof-type material, etc. So, airborne and spaceborne cameras tend to offer similar capabilities, commonly four channels for digital systems and three for film in the VIS (VISible) and NIR (Near-InfraRed) region. This convergence is natural as digital systems, whether on aircraft or satellite, use common CCD technology, mostly linear CCDs. LIDAR, being monochromatic with a very narrow spectral range, offers the least spectral capabilities.

3.3.4   *Spatial resolution*

Regarding spatial resolution, the requirements again vary greatly from, for example, sub-dm-level for cadastral applications to 20–100 m for coarse land use/land cover mapping. High spatial resolution satellite imagery with 1-m resolution, and even more so 0.61 m, can tackle the requirements of many applications, but clearly not all. Through the resolution convergence of airborne and spaceborne systems, an increased competition in the applications requiring 0.5-m to 1-m spatial resolution will be observed, with each sensor category preserving its particular strengths. Therefore, it is envisaged that in developed countries airborne data will be more widely used compared with developing countries or small and remote urban areas, where spaceborne data may be more suitable. Jensen and Cowen (1999) discuss in detail different urban objects and applications and the associated requirements regarding temporal, spectral and spatial resolutions. Despite the emergence of new applications and higher requirements their studies still have validity and value today.

3.4   Suitability and comparison of sensors for various urban applications and products

3.4.1   *DSM and DTM generation*

DSM and DTM generation from aerial film imagery is well established. Manual measurement of film or digital imagery can be very accurate but is time-consuming and laborious. Automatic image matching for DSM generation is much faster but results include a significant number of blunders that require manual editing. Matching results degrade and manual editing time rises with increasing scale, higher terrain relief and more three-dimensional above-ground objects, especially buildings and vegetation. Under good circumstances, matching delivers an RMS (Root Mean Square) accuracy of 0.5–1.5 pixels, but existing blunders can be tens of metres, especially when surface discontinuities, like building outlines, are smoothed. Matching methods of commercial systems have not advanced far during the 1990s, and all methods perform a final matching using only two images. A DSM produced by matching can be automatically reduced to a DTM in exceptional cases, when the terrain is flat and above-terrain objects are few and isolated. Urban areas are often imaged at a large scale and include many above-terrain objects. Therefore, very often instead of matching, manual DSM or DTM measurements are performed, very often including break line measurements. Digital cameras offer better opportunities for higher quality DSM/DTM generation. Regarding the DMC and ADS40 sensors, no results have been published so far. However, investigations with similar non-commercial systems show potential. Institut Géographique National (IGN) (France) has developed a 4K by 4K pixel area CCD camera, which has been employed especially for DSM and orthoimage generation in urban areas. Through use of more sophisticated matching algorithms, the amount of errors is reduced, surface discontinuities are preserved and the completeness of the results is improved (unmatched areas or occlusions are less). The HRSC has been used both by DLR and especially the firm ISTAR for generation of DSMs over many cities in Europe and North America with very impressive results. Thereby, matching exploits the fact that the various HRSC models provide up to five panchromatic stereo channels. This redundancy increases accuracy and reliability, reduces occlusions and increases completeness. Digital sensors that provide multispectral capability, especially if they include NIR, can provide a better reduction of DSM to DTM by easier detection and elimination of vegetation and buildings, via a combination of various cues, like DSM blobs, spectral characteristics, edge information, shadows, etc. So, although DSM generation by matching using digital airborne sensors has not been extensively investigated, there is enough evidence to suggest that it would be of higher quality than from digitized film imagery.

LIDAR systems are active, are not influenced by shadows, can be employed at night, preserve surface discontinuities, data processing is automated to a high degree and production times are relatively short. As such they provide dense and accurate urban measurements. However, they also have some disadvantages, including errors of secondary reflections close to vertical structures, a narrow flight swathe and longer flight time compared with aerial imagery. For a more detailed comparison between photogrammetry and LIDAR, also regarding DSM/DTM generation, see Baltsavias (1999a). In practice, LIDAR has been widely employed for urban DSMs, especially for urban planning, rooftop heights for communication antennas, etc. The vegetation penetration capability of LIDAR permits an easier direct DTM measurement or automatic reduction of DSM to DTM. In other cases, LIDAR is employed for DSM generation of certain objects that can be part of urban areas:

- mapping of long, narrow features such as road mapping, urban planning and design, powerline corridor planning and tower design, coastal erosion monitoring, coastal zone management, traffic and transport, riverways and water resources and traffic management, corridor planning, mapping of railway lines, fibre-optic corridors, pipelines, dykes, etc.;
- high point density, high-accuracy mapping applications such as the monitoring of open pits or dumps, flood mapping, mapping of local infrastructures (e.g., airports);
- fast-response applications. LIDAR provides digital range measurements, which can be quickly converted to three-dimensional coordinates, and applied, for instance, in cases involving natural disasters.

Investigations on DSM generation from high spatial resolution imagery include the following. Ridley *et al.* (1997) evaluated the potential of generating a national mapping database of maximum building heights of at least 5 m by 10 m in planimetry by using DSMs extracted by matching 1 m imagery. They reported that matching has a potential to provide the requested information if the DSM has a spacing of 1–3 m, but with lower accuracy (1.5-m to 3-m RMS) and completeness compared with manual measurements. Muller *et al.* (2001) used simulated 1-m resolution and IKONOS data for DSM generation and land use determination to estimate effective aerodynamic roughness for air pollution modelling and determine the position of trees close to buildings that may cause soil subsidence for insurance risk assessment. Up until now, the most extensive tests on DSM generation from IKONOS have been reported by Toutin *et al.* (2001) and Zhang *et al.* (2002). Toutin *et al.* (2001) used a stereo Geo IKONOS pair in an area with relatively low urban/residential percentage (15.5 per cent) with low and detached buildings. They examined the accuracy of the automatically generated DSM based on land cover type. The LE90 accuracy value (Linear Error, 90 per cent confidence interval) varied from 5 m to 18 m and the bias from 0.2 m to 2.5 m. The urban/residential land cover had an LE90 error of 5 m and a bias of 0.2 m (i.e., relatively good results due to the low density and height of buildings). Toutin *et al.* (2001) report that slopes facing the Sun have an error by 1 m smaller than that for slopes away from the Sun. Zhang *et al.* also noticed a similar effect, although the topic needs further investigation. Zhang *et al.* processed both a stereo Geo IKONOS pair in the city of Melbourne with dense buildings and high-rise buildings in the city centre, and a multitemporal pair of Geo images over the Greek island of Nisyros. In the first case, the LH Systems DPW770 and VirtuoZo

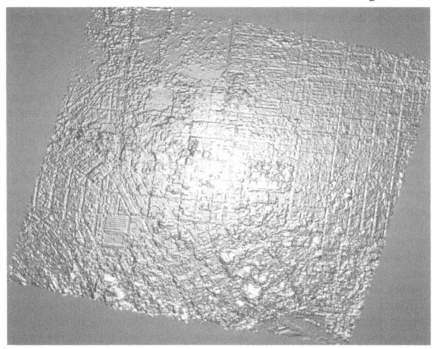

Figure 3.1 Shaded 2-m IKONOS-DSM (a part only) generated by VirtuoZo. The oval University of Melbourne campus is at the top left.

digital photogrammetric systems were used for automatic DSM generation. Comparison with manually and automatically measured points in aerial images gave a very low standard deviation of differences (0.9–1.2 m) and a higher bias (~2–2.5 m), with the DPW770 being slightly more accurate than the VirtuoZo. The results show the accuracy potential of IKONOS while at the same time they are too optimistic as the comparison points were selected in open-ground areas. The full DSM results showed visually many problems due to occlusions and shadows, repetitive patterns that lead to multiple solutions and the smoothing of surface discontinuities or the missing of small objects. The grey value coded result of VirtuoZo is shown in Figure 3.1. It should also be noted that for this stereopair the ground control points had very accurate (1–2 dm) object and image coordinates, the latter being generally quite difficult to achieve. For the multitemporal Geo images, and after manual exclusion of the different cloud regions, an RMS of as low as 3.2 m was achieved with a bias of 1.4 m. This is very encouraging and without the need of more expensive stereo IKONOS images and use of RPCs. As a comparison with the above results, the vertical accuracy specification of IKONOS for the Precision stereo product and use of GCPs is 3 m LE90 (Grodecki and Dial 2001). This is, however, for single, well-defined points, not arbitrary points on natural surfaces. In summary, the high DSM accuracy requirements in urban areas are barely satisfied by IKONOS imagery.

3.4.2 Orthoimage generation

Orthoimage generation from aerial film imagery is a simple and straightforward process. Generation of so-called true orthoimages (i.e., without radial displacement of above-terrain objects) is possible either with expensive manual measurements or automatic generation of a good-quality DSM (the latter usually requiring manual editing as explained in Section 3.4.1). However, with the advent of semi-automated commercial systems for three-dimensional city modelling, and especially building reconstruction, true orthoimages have become more common. Good colour balancing and automatic optimizing of seam lines during the mosaicking of many orthoimages may still be a problem with some commercial systems. The orthoimage accuracy depends on the accuracy of raw data (scanner influence), sensor interior and exterior orientation and primarily DSM/DTM quality. For high geometric accuracy, sometimes a LIDAR-derived DSM/DTM is used. Orthoimage generation from digital aerial cameras show similar characteristics. When a camera employs both panchromatic and multispectral channels, image sharpening can be performed quite easily, since the acquisition of all these channels is quasi-simultaneous. Image sharpening generates a new image by injecting in spectral images the usually higher spatial resolution of a panchromatic channel. With linear CCDs, image sharpening results in fewer differences due to occlusions among the used channels, if they are placed on a single focal plane, next to each other. An additional advantage of the linear CCDs is that there are no perspective displacements in the flight direction. Therefore, if the orthoimages are generated only from the central part of each strip and the nadir channel, a quasi-true orthoimage can be achieved even when using a DTM in orthoimage generation.

An increasing number of LIDAR systems record the intensity of the returned signal. However, LIDAR is rarely used for pure imaging and orthoimage generation for various reasons. The laser footprint is approximately circular and varies with the scan angle and the topography. The point spacing along and across track differs and the latter (often the along track too, depending on the scan pattern) also varies along the scan line with scan angle. Accordingly, it is impossible to image a whole area homogeneously and without gaps and overlaps. Moreover, during further visualization, processing of the image requires the interpolation of a regular grid. These problems are mainly due to the active nature of the laser (i.e., the image is formed on the ground, and not in the sensor focal plane as with passive optical sensors). The "lasels" have a much larger footprint (i.e., worse geometric resolution) than the pixels from the same flying height (a typical laser beam divergence of 1 mrad results in a 1-m footprint for

1,000-m flying height, while a 15-µm pixel with 15-cm camera constant has a footprint of just 10 cm). In addition, the radiometric quality is inferior to that of cameras (with LIDAR the signal can be very low especially for high flying heights and low-reflectivity targets, see range equation in Baltsavias 1999b), and even residues (interference patterns) that completely distort the image have been observed in some cases. Additional problems regarding the spectral properties have already been mentioned. However, with pulse lasers, used in commercial LIDAR systems, the recorded intensity is in most cases not the integration of the returned echo but just its maximum. One minor advantage of LIDAR images is that, because they are produced by active systems, they are insensitive to illumination shadows. Furthermore, laser images are already geocoded (i.e., no orthoimage generation is necessary). Concluding laser images cannot compete and substitute high-quality optical imagery. However, they can provide useful additional clues, which, together with the three-dimensional object description, can help the detection and classification of urban objects (see, e.g., Hug and Wehr 1997).

Orthoimage generation from IKONOS has been the main application. Results are reported in Davis and Wang (2001), Kersten *et al.* (2000), Toutin and Cheng (2000), Toutin (2001), Baltsavias *et al.* (2001a), Jacobsen (2001) and Vassilopoulou *et al.* (2002). The planimetric accuracy of the orthoimages depends on the accuracy of the GCPs and the DTM. For IKONOS, when compared with other spaceborne and airborne sensors, DTM accuracy is less important due to the small FOV, while GCP accuracy becomes more important due to the small pixel footprint. As shown by Fraser *et al.* (2002), the planimetric potential of IKONOS Geo lies in $\sim \frac{1}{3}$ pixel. So, the GCPs should be 1–2 dm accurate in both object and image space. While getting this accuracy with GPS in object space poses few if any problems, finding image points suitable for measurement by image analysis techniques with 0.1–0.2 pixel accuracy and accessibility in the scene for GPS measurement can be problematic. The best GCPs have good contrast, are preferably on the ground and are intersections of straight, long lines or centres of gravity of circular/elliptical features. Since such features are more common in urban areas, accurate orthoimage generation becomes easier. The planimetric accuracy of the orthoimage can be easily estimated by GCP accuracy (in Baltsavias *et al.* 2001a) using as input the DTM accuracy and the known sensor azimuth and elevation. For example, with 1–2-dm GCP accuracy, DTM accuracy of 2 m and elevation larger than 70 deg, a submetre planimetric accuracy can be achieved, similar to the much more expensive Precision Plus product. The method of Baltsavias *et al.* (2001a) for orthoimage generation is very simple and does not require knowledge of the sensor model or RPCs. As few as three GCPs are sufficient, while their spatial distribution does not have to be ideal. For object-to-image transformation, the simple terrain-corrected affined transformation is used. The authors present results from three varying scenes with achieved X- and

Figure 3.2 Geo IKONOS images of Canton Zug. Overlay of vector reference data on the PAN orthoimage (left) and multispectral image (MSI) (right).

Y-planimetric accuracy (RMS) of 1.5 m to 2.5 m, respectively. An example from the region of Zug in Switzerland is shown in Figure 3.2. If the quality of the used GCPs were better, accuracies of close or below 1 m could have been achieved. A very favourable characteristic of IKONOS is that for high sensor elevations, DTM errors have a small or no influence on orthoimage accuracy.

3.4.3   Extraction and three-dimensional reconstruction of urban objects

Aerial film or digitized imagery is currently the main source for mapping and extraction of urban objects using manual measurements. A collection of investigations, mainly from aerial imagery, but also increasingly from satellites, LIDAR and SAR can be found in Gruen *et al.* (1995), Gruen *et al.* (1977) and Baltsavias *et al.* (2001b). Some researchers have concentrated on the development of semi-automated systems, especially for simulation and three-dimensional city modelling (see Section 3.5). Currently, such stand-alone systems are commercially available and increasingly used, while commercial photogrammetric and remote sensing systems have almost nothing to offer in this respect. Another direction toward development of practical and operational object extraction is the use of a priori knowledge,

Figure 3.3 Automatically extracted roads using 1:15,000 aerial imagery and a priori information from 1:25,000 map vector data. Example in rural area in the region of Albis (near Zürich), Switzerland.

rules and models and combination of multiple information sources and clues that ease object detection and reconstruction. One such example is the project ATOMI (Eidenbenz *et al.* 2000) that uses vectorized 1:25,000 maps and 1:15,000 scale colour imagery to improve and update road centrelines and building outlines to determine their height. In particular, the work on road extraction has reached an operational stage (Zhang and Baltsavias 2001, 2002), and for rural areas a very high percentage of roads can be automatically extracted with high accuracy (over 90 per cent of the existing vector map data). Blind tests have been performed with several stereo pairs provided by the Swiss Federal Office of Topography and the Belgian National Geographic Institute covering several hundred kilometres of road. Quantitative analysis using manually measured data gave typical planimetric and vertical accuracies of 1 m or less. An example of extracted roads in rural areas is shown in Figure 3.3 (total road length ∼25 km).

In spite of the advantages of digital aerial sensors, automated object extraction has not yet reached an operational stage, at least not without manual editing. LIDAR has been used primarily for building extraction and less for vegetation mapping. In some cases, intensity information from the LIDAR system is included in this process (Hug and Wehr, 1997) or image information (Haala and Cramer 1999; Steinle and Voegtle 2001), or map information (Vosselman and Suveg 2001; Stilla *et al.* 2001), or just the raw laser data are processed (Weidner and Förstner 1995; Maas and Vosselman 1999). Clearly, the more additional information that exists and the higher its accuracy and completeness, the easier the object detection task becomes. The major current tendency is to combine LIDAR either with imagery or cadastral plans (see Barnsley *et al.*, Chapter 4 in this book). A special application on building change detection caused by earthquakes is reported by Murakami *et al.* (1999) and Steinle and Voegtle (2001). Regarding building extraction, the advantage in using LIDAR, over other imagery, is that the DSM blobs corresponding to buildings are usually better defined with sharper boundaries. However, this is true only with dense LIDAR data, often with an average point distance of 1 m or less. LIDAR data with coarser spacing (e.g., 2-m average point distance) hardly look any better than a DSM produced by a decent image matching method.

Object and feature extraction investigations using IKONOS imagery do not abound. Early research was performed using simulated data with empirical analysis (Aplin, Chapter 2 in this book). Ridley *et al.* (1997) evaluated the potential of 1-m resolution satellite imagery for building extraction, and findings were that only 73 per cent and 86 per cent of buildings could be interpreted correctly using monoscopic and stereoscopic imagery, respectively. More recently, Sohn and Dowman (2001) reported an investigation into the identification of buildings from high spatial resolution imagery. However, the study dealt with large detached buildings only and a comprehensive analysis of accuracy and completeness in the modelling of structure detail was not performed. In a broader feature extraction context, Hofmann (2001) reported on two-dimensional detection of buildings and roads in IKONOS imagery using spectral information, a DSM derived from laser scanning and image context and form via the commercial image processing package eCognition. Dial *et al.* (2001) presented the results of investigations into automated road extraction, with the focus on wide suburban roads in expanding US cities. Fraser *et al.* (2002) present the first ever investigations on the accuracy and completeness of three-dimensional building reconstruction with manual measurements in stereo IKONOS images. Using nineteen roof corner points measured with GPS as checkpoints, they generated results in planimetric and height accuracy of 0.6 m and 0.8 m, respectively. To provide a qualitative and more extensive quantitative assessment, the University of Melbourne campus was measured manually in stereo using an in-house-developed software tool for the IKONOS stereo images, and an analytical plotter for 1:15,000 colour aerial imagery. The resulting plots of extracted building features are shown in Figure 3.4.

The manual measurements of roof corners and points of detail were topologically structured automatically using CyberCity's software package CC-Modeler (Gruen and Wang 1998) (Figure 3.5). Comparison of the two models in Figure 3.4 reveals the following regarding the IKONOS stereo feature extraction: about 15% of the buildings measured in the aerial images

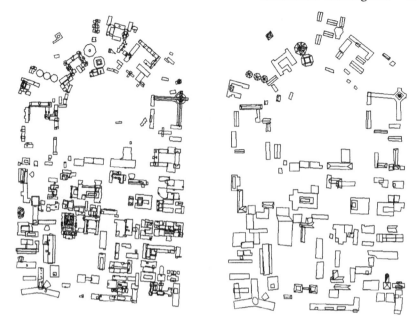

Figure 3.4 Buildings of University of Melbourne campus reconstructed from 1:15,000 aerial images (left) and stereo IKONOS imagery (right). To simplify visualization, first points and first lines have been omitted in the left figure.

could not be modelled, and it is interesting to note that this figure fits well with the findings of Ridley *et al.* (1997). Also, a number of both small and large buildings could not be identified and measured, though some new buildings could be reconstructed, even if small. Finally, as indicated in Figure 3.6, buildings could often only be generalized with a simplified roof structure and variations to their form and size. Measurement and interpretation in stereo proved to be a considerable advantage, and we expect that colour imagery would also have been very advantageous if available. Other factors influencing the feature extraction process are shadows, occlusions, edge definition (also related to noise and residues), saturation of bright surfaces, Sun and sensor elevation and azimuth, and atmospheric conditions. The 1-m resolution of IKONOS also leads to certain interpretation restrictions. However, additional tests with different IKONOS stereo imagery are needed in order to draw more conclusive results.

3.5 Three-dimensional city models

3.5.1 Introduction

It is only since around 1990 that photogrammetric approaches to building identification and modelling have evolved. What started out as a pure

Figure 3.5 Visualization from stereo IKONOS images of extracted buildings and trees of University of Melbourne campus.

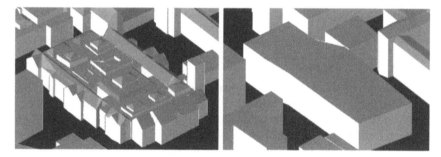

Figure 3.6 Building with complicated roof structure as extracted from stereo aerial (left) and IKONOS images (right).

research issue has now found firm ground in professional practice. After the first phase of efforts to extract buildings fully automatically, the tight specifications of users have led to the development of efficient manual and semi-automated procedures. Actually, the need to extend modelling from simple to much more complex buildings and full ensembles and even to generate complete city models (including DTMs, roads, bridges, parking lots, pedestrian walkways, traffic elements, waterways, vegetation objects, etc.) puts fully automated methods even further back in the waiting line of technologies for practical use. In a sense, user requirements have outpaced the capabilities and performance of automated methods. However, to clarify, automation in object extraction from images is still and will continue to be a key research topic. There are many fully automated approaches to

Figure 3.7 Modelling of a chemical plant (combination of vector and raster image data) (courtesy CyberCity AG, Bellikon, Switzerland).

building extraction, but only very few were designed to be semi-automated from the very beginning. Very often, procedures are declared as automatic but require so much post-editing that their status as automatic methods becomes questionable. For details on fully automated approaches see Section 3.3.

Applications of city models are manyfold. Currently the major users in Europe are in city planning, facility mapping (especially chemical plants and car manufacturers, see Figure 3.7), telecommunication, construction of sports facilities and other infrastructure. Others include environmental studies and simulations, Location-Based Services (LBS), transport risk assessment, car navigation, simulated training (airplanes, trains, trams, etc.), energy providers (placement of solar panels), real estate business, virtual tourism and microclimate studies. Interesting markets are expected in the entertainment and "infotainment industries (e.g., for video games, movies for TV and cinema, news broadcasting, sports events, animation for traffic and crowd behaviour and many more). When designing an efficient method for object extraction and modelling the following requirements should be observed;

- extract not only buildings, but also other objects;
- generate three-dimensional geometry and, if a GIS platform is used, topology as well;
- integrate natural image texture (for DTMs, roofs, façades and special objects);
- allow for object attributes;
- keep the level of detail flexible;

- allow for a wide spectrum of accuracy levels in the cm and dm ranges;
- produce structured data, compatible with major computer aided design (CAD) and visualization software;
- provide for internal quality-control procedures, leading to absolutely reliable results.

We currently see three major techniques used in city model generation.

DIGITIZING OF MAPS

This only gives two-dimensional information. The height of objects has to be approximated or derived with a lot of additional effort. It does not provide detailed modelling of the roof landscape. The roof landscape is usually very important because city models are mostly shown from an aerial perspective. Also, map data are often outdated.

EXTRACTION FROM AERIAL LASER SCANS

Laser scans produce regular sampling patterns over the terrain. Most objects in city models are best described by their edges, which are not easily accessible in laser scans and often cannot be derived unambiguously. Some objects of interest that do not distinguish themselves through height differences from their neighbourhood can thus not be found in laser data. Finally, the resolution of current laser scan data is not sufficient for detailed models.

PHOTOGRAMMETRIC GENERATION

Aerial and terrestrial images are appropriate data sources for the generation of city models. They allow the construction of both the geometrical and the texture models from one unique data set. The photogrammetric technique is highly scale-sensitive, and can adapt to required changes in resolution and accuracy. The processing of new images guarantees an up-to-date model. Images are a multipurpose data source and can be used for many other purposes as well.

In considering the above, the photogrammetric approach must be considered the most relevant technique. A scheme for image-based reconstruction of a hybrid city model is shown in Figure 3.8. Hybrid refers to the fact that both vector and raster data can be represented by the model. According to this scheme, roof landscapes, DTMs, transportation elements, land use information, etc. can be extracted from aerial images. Combining roofs and DTMs will result in the building of vector models, which, in turn, can be refined by using terrestrial images, taken with camcorders or still video cameras. Aerial images, terrestrial images and digitized maps can all

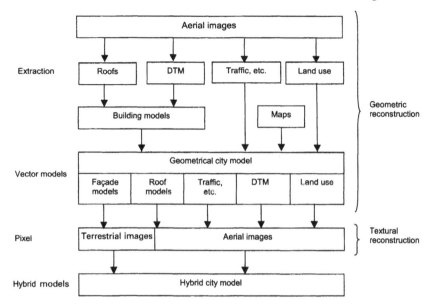

Figure 3.8 Image-based reconstruction of a hybrid city model.

contribute to the texture requirements of the hybrid model. It is also well known that to a certain extent texture information can compensate for missing vector data.

As fully automated extraction methods cannot cope with most of the aforementioned requirements, semi-automated photogrammetric methods are currently the only practical solution. For a review on semi-automated methods for site recording see Gruen (2000). There are two semi-automated approaches that have made it into the commercial domain so far:

- InJECT, a product of INPHO GmbH, Stuttgart. This approach is based on the fitting of elementary, volumetric building models or, in the case of complex buildings, building component models to image data. This concept, originally introduced at the Stanford Research Institute, Menlo Park, CA was refined and extended at the Institute of Photogrammetry, University of Bonn and is now available as a commercial software package.
- CyberCity Modeler (CC-Modeler) is a method and software package that fits planar surfaces to measured and weakly structured point clouds, thus generating CAD-compatible objects like buildings, trees, waterways, roads, etc. Usually these point clouds are taken from aerial images, but it is also possible to digitize them from existing building plans. This product is marketed by CyberCity AG, Bellikon, Switzerland, a spin-off company of ETH Zürich.

We will focus on CyberCity Modeler.

3.5.2   CyberCity Modeler

For the generation of three-dimensional descriptions of objects from aerial images, two major components are involved: photogrammetric measurements and structuring. In CyberCity Modeler (CC-Modeler) object identification and measurement is performed in manual mode by an operator within a stereoscopic model on an analytical plotter or a digital station. According to our experiences stereoscopy is very crucial for object identification especially as in many complex urban conditions monoscopic image interpretation inevitably fails. So, the human operator defines the level of detail.

The structuring of the point clouds is done automatically with the CC-Modeler software. Structuring involves essentially the intelligent assignment of planar faces to the given cloud of points or, in other words, the decision about which points belong to which planar faces. This problem is formulated as a consistent labelling problem and solved via a modified technique of probabilistic relaxation. Then, a least-squares adjustment is performed for all faces simultaneously, fitting the individual faces in an optimal way to the measured points considering the fact that individual points are usually members of more than one face. This adjustment is amended by observation equations that model orthogonality constraints of pairs of straight lines. For the purpose of visualization the system can also triangulate the faces into a Triangulated Irregular Network (TIN) structure. Figure 3.9 shows the data flow and the procedures involved in

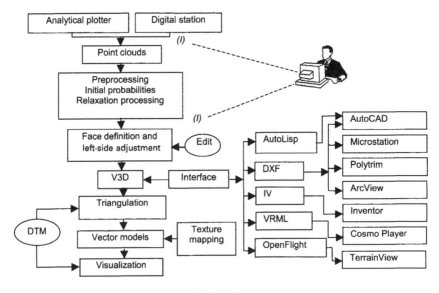

(I) – Interactive functions

Figure 3.9 Data flow of CC-Modeler.

Figure 3.10 City model Zürich Oerlikon. Note the modelling of non-planar surface objects.

CC-Modeler. A detailed description can be found at our homepage www.photogrammetry.ethz.ch/Research/Projects/CC-Modeler and in Gruen and Wang (1998). With this technique, hundreds of objects can be measured in a day. Although CC-Modeler generates a polyhedral world, objects with non-planar surfaces can also be modelled, given the resolution (compare Figure 3.10).

A digital terrain model can also be measured and integrated. Texture from aerial images is mapped automatically onto the terrain and the roofs, since the geometrical relationship between object faces and image patches has been established. Façade texture is produced semi-automatically via projective transformation from terrestrial images usually taken by camcorders or still video cameras (see also Section 3.5.4). Figure 3.11 shows the user interface of CC-Modeler.

The system produces its own internal data structure (V3D), thereby accommodating vector and raster data, and interfaces to major public data formats, including OpenFlight, are available. V3D is a self-developed vector-based data structure that builds the facet model of objects (Figure 3.12). The basic geometrical element is the point. Points are used to express faces and line segments. An object consists of faces and a polyline consists of line segments. Once some attributes are attached, an object or a polyline becomes an entity class. The polyline representation is used to

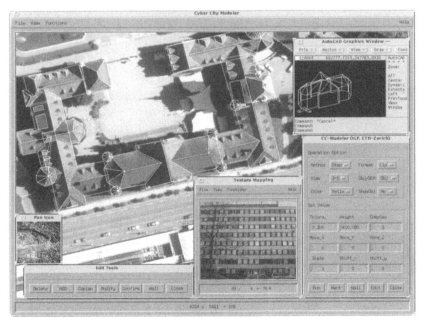

Figure 3.11 User interface of CyberCity Modeler.

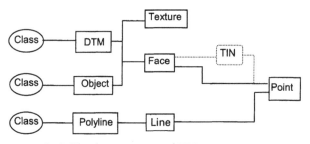

Figure 3.12 The data structure of V3D.

describe one-dimensional objects, such as property boundaries, sidewalk borders, line features on roads, etc. For more details on the data model and on GIS-related aspects see Section 3.5.5. The system and software is fully operational. Over 200,000 buildings at very high spatial resolution have been generated already in cities and towns like Amsterdam, Chur (Switzerland), Florence, Giessen (Germany), Hamburg, Melbourne, Tokyo, Zürich, in chemical plants and car manufacturer production sites. Figure 3.13 shows the integration of vector and image raster data into a joint model (images are mapped onto DTMs, roofs and façades).

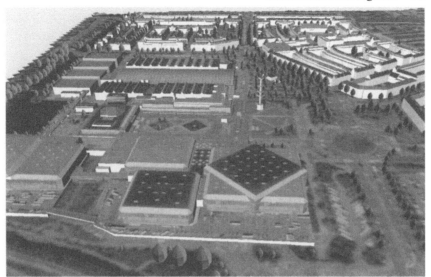

Figure 3.13 Three-dimensional model of the Congress Centre RAI, Amsterdam. Vector data, overlaid with natural texture.

3.5.3 Recent extensions to CC-Modeler

A modification to CC-Modeler is geometric regularization. The requirement is to make straight lines parallel and perpendicular where they are not, or to have all points of a group (e.g., eaves or ridge points) at a unique height. Another problem is that CC-Modeler was originally designed to handle individual buildings sequentially and independent of each other. Building neighbourhood conditions were not considered. The resulting geometrical inconsistencies, like small gaps or overlaps between adjacent buildings (in the cm/dm range), are not dramatic and are certainly tolerable in many applications, especially those which are purely related to visualization. However, the topological errors constitute a serious problem in projects where the three-dimensional model is subject to legal considerations or some other kind of analysis which requires topologically correct data.

Another significant extension refers to the precise modelling of building façades. Façades are usually not visible in aerial images, but are available in cadastral maps. We combine façade information with the roof landscape modelled by CC-Modeler in order to be able to represent roof overhangs. We also show that we can model other vertical walls explicitly. In the following we will describe all these extensions in more detail. The flowchart in Figure 3.14 represents the processes described above. These are executed after the face definition by probabilistic relaxation is completed. The dashed connection between "Vertical wall integration" and "V3D corrected" is under development. For a more detailed description of these extensions see Gruen and Wang (2001).

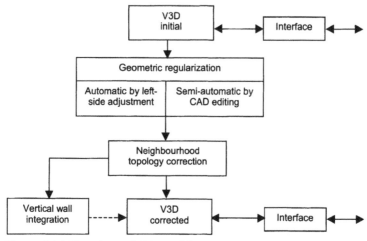

Figure 3.14 Flowchart of CC-Modeler extensions.

3.5.4 *Texture mapping and visualization/simulation*

Photorealistic texturing applied to three-dimensional objects gives the most natural representation of the real world. Texture supplies information on material properties and represents details that are not modelled in the vector data set.

Generally, there are two types of data source: aerial images, and terrestrial images taken from street level. The former are usually employed in the mapping of terrain surfaces and roofs of buildings, the latter are for building façades and other vertical faces. From a data structure point of view, both kinds of image are expressed as two-dimensional raster data, which can be stored or manipulated as a special layer in our three-dimensional system. Visualization of three-dimensional city models becomes a key issue when dealing with user requests. The best model is limited by speed. It is important to distinguish between real-time and snail-time visualization requirements. Snail-time performance is acceptable if images are produced for, say, publications, but in most applications real-time capabilities are essential. Although there are visualization programs available on the international market (compare www.tec.army.mil/TD/tvd/survey/survey_toc.html), only very few are real-time.

For high-end performance, Level-of-Detail (LoD) capabilities for both vector and image data is indispensable. LoD provides for on-the-fly switching between several resolution levels (three are mostly sufficient), depending on the viewing distance. With this functionality and sufficient host and graphics memory and an appropriate, but still standard graphics board, even laptops can handle very large data sets in real time.

In our group we are currently using packages like Cosmo Player (for very small data sets), AutoCAD, Microstation, Inventor/Explorer (SGI), Maya

(Alias Wavefront), Terrainview (Viewtec), Skyline (IDC AG, Switzerland) and a variety of self-developed software. Modern visualization software not only shows the "naked" model, but in addition allows for features like import of various standard data formats, preparation of interactive and/or batchmode flyovers and walkthroughs, generation of videos, integration of text information, definition of various layer systems above terrain, search functions for objects, coupling of information in different windows, import of synthetic textures, integration and manipulation of active objects (e.g., clouds, fog, multiple light sources, cars, people, etc.), hyperlink functions for the integration of object properties, export via Internet/Intranet/CD/DVD, etc.

3.5.5   GIS aspects

Although many users are currently interested in the visualization of city models there is also a clear desire to integrate the data into a GIS platform in order to utilize the GIS data administration and analysis functions. The commercial GIS technology is still primarily two-dimensional-oriented and is thus not really prepared to handle three-dimensional objects efficiently. Therefore we have developed in a pilot project a laboratory version of a hybrid three-dimensional spatial information system, whose major aspects we will demonstrate in the following.

3.5.5.1   Data structure

Our V3D is a hybrid data structure. It not only models three-dimensional objects, but also combines raster images and attribute information for each object. The terrain objects are grouped into four different geometric object types:

- point objects – zero-dimensional objects that have a position but no spatial extension;
- line objects – one-dimensional objects with length as the only measurable spatial extension, which means that line objects are built up of connected line segments;
- surface objects – two-dimensional or two-and-a-half-dimensional objects with area and perimeter as measurable spatial extension, which are composed of facet patches;
- body objects – three-dimensional objects with volume and surface area as measurable spatial extension which are bordered by facets.

In V3D, each special object is identified by Type Identifier Code (TIC), referred to as PIC, LIC, SIC and BIC, respectively. Two data sets are attached to each object type: thematic data and geometric data. Image data can be attached to the surface object, body object and DTM object. In

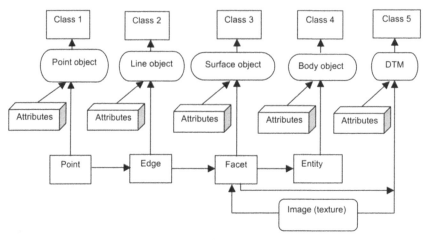

Figure 3.15 The logical data structure of V3D.

fact, the thematic data attributes are built up in a separate data table. It is linked to the object type with a related class label. The definition of the thematic data is user-dependent. The geometric data set contains the geometric information of three-dimensional objects (i.e., the information of position, shape, size, structure definition and image index). The diagram in Figure 3.15 shows the logical data structure.

For the four object types, four geometrical elements are designed (i.e., *Point*, *Edge*, *Facet* and *Entity*). *Point* is the basic geometric element in this diagram. The *Point* can present a point object. It also can be the start or end point of an *Edge*. The *Edge* is a line segment, which is an ordered connection between two points: begin point and end point. Further, it can be a straight part of a line object or lie on a facet. The *Facet* is the intermediate geometrical element. It is completely described by the ordered edges that define the border of the facet. One or more facets can be related to a surface object or *Entity* geometrical element. Moreover, *Facet* is related to an image patch. *Entity* is the highest level geometrical element, and carries shape information. An entity is completely defined by its bordering facets. Image data and thematic data are two special data sets, which are built up in two separate data tables. Each facet is always related to an image patch through a corresponding link.

Once the attribute table is attached and the TIC is labelled, a geometrical element becomes an object type. The DTM is treated as a special data type, which is described by a series of facets. Obviously, the topological relationships between geometrical elements are implicitly defined by the data structure. A point object is presented by a distinct *Point* element. The line object is described by ordered *Edges*. The surface object is described by the *Facet* with the information of image patches. Similarly, the body object is described by *Entity* defined by the facets. Therefore, the topological relationships between *Point* and *Edge*, *Edge* and *Facet*, *Facet* and *Entity* are registered by the links between the geometrical elements.

3.5.5.2 *Implementation in a relational database*

In a relational database the most common object to be manipulated is the relation table. Other objects such as index, views, sequence, synonyms and data dictionary are usually used for query and data access. "Table" is the basic storage structure, which is a two-dimensional matrix consisting of columns and rows of data elements. Each row in a table contains the information needed to describe one instance of the entity; each column represents an attribute of the entity.

The data model shown in Figure 3.15 is a relational model, which can be implemented by relational data base technology. Figure 3.16 shows the relational model of the V3D data structure.

Each object type is defined as a table, shown as the upper row. A point-type table includes three terms. The Point Identification Code (PIC) is an identification code for a point-type object. The Attribute Identification (AID) is coded to relate to an attribute table. Different types of object may have different attribute tables. For example, the attribute tables of "tree" may have different thematic definitions than "pole". The Name of Point (NP) is the identification of a geometric point that is used to relate it to a distinct element in a point-geometric element table. The *Point* table is the most basic geometrical element table, which defines the coordinate position of the geometrical points.

The line-type table has similar content as the point-type table. The difference is that a line-type object is identified by the Line Identification

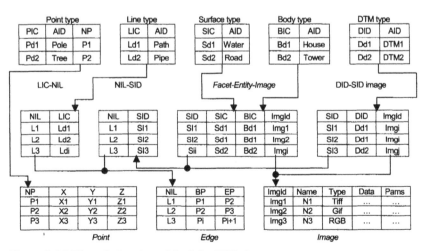

Figure 3.16 The relational model of the V3D data structure.

Code (LIC), which is not directly linked with the geometric element table *Edge*, but linked with an intermediate relational table LIC-NIL and then indexed to the *Edge* table. The *Edge* table defines the geometrical element *Edge*, in which each edge (NIL) is described by the beginning point (BP) and the end point (EP). The surface-type table and body-type table have similar terms as the line-type table. For each type of object a distinct identification code (SIC [Surface Identification Code] or BIC [Body Identification Code]) is labelled. Both SIC and BIC are linked with a merging geometrical element table, *Facet-Entity-Image*, in which the topological relationships between *Facet* and *Surface*, *Facet* and *Entity*, *Facet* and *Image* are defined. The *Facet-Entity-Image* table has two links: one is related to the *Image* table; the other is related to the NIL-SID table. Image table is a basic table that describes all attributes of images, such as the image name, format, pixels, camera parameters, orientation parameters, etc. The NIL-SID table is another intermediate table that defines the corresponding relationships between *Facet* and *Edge*. Its NIL column is related to the *Edge* table. The DTM is treated as a special class, which is related to the NIL-SID table through an intermediate table DID-SID-Image.

Based on the relational structure shown on the diagram in Figure 3.16, the query of a geometrical description of a distinct object type is easily realized. For example, the query "Select the geometrical description of an object with the identification code BIC = 202", will first index all *Facet* identification in the *Facet-Entity-Image* table by its BIC, and then retrieve all edge name identifications (NIL), and finally index the position information of structure points with the help of *Edge* table and *Point* table.

The queries of topological relationships are divided into two types: relationships between the geometrical elements of an object and those between objects themselves. The relationships between the geometrical elements are implicitly defined in the above data structure. Though the internal topology is not directly supplied, users can flexibly deduce the relationships, such as joint, adjacency, left or right, etc. The queries of topological relationships between objects are not considered in the above data structure because they are application-dependent.

3.5.5.3   *A prototype system*

Based on the above data model and structure, a special information system, CC-SIS, has been successfully developed and implemented on a workstation (Sun SPARC) under X-Windows, OSF/Motif, OpenGL as well as ORACLE database. Although its main purpose is scientific investigations, it can also be used in applications like photorealistic representation with possibilities for navigation through the three-dimensional city model, creation, storage, analysis and query of a city object. In combination with our topology

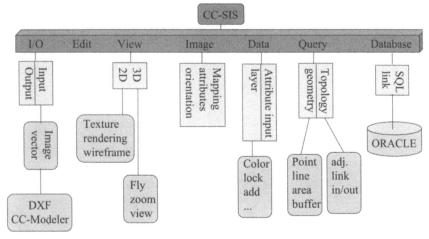

Figure 3.17 The function units of CC-SIS.

generator CC-Modeler, it builds up a system for geometrical information generation, storage and manipulation.

There are currently seven function units in CC-SIS, as shown in Figure 3.17. CC-SIS can directly read the data file generated by CC-Modeler and output results in the format DXF, Autolisp, Inventor and V3D. The Edit function is used for graphic editing, which is to be developed in the future. The View function supplies the tools for two-dimensional or three-dimensional viewing, such as dynamically selecting a view port, zooming, etc. Further, three types of rendering are available, wireframe, shading and texture mapping. The Image function supplies the tools for interior orientation of the images in order to map natural texture from images. Also, artificial texture can be mapped. An example of this function is shown in Figure 3.18. The data function manipulates the data. It includes two submodules: one is used for operation on layers (objects are defined as different layers in CC-SIS, such as building, DTM, waterway, pathway, tree, etc.); the other is to input the attributes for the selected object. The Geo-query function includes two tools: geometry query and topology query. The former is used to query the separated object by the point, line or entity selection; the latter is employed to query topological relationships between different objects. Figure 3.18 shows the geometric query of CC-SIS. The user can mark an object (e.g., building) with a cursor. This triggers and displays the corresponding attributes and geometrical/topological information. The operations on a database are defined in the Database function, including database link and structured query language (SQL)-query. SQL-query is a submenu, in which standard SQL queries are supplied.

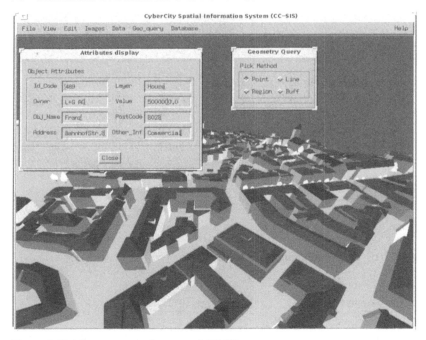

Figure 3.18 The geometrical query of CC-SIS.

3.6   Conclusions

The demand for three-dimensional city models has matured from a narrow research niche into an expanding market of applications. In satisfying the broadening of the application base there is a need for higher resolution data, both in terms of geometrical and textural detail. Aerial photogrammetry plays an important role as a flexible and economic technique for data generation. Semi-automated photogrammetric techniques are available for efficient data production. However, satellite sensor images, even in high spatial resolution mode, do not show enough promise for detailed and reliable modelling. Terrestrial images, on the other hand, are relevant already for façade texture generation, and in the future could possibly be used for façade geometry modelling and the recording of other objects that are not accessible from the air.

It remains to be seen to what extent aerial laser scan data can be integrated into the data generation process. Terrestrial laser scanners already play a significant role in the modelling of indoor scenes. The need to combine outdoor and indoor data will thus lead to hybrid sensor approaches in the future. Much of the current discussion is centred on the generation of virgin databases, and methods for data maintenance. The fast pace with which our man-made environment is changing will also require innovative techniques for the updating of three-dimensional city models.

3.7 Acknowledgements

The provision of imagery and ground control information by the Department of Geomatics, University of Melbourne and the cooperation of C. Fraser, H. Hanley and T. Yamakawa are gratefully acknowledged. The IKONOS image of Lucerne was kindly provided by the National Point of Contact, Swiss Federal Office of Topography, Wabern, the images of Zug by the company Swissphoto AG, while the IKONOS image of Nisyros was made available by Prof. E. Lagios, University of Athens, Greece within the EU project Geowarn (www.geowarn.org). Gene Dial and Laurie Gibson (Space Imaging) are acknowledged for valuable discussions and suggestions and provision of information. We also thank CyberCity AG, Bellikon, Switzerland for providing some views on three-dimensional city models.

3.8 References

Airborne Laser Mapping, 2002, Airbornelasermapping.com [reference site on an emerging technology].

Baltsavias, E. P., 1999a, A comparison between photogrammetry and laser scanning. *ISPRS Journal of Photogrammetry and Remote Sensing*, 54, 83–94.

Baltsavias, E. P., 1999b, Airborne laser scanning: Basic relations and formulas. *ISPRS Journal of Photogrammetry and Remote Sensing*, 54, 199–214.

Baltsavias, E., Pateraki, M. and Zhang, L., 2001a, Radiometric and geometric evaluation of IKONOS GEO images and their use for 3D building modelling. Paper given at Conference of the Joint ISPRS Workshop on High Resolution Mapping from Space (CD-ROM).

Baltsavias, E. P., Gruen, A. and Van Gool, L. (eds), 2001b, *Automatic Extraction of Man-Made Objects from Aerial and Space Images III* (Lisse, The Netherlands: Balkema).

Davis, C. H. and Wang, W., 2001, Planimetric accuracy of IKONOS 1-m panchromatic image products. Paper given at Conference of the American Society of Photogrammetry and Remote Sensing (CD-ROM).

Dial, G., Gibson, L. and Poulsen, R., 2001. IKONOS satellite imagery and its use in automated road extraction. In E. P. Baltsavias, A. Gruen and L. Van Gool (eds) *Automatic Extraction of Man-Made Objects from Aerial and Space Images III* (Lisse, The Netherlands: Balkema), pp. 49–58.

Flood, M., 2001, Laser altimetry: From science to commercial LIDAR mapping. *Photogrammetric Engineering and Remote Sensing*, 67, 1209–17.

Fraser, C., Baltsavias, E. and Gruen, A., 2002, Processing of IKONOS imagery for sub-metre 3D positioning and building extraction. *ISPRS Journal of Photogrammetry and Remote Sensing*, 56, 174–94.

Fricker, P., 2001, ADS40 – Progress in digital aerial data collection. In D. Fritsch and R. Spiller (eds) *Photogrammetric Week '01* (Heidelberg, Germany: Wichmann), pp. 105–16.

Fuchs, F., 2001, Building reconstruction in urban environment: Graph-based approach. In E. P. Baltsavias, A. Gruen and L. Van Gool (eds) *Automatic Extraction of Man-Made Objects from Aerial and Space Images III* (Lisse, The Netherlands: Balkema), pp. 205–16.

Grodecki, J. and Dial, G., 2001, IKONOS geometric accuracy. Paper given at Conference of the Joint ISPRS Workshop High Resolution Mapping from Space 2001 (CD-ROM).

Gruen, A., 2000, Semi-automated approaches to site modeling. *ISPRS International Archives of Photogrammetry and Remote Sensing*, **XXXIII**, 309–18.

Gruen, A. and Wang, X., 1998, CC-Modeler: A topology generator for 3-D city models. *ISPRS Journal of Photogrammetry and Remote Sensing*, 53, 286–95.

Gruen, A. and Wang, X., 2000. A hybrid GIS for 3D city models. *International Archives of Photogrammetry and Remote Sensing*, 33(Part 4/3), 1165–72.

Gruen, A. and Wang, X., 2001, News from CyberCity – Modeler. In E. P. Baltsavias, A. Gruen and L. Van Gool (eds) *Automatic Extraction of Man-Made Objects from Aerial and Space Images III* (Lisse, The Netherlands: Balkema), pp. 93–101.

Gruen, A., Kuebler, O. and Agouris, P. (eds), 1995, *Automatic Extraction of Man-Made Objects from Aerial and Space Images* (Basel, Switzerland: Birkhäuser Verlag).

Gruen, A., Baltsavias, E. P. and Henricsson, O. (eds), 1997, *Automatic Extraction of Man-Made Objects from Aerial and Space Images II* (Basel, Switzerland: Birkhäuser-Verlag).

Haala, N. and Cramer, C., 1999, Extraction of buildings and trees in urban environments. *ISPRS Journal of Photogrammetry and Remote Sensing*, 54, 130–7.

Hinz, A., Doerstel, C. and Meier, H., 2001, DMC – the digital sensor technology of Z/I Imaging. In D. Fritsch and R. Spiller (eds) *Photogrammetric Week '01* (Heidelberg, Germany: Wichmann), pp. 93–103.

Hofmann, P., 2001, Detecting buildings and roads from IKONOS data using additional elevation information. *GIS*, 6, 28–33.

Hug, Ch. and Wehr, A., 1997, Detecting and identifying topographic objects in imaging laser altimeter data. *International Archives of Photogrammetry and Remote Sensing*, 32(Part 3-4W2), 19–26.

Jacobsen, K., 2001, Automatic matching and generation of orthophotos from airborne and spaceborne line scanner images. Paper given at Conference of the Joint ISPRS Workshop High Resolution Mapping from Space 2001 (CD-ROM).

Jensen, J. R. and Cowen, D. C., 1999, Remote sensing of urban/suburban infrastructure and socio-economic attributes. *Photogrammetric Engineering and Remote Sensing*, 65, 611–22.

Kersten, T., Baltsavias, E., Schwarz, M. and Leiss, I., 2000, IKONOS-2 Carterra Geo – Erste geometrische Genauigkeitsuntersuchungen in der Schweiz mit hochaufgeloesten Satellitendaten. *Vermessung, Photogrammetrie, Kulturtechnik*, 8, 490–7 [in German].

Lehmann, F., 2001, The HRSC digital airborne imager – a review of two years of operational production. *Geoinformatics*, 4, 22–5.

Maas, H-G. and Vosselman, G., 1999, Two algorithms for extracting building models from raw laser altimetry data. *ISPRS Journal of Photogrammetry and Remote Sensing*, 54, 153–63.

Muller, J-P., Kim, J. R. and Tong, L., 2001, Automated mapping of surface roughness and landuse from simulated and spaceborne 1m data. In E. P. Baltsavias, A. Gruen and L. Van Gool (eds) *Automated Extraction of Man-Made Objects from Aerial and Space Images III* (Lisse, The Netherlands: Balkema), pp. 359–69.

Murakami, H., Nakagawa, K., Hasegawa, H., Shibata, T. and Iwanami, E., 1999, Change detection of buildings using an airborne laser scanner. *ISPRS Journal of Photogrammetry and Remote Sensing*, 54, 148–52.

Neukum, G., 1999, The airborne HRSC-A: Performance results and application potential. In D. Fritsch and R. Spiller (eds) *Photogrammetric Week '99* (Heidelberg, Germany: Wichmann), pp. 83–8.

Neukum, G. and HRSC Team, 2001, The airborne HRSC-AX cameras: Evaluation of the technical concept and presentation of application results after one year of operations. In D. Fritsch and R. Spiller (eds) *Photogrammetric Week '01* (Heidelberg, Germany: Wichmann), pp. 117–30.

Petrie, G., 2002, Optical imagery from airborne and spaceborne platforms. *Geoinformatics*, 5, 28–35.

Ridley, H. M., Atkinson, P. M., Aplin, P., Muller, J. P. and Dowman, I., 1997, Evaluating the potential of the forthcoming commercial US high-resolution satellite sensor imagery at the Ordnance Survey. *Photogrammetric Engineering and Remote Sensing*, 63, 997–1005.

Sandau, R., Braunecker, B., Driescher, H., Eckardt, A., Hilbert, S., Hutton, J., Kirchhofer, W., Lithopoulos, E., Reulke, R. and Wicki, S., 2000, Design principles of the LH systems ADS40 airborne digital sensor. *International Archives of Photogrammetry, Remote Sensing and Spatial Information Sciences*, **XXXIII**(Part B1), 258–65.

Starlabo, 2002, High-definition digital airborne sensors (http://www.starlabo.co.jp/en/business/index.html).

Steinle, E. and Voegtle, T., 2001, Automated extraction and reconstruction of buildings in laser scanning data for disaster management. In E. P. Baltsavias, A. Gruen and L. Van Gool (eds) *Automatic Extraction of Man-Made Objects from Aerial and Space Images III* (Lisse, The Netherlands: Balkema), pp. 309–18.

Thom, C. and Souchon, J-P., 1999, The IGN digital camera system in progress. In D. Fritsch and R. Spiller (eds) *Photogrammetric Week '99* (Heidelberg, Germany: Wichmann), pp. 89–94.

Toth, C., 1999, Experiences with frame CCD arrays and direct georeferencing. In D. Fritsch and R. Spiller (eds) *Photogrammetric Week '99* (Heidelberg, Germany: Wichmann), pp. 95–107.

Toutin, T., 2001, Geometric processing of IKONOS Geo images with DEM. Paper given at Proceedings of Joint the ISPRS Workshop on High Resolution Mapping from Space 2001 (CD-ROM).

Toutin, Th. and Cheng, P., 2000, Demystification of IKONOS! *EOM*, 9, 17–21.

Toutin, Th., Chénier, R. and Carbonneau, Y., 2001, 3D geometric modelling of IKONOS GEO images. Paper given at Conference of the Joint ISPRS Workshop on High Resolution Mapping from Space 2001 (CD-ROM).

Vassilopoulou, S., Hurni, L., Dietrich, V., Baltsavias, E., Pateraki, M., Lagios, E. and Parcharidis, I., 2002, Orthophoto generation using IKONOS imagery and high resolution DEM: A case study on monitoring of volcanic hazard on Nisyros island, Greece. *ISPRS Journal of Photogrammetry and Remote Sensing* (in press).

Weidner, U. and Förstner, W., 1995, Towards automatic building extraction from high resolution digital elevation models. *ISPRS Journal of Photogrammetry and Remote Sensing*, 50, 38–49.

Zhang, C. and Baltsavias, E. P., 2001, Improvement and updating of a cartographic road database by image analysis techniques using multiple knowledge sources and cues. In T. Blaschke (ed.) *Fernerkundung und GIS: Neue Sensoren – Innovative Methoden* (Heidelberg, Germany: Wichmann Verlag), pp. 65–77 [in German].

Zhang, C and Baltsavias, E. P., 2002, Improving cartographic road databases by image analysis. *International Archives of Photogrammetry, Remote Sensing and Spatial Information Sciences*, **XXXIV** (Part 3A) 400–5.

Zhang, L., Pateraki, M. and Baltsavias, E., 2002, Matching of IKONOS stereo and multitemporal Geo images for DSM generation. *Proc. Map Asia 2002*, 7–9 August, Bangkok, Thailand (on CD-ROM). Also available at http://www.gis.development.net/technology/ip/techip012.htm (accessed 21 October, 2002).

4 Determining urban land use through an analysis of the spatial composition of buildings identified in LIDAR and multispectral image data

Michael J. Barnsley, Alan M. Steel and Stuart L. Barr

4.1 Introduction

It has been suggested that different categories of urban land use may be identified in high spatial resolution remotely sensed images by analysing the structural characteristics of their constituent land cover parcels (e.g., buildings, roads and open spaces), in terms of their size, shape and spatial distribution within the scene (Wharton 1982; Møller-Jensen 1990; Gong and Howarth 1992; Eyton 1993; Barnsley and Barr 1996; Barnsley and Barr 2000). The implicit assumption is that each land use exhibits distinct, broadly consistent patterns of buildings and other land cover objects by which it may be recognized. This hypothesis has yet to be fully and rigorously tested, although encouraging results have been obtained in studies making use of multispectral scanner images (Barnsley and Barr 1997; Barr and Barnsley 2000; Bauer and Steinnocher 2001) and, separately, rasterized digital map data (Barr and Barnsley 1998a). These studies have employed a graph-based data-processing system, known as SAMS (Structural Analysis and Mapping System), that takes as input rasterized, thematic (land cover) maps (Barr and Barnsley 1995; Barr and Barnsley 1997; Barr and Barnsley 1998b). The system aggregates contiguous pixels sharing the same thematic label into discrete regions and derives a number of morphological properties (e.g., area, perimeter, compactness) and structural relations (e.g., adjacency, containment, distance and direction) from them. This information is stored in a graph-theoretic data model, known as XRAG (eXtended Relational Attribute Graph). In this model, each building (or other land cover parcel) is represented by a node in one or more graphs, and for which morphological properties, such as the area and compactness of the corresponding region, may be stored. Structural relations, such as adjacency, are represented by edges connecting these nodes in the appropriate graph, while structural properties, such as the distance between buildings, are represented as properties of these edges.

Previous studies using SAMS/XRAG have typically employed data acquired by multispectral scanning systems to divide the observed scene into its constituent land cover parcels, but images generated by LIDAR (Laser-Induced Detection And Ranging) scanning altimetry represent a further source of appropriate information. In this context, LIDAR data offer the advantage of providing additional information on building height and, in some instances, roof form (i.e., pitched or flat). In this study, therefore, we use a combination of airborne LIDAR and satellite multi-spectral scanner data to derive information on the buildings present in two residential and two industrial districts of the city of Cardiff, Wales, UK. We employ SAMS/XRAG to derive from these data information on the structural composition of land use. This is used to analyse the extent to which these urban land use categories can be identified and distinguished on the basis of their structural composition. The results are compared with those obtained from reference data of the same areas, in the form of fine spatial scale Ordnance Survey digital map data. The level of agreement between the image and reference (map) data is indicative of the degree to which information on the detail of the urban structures is preserved or lost in the image formation process.

4.2   LIDAR – systems, data and applications

4.2.1   *LIDAR data quality and cost-effectiveness*

LIDAR devices are active systems that emit laser pulses and receive the signals returning from the ground. They also record information on the angle of the emitted pulse and the time elapsed before the return signal is received at the sensor. By combining these data with precise measurements of the position and attitude of the sensor platform, using Global Positioning System (GPS) technology, it is possible to generate very detailed, digital models of the land surface: horizontal sampling intervals of <3 m on the ground and a vertical accuracy of 10 cm to 15 cm are typical (Environment Agency 1997; Environment Agency 1998), although greater levels of detail and accuracy can be achieved depending on the sensor used and the altitude of the platform. Unlike Digital Elevation Models (DEMs) produced from topographic (contour) maps, however, LIDAR-derived Digital Surface Models (DSMs) also record the details of localized surface protrusions, such as individual buildings and trees.

LIDAR systems represent a cost-effective means of acquiring digital elevation data over large areas (Flood and Gutelius 1997). For example, a feasibility study commissioned by the UK's Environment Agency in 1998 estimated the cost of a large-area LIDAR survey to be about £200 per square kilometre, based on a point sampling density of ~0.4 per square metre and a data capture rate of 90 km² to 100 km² per hour (Environment Agency 1998). This compares favourably with the cost of conducting a similar survey using ground-based techniques (roughly £1,000 per square kilometre) (Environment Agency 1998). These qualities have encouraged the use of LIDAR data in a diverse set of applications, ranging from studies of vegetation canopy structure (Blair *et al.* 1999), surface hydrology (Ritchie 1996) and ice sheet dynamics (Krabill *et al.* 1995) through to detailed topographic survey (Krabill *et al.* 1994). Refer to Chapter 3 in this book by Baltsavias and Gruen for more details on LIDAR.

4.2.2   LIDAR system parameter issues

The point sampling density of LIDAR systems, and hence the level of detail with which information on surface elevation can be recorded, is dependent on a number of factors. These include the altitude and velocity of the platform, and the field-of-view and sampling frequency of the sensor. There is, however, a trade-off to be made between the sampling frequency of the sensor, its ranging precision and its ability to detect weak signals. Specifically, increasing the sampling frequency of the LIDAR system in an attempt to produce a more densely sampled DSM is typically achieved at the cost of a reduction in the power of the laser pulse and a widening of the angle over which it is emitted (DeLoach 1998). The effect of the former is to reduce the potential for detecting weak return signals, while the latter compromises the ranging precision of the sensor. This interdependence influences the selection of the most suitable system parameters for a given application, which may also be scale-specific (i.e., the requirements for generating high spatial resolution models of individual buildings are different from those of forest biomass monitoring: DeLoach 1998).

There are two other attributes that can be retrieved from some scanning LIDAR systems, namely the magnitude/amplitude of the return pulse and the timings of the first and last returns (Axelsson 1999). Amplitude information can, in principle, assist in distinguishing different surface cover types, depending on their spectral reflectance properties and the wavelength of the laser. Similarly, a dual (first and last) return capability provides information that may help to separate "solid" objects (e.g., buildings) from "light-permeable" ones (e.g., vegetation canopies). Therefore, solid objects tend to produce a strong, single return signal, since reflection of the laser pulse occurs predominantly from the surface of the object (i.e., at a single, fixed distance from the LIDAR). Light-permeable objects, on the other hand, will typically produce a more complex return signal. In the case of a tree canopy, for example, the first return will be produced by the tree crowns, since these are closest to the LIDAR. Depending on the density of the tree canopy, subsequent – typically weaker – returns may be produced by the understorey and the soil substrate.

Taking the tree canopy as a whole, variations in the returns from the tree crowns may give an indication of canopy-top roughness and density, while height differences between the tree crowns (first return) and the soil substrate (last return) may offer information useful to the estimation of canopy height and above-ground biomass (Dubayah *et al.* 1997).

4.2.3 *Applications of LIDAR data in urban studies*

LIDAR systems have also been employed in studies of urban areas. Typically, the aim of such studies is to segment the scene into discrete land cover classes or to generate very detailed models of building elevation. The accuracy with which these objectives can be achieved is dependent on the relationship between the LIDAR point sampling density and the size (scale) of the objects to be classified or modelled. Maas (1999b), for example, has shown that LIDAR data can be used to divide urban scenes into four broad land cover classes (i.e., buildings with pitched roofs, buildings with flat roofs, trees, and flat terrain or roads) using measures of height and height variation as inputs to a standard, per-pixel, maximum likelihood (ML) classification algorithm. Applying this approach to LIDAR data recorded at a sampling density of five points per square metre achieves an overall accuracy of 98 per cent and shows that this can be improved to 99 per cent through subsequent use of morphological filtering and connected component labelling techniques. A building elevation model is also derived from the resultant scene segmentation, via analysis of invariant moments in the height data (Maas 1999a; Maas and Vosselman 1999).

LIDAR data acquired at the point sampling densities reported above are not yet widely available. Consequently, Maas (1999a) also tests the effect of a reduction in point sampling density on the retrieval of selected scene properties, such as building centroid location, building dimensions (orientation, length, width and height) and roof slope. In this context, stepwise thinning of the original data set suggests that it may not be possible to estimate accurately some of these properties, particularly roof slope, at point sampling densities of much less than one per square metre (Maas 1999a). This has potentially important implications for the retrieval of building area, perimeter and compactness in the present study, in which the LIDAR data are sampled at ∼0.4 points per square metre. Ancillary data, in the form of a multispectral IKONOS image, are available, however, to augment the information provided by the LIDAR image used here.

4.3 Derivation of building information

4.3.1 *Data sets*

The data used in this investigation cover both residential and industrial areas in the city of Cardiff, Wales, UK (Figure 4.1). LIDAR data were

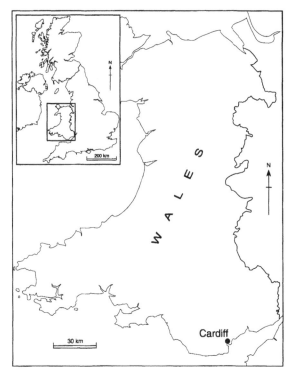

Figure 4.1 Location of the city of Cardiff, Wales, UK.

acquired by the UK Environment Agency using an ALTM 1020 device mounted on a light aircraft. These were recorded on 28 March 1998 and were subsequently interpolated by the Environment Agency onto a 2-m resolution grid (Figure 4.2a). The accuracy of this product is given as 9 cm to 15 cm in the vertical and 10.8 cm to 14.5 cm in the horizontal. The second image data set used in this study is a multispectral IKONOS scene acquired close to midday on 14 August 2001. These data have a nominal spatial resolution of 4 m and consist of four spectral channels (blue, green, red and near-infrared) in the range 450 nm to 880 nm. They were subsequently resampled onto a 2-m grid, to match the spatial resolution of the LIDAR data (Figure 4.2b–d).

Reference data on the size, shape and spatial location of building objects in the study area are provided by digital map data, supplied by the UK Ordnance Survey (OS). These are taken from the Land-Line.93+ series and consist of vector-format 1:1,250 scale data, containing information on various types of point, line and area features. The majority of the Land-Line.93+ tiles used in this study were digitized between 1997 and 2000, such that they can be considered to have the same effective acquisition date as the LIDAR data. Building features were extracted from these data and

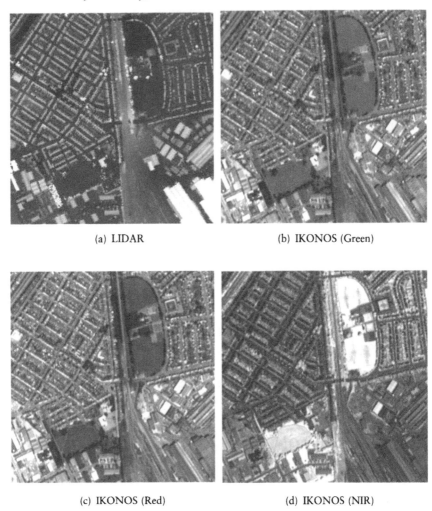

(a) LIDAR   (b) IKONOS (Green)

(c) IKONOS (Red)   (d) IKONOS (NIR)

Figure 4.2 1-km^2 extract from (a) the LIDAR and (b–d) the IKONOS image data
sets.

were topologically structured to generate a series of polygon coverages.
These were then exported as a binary (building vs.non-building) raster, with
a 2-m spatial resolution, suitable for input into SAMS/XRAG. A second
source of reference data is also used, namely OS Land-Form Profile data.
This is a 1 : 10,000 scale digital elevation model, with a vertical accuracy of
±1.0-m root mean square error (RMSE). These data are used to remove
variations in the underlying terrain elevation from the LIDAR-derived DSM.

The LIDAR and IKONOS images were geometrically rectified so that they could be co-registered with the reference data. The resultant registration accuracy is estimated to be better than 0.46 pixels (i.e., 0.92-m) RMSE.

4.3.2 Data processing stages

4.3.2.1 Removal of variations in the underlying terrain

The LIDAR image data represent the height of each pixel in metres above a reference datum. Consequently, the first stage in extracting buildings from these data involves removal of the underlying (baseline) terrain to produce estimates of height above local ground level (i.e., to produce a topographically "flat" scene). This was achieved by resampling the Land-Form Profile data onto a 2-m grid and subtracting the resultant terrain elevation values from the heights reported by the equivalent pixels in the LIDAR image.

4.3.2.2 Identification of buildings in the terrain-adjusted LIDAR data

Having removed the underlying terrain from the LIDAR data, the next stage was to identify and delineate each of the buildings present within the scene. This seemingly simple task is complicated somewhat for two reasons: first, a number of features other than buildings protrude above local ground level, most notably trees, and these must be excluded from further analysis; second, the 2-m spatial sampling of the gridded LIDAR data is such that the edges and height profiles of buildings cannot be determined exactly. The latter is essentially a "mixed pixel" problem, broadly analogous to that encountered in images acquired by multispectral scanners (Ichoku and Karnieli 1996). Therefore, some pixels in the LIDAR image only partially overlap a building, such that the reported elevation is a function of the height of the building, the fraction of the pixel that it occupies, and the spatial interpolation procedure employed to generate the DSM from the LIDAR point samples. This typically produces an elevation value lower than the actual height of the building and affects the delineation of the building outline. We have, therefore, explored two methods for distinguishing "building" from "non-building" pixels in the terrain-adjusted LIDAR data. Each employs a combination of the LIDAR and IKONOS images, in which the latter are used to help distinguish trees from buildings. The performance of the two methods was evaluated for the 1-km^2 test area shown in Figure 4.2, with respect to the reference data set generated from the OS digital map data.

IMAGE CLASSIFICATION APPROACH

In the first method, a supervised, per-pixel, ML classifier was used to identify building and non-building pixels in the combined LIDAR and IKONOS images. The resultant data exhibited very good (91.53 per cent) agreement, measured on a pixel-by-pixel basis, with the OS reference data. Nevertheless, a number of very small "buildings" were found in the classified image data that were not present in the reference data. There are several possible reasons for this, including, in order of likely significance, (i) errors in the image classification, (ii)the effect of the spatial resampling procedures performed on both the LIDAR and IKONOS images, (iii) changes on the ground between the acquisition dates of the image and reference data sets, (iv) errors introduced by removing the underlying terrain using the Land-Form Profile data and (v) omission of small building objects from the Land-Line.93+ data or the products that we have derived from them. Since it was not possible to establish the true cause of these discrepancies, it was decided to exclude the erroneous buildings from further analysis by relabelling all building objects with an area less than 24 m² (the minimum expected size of buildings in this region). This resulted in a similar number of building objects being reported in both the image and reference data sets.

ELEVATION AND NDVI THRESHOLD APPROACH

The second approach involved application of a simple height threshold to the terrain-adjusted LIDAR data and an NDVI (Normalized Difference Vegetation Index; NDVI = (NIR−Red)/(NIR + Red)) threshold to the IKONOS data. A threshold value of NDVI = 0.24 proved effective in terms of distinguishing vegetation from non-vegetation, but selection of an objective height threshold to distinguish buildings from non-buildings in the LIDAR data proved more problematic. While the average height of a one-storey (i.e., the shortest expected) building can be established very easily through field inspection, the "mixed pixel" problem, described above, typically produces somewhat lower height values for many "building" pixels in the resultant LIDAR data. The choice of appropriate threshold is, therefore, not straightforward: selection of a threshold that is too high may result in "erosion" of the derived building objects (i.e., errors of omission from the building class), while selection of one that is too low may result in their overestimation (i.e., errors of commission). In the circumstances, the solution adopted in this study is one based on "trial-and-error" – assessing the accuracy of building/non-building identification relative to the reference data, based on a number of different height thresholds (Figure 4.3). On this basis, a height threshold of 2.62 m was selected, since this is the point at which the curves for producer's and user's accuracy cross for both the building and non-building classes. It therefore represents a compromise between errors of commission and omission for each class. The agreement between the combined LIDAR and IKONOS image data and the reference data set was 91.65 per cent using this threshold. Finally, buildings with an area less than 24 m² were removed, as per the image classification approach.

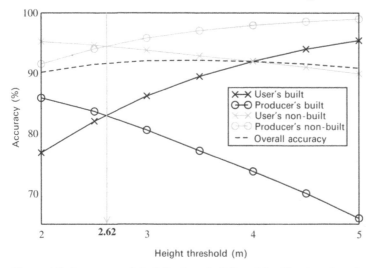

Figure 4.3 Accuracy of building/non-building identification based on different height thresholds applied to the terrain-corrected LIDAR data.

COMMENT ON THE TWO APPROACHES

Although there is little difference in terms of overall performance between the two methods of building identification, the simple threshold approach is adopted here in all subsequent analyses. There are two reasons for this: first, and most importantly, because of its simplicity and transparency; second, because it performed very slightly better in terms of the number of building pixels identified in the image data compared with the reference data set.

4.3.2.3   Assessment of building extraction by land use category

Although it is instructive to examine the overall agreement between the building pixels derived from the combined LIDAR and IKONOS images and those identified in the OS reference data for the 1-km^2 test area, it is possible that this masks variations in performance as a function of land use. These are expected to occur because the typical size, shape, orientation and spacing of buildings differs between urban land use categories. Consequently, we have examined in greater detail the ability of the height, NDVI and minimum area thresholds, described above, to identify and delineate building pixels in four discrete sample areas within the full study area – two drawn from residential districts of late nineteenth century construction (hereafter referred to as Residential #1 and #2) and two from twentieth century industrial zones (Industrial #1 and #2). Each covers an area of 400 m × 440 m on the ground. It should be noted, however, that the sample areas are not exclusively filled with buildings of

Table 4.1 Agreement between building (B) and non-building (!B) pixels identified in the LIDAR and IKONOS image data and in the reference data for four test extracts

Land use extract	Agreement (%)					
	Producer's		User's		Overall	Ratio building pixels LIDAR/reference
	B	!B	B	!B		
Industrial #1	90.50	96.40	89.90	96.60	94.85	1.01
Industrial #2	91.60	92.30	80.40	96.90	92.10	1.14
Residential #1	76.40	92.90	84.20	88.80	87.40	0.91
Residential #2	77.80	93.80	86.10	89.50	88.50	0.90

the designated land use, but are taken to represent the diversity present within it: for example, the residential areas may contain some buildings dedicated to commercial, educational or recreational activity. The results are presented in Table 4.1 and Figure 4.4.

Table 4.1 indicates that the overall agreement, measured on a pixel-by-pixel basis, between the reference data and the image data is slightly higher for the two industrial subscenes (92 per cent to 95 per cent) than for their residential counterparts (87 per cent to 89 per cent), although in each case the results are very good. The producer's accuracy for buildings in the residential subscenes is comparatively low, however, and this results in an underestimation of the number of building pixels (i.e., the ratio of image-to-reference building pixels is less than one). Figure 4.4 indicates that this manifests itself as erosion of the building objects derived from the LIDAR and IKONOS images relative to those in the OS reference data. It is possible that this could have been partly overcome by reducing the height threshold used to identify buildings in the LIDAR data, but probably only at the expense of greater errors of commission into the building class (i.e., a reduction in the user's accuracy), given the findings presented in Figure 4.3.

The differences between the image-derived and reference building data sets are perhaps easier to visualize in Figure 4.5. In this figure, black pixels represent areas where both the image and the reference data identify part of a building, while white pixels represent areas where neither data set indicates that a building is present. Similarly, light-grey pixels represent areas identified as being part of a building in the reference data, but not in the image (i.e., an error of omission), while dark-grey pixels represent areas identified as being part of a building in the image, but not in the reference data (i.e., an error of commission). Clearly, both types of error are present in each of the four subscenes. Although the majority of these differences are likely to result from the limitations of the height, NDVI and minimum area thresholds used to identify the building pixels, other differences may be due to the accumulation and removal of stockpiles of freight or temporary

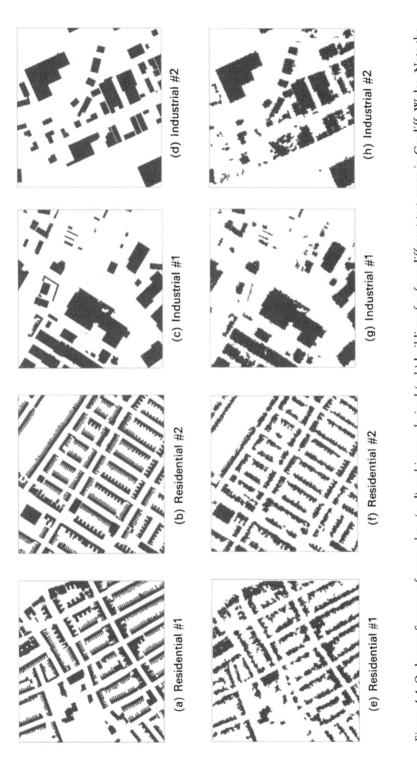

(a) Residential #1   (b) Residential #2   (c) Industrial #1   (d) Industrial #2

(e) Residential #1   (f) Residential #2   (g) Industrial #1   (h) Industrial #2

Figure 4.4 Ordnance Survey reference data (a–d) and image-derived (e–h) buildings for four different test areas in Cardiff, Wales. Note the erosion of the building areas in the image-derived data sets.

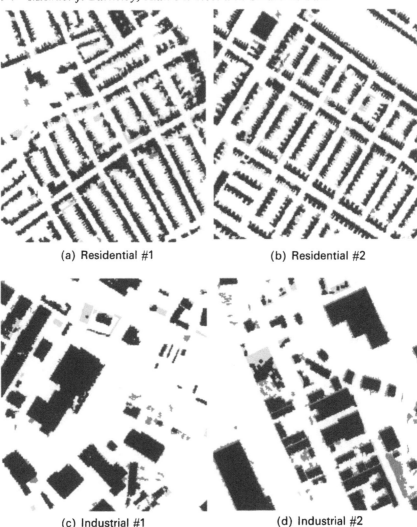

(a) Residential #1   (b) Residential #2

(c) Industrial #1   (d) Industrial #2

Figure 4.5 Comparison between the "building" pixels identified in the image data and the Ordnance Survey (OS) reference data for two residential and two industrial sample areas: black = building in both; white = non-building in both; light grey = building in the OS data but not in the images; dark grey = building in the images but not in the OS data.

constructions, particularly in the industrial areas. The residential subscenes, on the other hand, appear to be most significantly affected by errors of omission (i.e., underestimation, or erosion, of building extent). This is most likely an artefact of the relatively low point sampling density (∼0.4 per square metre) of the LIDAR data used in this study and its subsequent interpolation onto a 2-m grid (Maas 1999a; Maas and Vosselman 1999).

The resultant smoothing of building edges inhibits precise detection and delineation of the narrow, densely spaced extensions at the back of the Victorian terraced houses in this area.

4.3.2.4 Analysis of the derived building height profiles

Underestimation of the physical extent of certain buildings in the image data can be explored further through an analysis of height profiles along a number of sample transects running across the residential and industrial test areas, the results of which are shown in Figure 4.6. In this figure, solid lines represent the LIDAR-derived building profiles, vertical dashed lines indicate the limits of the corresponding buildings identified in the OS reference data, while dotted lines denote the building extent produced by applying the

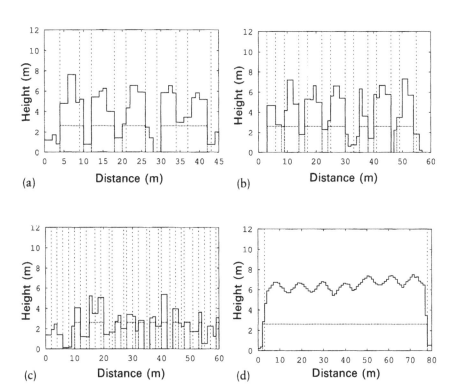

Figure 4.6 Effectiveness of the height threshold in deriving building profiles. The solid lines show the LIDAR-derived building profiles, the vertical dashed lines indicate the limits of the corresponding buildings identified in the Ordnance Survey reference data and the dotted lines denote the inferred building extent resulting from the application of the height, NDVI and minimum area thresholds to the combined LIDAR and IKONOS image data.

height, NDVI and minimum area thresholds to the combined LIDAR and IKONOS images. Figures 4.6a–c show transects across the back of three residential terraces. It is evident that the processing chain is quite effective in delineating the buildings where they are relatively widely spaced (Figures 4.6a–b) but, not surprisingly, is less effective where the spacing between the buildings approaches the grid resolution of the image data (Figure 4.6c). The smoothing effect of the LIDAR data interpolation on the building edges is also evident in these figures, again partly as a result of resolution/gridding issues, but also because of projection and occlusion effects resulting from the scanning operation of the LIDAR. Figure 4.6d illustrates that the larger, more widely spaced industrial buildings are typically much more accurately delineated.

4.4   Analysis of the structural composition of land use

The previous section examined the potential to identify building objects in a combination of LIDAR and IKONOS images, assessing the success with which this can be achieved by comparing the results with Ordnance Survey digital map data. Overall, the degree of correspondence between the two data sets was found to be very high, when measured on a per-pixel basis, although there was evidence of variations in performance as a function of urban land use. In particular, the identification and delineation of small, closely spaced buildings, typical of the residential areas studied, was found to be slightly problematic, primarily as a result of the relatively low point sampling density of the LIDAR data. In this section, we explore a number of related issues, namely: (i)whether there are significant differences between the structural composition of buildings in the residential and industrial areas, determined using the reference data, (ii)whether these differences are greater *between* or *within* the two land use categories studied and (iii) whether the combined LIDAR and IKONOS images preserve the morphology and spatial composition of buildings sufficiently well to allow the two land use categories to be distinguished in these terms. The SAMS/XRAG system developed by Barr and Barnsley (1998b) is used for this purpose.

4.4.1   *Analysis of building morphology*

Three morphological properties of buildings are examined here, namely their area, compactness and height. Area and compactness were selected on the basis of several previous studies, which suggest that these properties may help to distinguish different types of urban land use (Barr and Barnsley

1998a; Barr and Barnsley 2000). Various measures of object compactness are available. In this study we use a scale-invariant measure of the object's shape relative to a perfect circle of the same area, for which compactness equals 1.0 (Sonka *et al.* 1993),

$$C = \frac{P^2}{4\pi} \times A \qquad\qquad (4.1)$$

where C = compactness, P = perimeter and A = area. As the shape of a building deviates from this norm, its compactness value increases (i.e., it becomes less compact). In terms of building height, the maximum elevation of each building is employed. This is intended to avoid problems associated with "mixed pixels" at the edges of buildings. Obviously, information on building height should help to separate high-rise from low-rise residential districts and, hopefully, residential from industrial land.

Table 4.2 presents descriptive statistics on building morphology for each of the four sample areas, based on an analysis of the discrete building objects (cf. building pixels) identified in the image data and the building polygons recorded in the corresponding OS reference data. Note that the number of buildings reported by the two data sets is quite similar for all four sample areas, although, with one exception, the image data tend to overestimate the number of buildings present. As a result, the mean building size in the image data is generally smaller than that in the reference data. The effect is most pronounced for the Residential #1 sample area. In the other three sample areas, the mean area of buildings reported by the image data lies within ± 7 per cent of that reported by the reference data.

In terms of building compactness, the values derived from the image data tend to underestimate those reported by the reference data in the residential sample areas, but are quite similar to the reference values for the industrial subscenes (Table 4.2). The former indicates that some of the complex building shape present in the digital map data has been lost, probably as a result of the limited point sampling density of the LIDAR and the effects of the gridding process used to generate an image from LIDAR point samples (Maas 1999a; Maas and Vosselman 1999).

One of the most surprising features of Table 4.2 is that the mean building height is greater in the residential subscenes than in the industrial ones, even though the maximum building height is larger in the latter (24.1 m and 31.1 m in the industrial sample areas, compared with 20.0 m and 15.9 m in the residential areas). Otherwise, the results reported in Table 4.2 indicate that building area and compactness are captured reasonably well by the image data, despite some loss of spatial detail in the residential areas. More significantly, however, Table 4.2 suggests that there are varying degrees of overlap in the statistical distributions of all three morphological properties between the residential and industrial sample areas.

Table 4.2 Descriptive statistics on the morphological properties of building objects identified in the combined LIDAR and IKONOS images and the Ordnance Survey (OS) digital map data

Land use	Data	Number of buildings	Area (m²)			Compactness (m²/m²)			Heights (m)		
			Mean	σ	Median	Mean	σ	Median	Mean	σ	Median
Residential #1	LIDAR	97	547	701	334	5.53	4.55	3.85	10.75	3.12	10.24
	OS	89	657	628	520	9.36	9.19	5.37	–	–	–
Residential #2	LIDAR	96	547	443	497	5.45	3.30	4.63	9.22	1.94	9.40
	OS	109	534	521	518	8.54	8.02	7.74	–	–	–
Industrial #1	LIDAR	64	725	1651	141	3.18	1.35	2.88	7.60	3.55	6.87
	OS	58	796	1679	212	3.18	1.86	2.63	–	–	–
Industrial #2	LIDAR	60	855	1727	123	3.64	2.37	2.94	8.41	4.75	8.27
	OS	51	883	1547	373	2.68	1.17	2.41	–	–	–

4.4.2 *Analysis of structural composition*

To represent more fully the structural complexity of land use, we need to extend our analysis beyond merely the morphological properties of individual buildings to consider their spatial arrangement within the scene. Numerous techniques and measures exist for this purpose (Barr and Barnsley 1998b). In principle, we might expect that different urban land uses may be distinguished in terms of the number of buildings per unit area (i.e., building density or spacing) – which is likely to be partly correlated with building size – and their location with respect to one another (i.e., the spatial pattern of buildings). This information can be captured effectively using a single, graph-theoretic measure, known as a Gabriel graph (Jaromczyk and Toussaint 1992; Mutula and Sokal 1980).

Figure 4.7 illustrates the process by which Gabriel graphs are constructed. The light-grey boxes in Figure 4.7a represent individual buildings located within the scene, while the small, dark-grey circles represent their centroids. The latter form the nodes in the resultant Gabriel graph. A proximity relation between a node pair is defined by a circle passing both nodes (Figure 4.7a). If no other node lies within this circle, the two nodes are considered to be relative neighbours and an edge is created connecting them (Figure 4.7b). All possible node pairs are evaluated in this way.

Having identified relative neighbours, two properties may be derived from the resultant graph: first, the distances between connected graph nodes (i.e., building centroids); second, the number of graph edges incident on each node (i.e., the number of relative neighbours for each building and, hence, the spatial arrangement of buildings that surround it). The second property is known as node degree or, sometimes, node valency ($\delta(v)$) (Biggs 1989; Piff 1992). Note that node degree is theoretically unbounded in value. Therefore, if the neighbours are distributed radially at an equal distance around a building (e.g., as the hour marks are from the centre of a clock face), no other centroids will fall within the connecting circles and graph

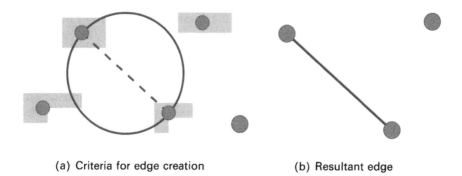

(a) Criteria for edge creation   (b) Resultant edge

Figure 4.7 Visual representation of the process leading to the construction of a Gabriel graph (see text for explanation).

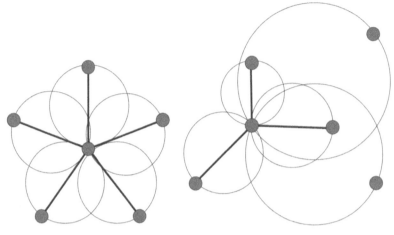

(a) Radial pattern of building nodes   (b) More typical pattern of building nodes

Figure 4.8 Effect of the spatial pattern of building centroids on the resultant node degree (see text for explanation).

edges will be created connecting all of them to the node concerned. This is illustrated in Figure 4.8a for which $\delta(v) = 5$. In reality, a perfectly symmetrical distribution of building centroids, such as this, is rare. The angles and distances between a given building centroid and each of its neighbours will differ, effectively limiting the range of possible values for $\delta(v)$ (see, e.g., Figure 4.8b for which $\delta(v) = 3$). For the data used in the present study, the typical range of values is $\leq \delta(v) \leq 6$.

Visualizations of the Gabriel graphs for each of the four sample areas described in the previous section are presented in Figure 4.9. This shows the results for both the OS reference data (Figures 4.9a–d) and the combined LIDAR and IKONOS images (Figures 4.9e–h). As each node in these subgraphs represents a building centroid, the number and spatial distribution of nodes in these figures indicate the density and locations of buildings in the corresponding sample areas (Figure 4.4). Likewise, the number, orientation and length of the graph edges indicate the spatial pattern of these buildings. For each sample area, the Gabriel graphs derived from the image data and from the OS reference data appear to be similar in terms of overall node location, node degree, edge length and edge orientation. Moreover, there appear to be qualitative differences between the Gabriel graphs derived from the residential and industrial areas. Quantitative analysis of the graph structures is, however, required to establish whether the differences between these two land uses are statistically significant; in other words, whether between-class variances exceed within-class variances. To this end, we explore the statistical distributions of two properties of the Gabriel graphs, namely edge length and node degree.

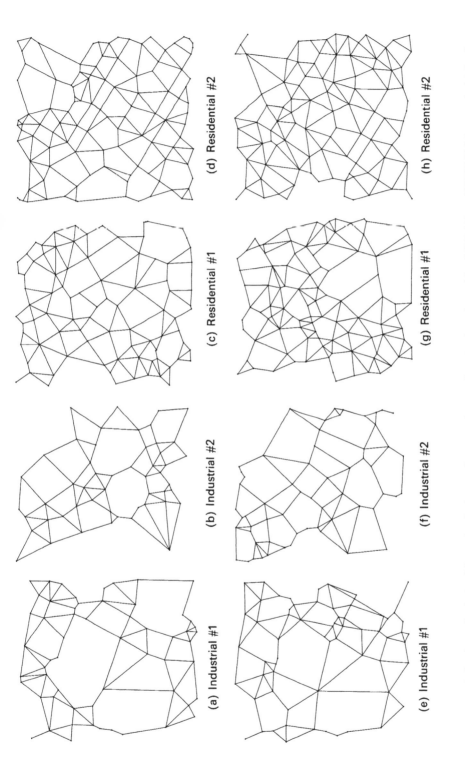

(a) Industrial #1   (b) Industrial #2   (c) Residential #1   (d) Residential #2

(e) Industrial #1   (f) Industrial #2   (g) Residential #1   (h) Residential #2

Figure 4.9 Gabriel graphs derived from the Ordnance Survey reference data (a–d) and combined LIDAR and IKONOS images (e–h) for four sample areas in Cardiff, Wales.

Table 4.3 Descriptive statistics on node degree for the Gabriel graphs of each sample area

Land use	Data	Number of nodes	Node degree, $\delta(v)$		
			Mean	σ	Median
Residential #1	LIDAR	97	4.31	1.09	4
	OS	89	3.54	1.10	3
Residential #2	LIDAR	96	4.34	1.14	4
	OS	109	3.71	1.12	4
Industrial #1	LIDAR	64	3.74	1.23	4
	OS	58	3.61	0.94	4
Industrial #2	LIDAR	60	3.42	1.18	3
	OS	51	3.36	1.10	3

Table 4.3 presents descriptive statistics on the total number of nodes and on node degree in the Gabriel graphs for each of the four sample areas. With one exception, the image data overestimate the number of building nodes present in the scene by about 10 per cent with respect to the reference data. This is largely due to the erosion of the building outlines and the resultant fragmentation of building structures in the image data. As a result, the mean node degree is similarly overestimated – by approximately 20 per cent in the residential zones where the effects of building erosion and fragmentation are greatest, but only approximately 3 per cent in the industrial zones. More importantly, given the similar mean values and relatively large standard deviations of node degree for all four land use samples, it seems unlikely that it would be possible to distinguish between these two land uses based on an analysis of this structural property alone.

Table 4.4 presents descriptive statistics on the total number of edges and on edge length in the Gabriel graphs for each of the four sample areas. This shows that, with one exception, the number of edges in the Gabriel graphs derived from the image data is broadly similar (±10 per cent) to that of the corresponding areas in the OS reference data. The exception is the Residential #1 sample, which has approximately 18 per cent more edges in the image data than in the reference data. This reflects a more fragmented building structure in the images, resulting from erosion of the building outlines, as previously discussed. Table 4.4 also indicates that there is a significant difference between the residential and industrial extracts in terms of the total number of edges present in each, reflecting the fact that industrial areas generally have fewer, but larger buildings. The mean edge length is slightly higher for the industrial sample areas, although there is overlap between the statistical distributions of this property for the two land uses. Consequently, it is unlikely that they could be distinguished unambiguously, solely on the basis of the distances between individual buildings.

Table 4.4 Descriptive statistics on edge length (m) for the Gabriel graphs of each sample area

Land use	Data	Number of graph edges	Graph edge length		
			Mean	σ	Median
Residential #1	LIDAR	194	42.96	19.45	39.47
	OS	164	46.21	19.28	43.88
Residential #2	LIDAR	195	46.14	17.37	43.09
	OS	202	42.65	19.38	39.35
Industrial #1	LIDAR	112	51.00	28.87	45.80
	OS	101	55.77	29.14	51.43
Industrial #2	LIDAR	97	49.07	27.80	40.92
	OS	92	51.56	24.79	44.75

4.4.3 Statistical separation of land use patterns

It should be evident from the preceding discussion that no single morphological property or structural relation enables us to distinguish unambiguously between the two categories of urban land use examined here, although their statistical distributions appear to differ as a function of land use in most instances. In practice, we must employ some combination of two or more morphological properties and structural relations, in much the same way that we use data from several spectral wavebands in per-pixel image classification for land cover mapping. The key issues that arise from this are: (i) what combination of properties and relations provides the best separation of the two land use categories, and (ii) is this sufficient to identify areas of each land use unambiguously?

To address these issues, we have used two tests of statistical separability – one of which is non-parametric (Kruskall–Wallace H-test), the other parametric (transformed divergence). The former is used to perform a comparison between the statistical distributions of a single morphological property or structural relation measured for each of the four land use samples. In this study, we ran the H-test on building area, compactness, height and Gabriel-graph edge length. Where the calculated value of this statistic (H_{calc}) falls below the critical value (H_{crit}) at the chosen significance level, we may conclude that there is no statistically significant difference between the sample areas in terms of that property or relation (Norcliffe 1982). Transformed divergence (D), on the other hand, allows us to combine two or more properties and relations when measuring the similarity (or difference) between the structural composition of the land use samples. Here, values of D are scaled between 0 (identical distributions) and 100 (non-overlapping distributions) (Mather 1987).

Table 4.5 Kruskall–Wallis H tests of statistical separability between urban land use categories

Properties/Relations	Land use comparison		
	Industrial vs. Industrial	Industrial vs. Residential*	Residential vs. Residential
Area	0.18	12.38	3.37
Compactness	0.25	31.48	0.97
Height	0.68	39.91	7.81
Edge length	0.29	4.05	4.17

* $H_{crit} = 3.84$ at the 5% significance level.

Table 4.6 Transformed divergence analysis of statistical separability between urban land use categories

Properties/Relations	Land use comparison		
	Industrial vs. Industrial	Industrial vs. Residential*	Residential vs. Residential
A/C/H/E/N	17.79	97.41	25.40
A/C/H/E	16.20	97.00	21.39
A/C/E/N	13.98	96.74	12.09
A/C/E	11.96	96.17	7.47
C/H/E/N	16.39	57.47	20.29

A = Area, C = Compactness, H = Height, E = Graph edge length, N = Node degree.
* These values are averages of the four possible extract comparisons.

The results of the Kruskall–Wallis H-tests are reported in Table 4.5. This suggests that the sample areas of industrial and residential land differ significantly in terms of all four properties (i.e., $H_{calc} > H_{crit}$), although the result for edge length is marginal. Otherwise, building compactness and building height yield the largest values of H_{calc}. The two industrial sample areas exhibit no significant differences in terms of their morphological or structural properties, but there are significant differences in terms of building height and edge length between the residential samples. This may reflect genuine variability in different areas of residential land, although it is not immediately apparent on visual inspection of Figure 4.4.

The results of the transformed divergence tests are given in Table 4.6. It is evident, once again, that the differences between the two urban land use categories (i.e., industrial vs. residential) are greater than the differences within them (i.e., industrial vs. industrial or residential vs. residential). Greatest separation of residential and industrial land is achieved by using all five morphological and structural properties simultaneously (i.e., A/C/H/

E/N), but it is worth noting that this is at the expense of a slight increase in within-class variability. It could, in fact, be argued that a combination of building area, building compactness and Gabriel graph edge length (i.e., A/C/E) offers the best compromise between within-class variability and between-class separability. Notice, however, the importance of building area: removing this property results in a dramatic reduction in between-class separability (i.e., C/H/E/N).

4.5   Discussion

The results outlined above suggest that it may be possible to distinguish and, hence, map different categories of urban land use based on an analysis of the structural composition of their constituent buildings. This requires that we are able to determine the morphological properties of the buildings (e.g., their area, compactness and height) and the structural relations between them (i.e., their spatial pattern and interrelationships), and that we have some means of analysing and interrogating these properties and relations.

It has been shown that building objects can be identified with a high level of accuracy (88 per cent to 95 per cent) from a combination of high spatial resolution LIDAR and multispectral images, compared with reference (digital map) data of the same area, albeit with some loss of fidelity in terms of the retrieved building area and compactness values. It has also been possible to capture important aspects of the spatial pattern of buildings in each land use by means of two structural properties (edge length and node degree) derived from a single graph-theoretic transformation (i.e., Gabriel graph), using the XRAG/SAMS system (Barr and Barnsley 1997; Barr and Barnsley 1998b).

Despite this, a number of uncertainties and concerns remain. First, the present study has considered only two categories of urban land use – namely, Victorian terraced housing and late twentieth century industrial/commercial developments. These might reasonably be expected to exhibit very different spatial/structural characteristics, and it remains to be seen whether it is possible to distinguish more subtle differences in urban land use. Second, only two sample areas have been used for each land use, and all four samples have been drawn from the city of Cardiff. It is entirely possible, therefore, that we have not captured the full variability typical of even these two land uses. Third, the expected statistical distributions of some of the morphological and structural properties used in this study are not yet known, so that it is unclear as to whether parametric or non-parametric tests of statistical similarity/difference are most appropriate or, indeed, whether the different properties may be used together untransformed in a single test. Combined with the very limited number of sample areas available, this implies that the results of this study are, at best, indicative.

Certainly, they cannot be regarded as conclusive. More generally, we are in the process of examining all three of these important issues.

Notwithstanding these caveats, the overall picture is positive. The potential for identifying and delineating discrete zones of land use in urban areas from high spatial resolution remotely sensed images seems clear, assuming that further analyses of other sample areas yield consistent results. What remains then is to develop a method for "classifying" urban land use based on remotely sensed images, through an analysis of building morphology and graph structure. In this context, the alternatives include standard image classification procedures (e.g., maximum likelihood or artificial neural network algorithms), to which a set of morphological and structural properties are presented as input data channels, or more formal graph searching techniques and graph similarity measures.

4.6   Conclusions

This study has provided a preliminary confirmation of the hypothesis that different types of urban land use can be identified in high spatial resolution remotely sensed images through an analysis of the morphological properties of their constituent buildings and the spatial/structural relations between them. More specifically, it has been shown that there are statistically significant differences between the area, compactness, height and spatial arrangement of buildings in areas of Victorian terraced housing and areas of twentieth century industrial/commercial land, and that these can be captured effectively using a graph-based spatial analysis system. We are currently extending the research reported in this study through analysis of the structural composition of a much wider range of urban land uses in Cardiff, using multiple extracts for each land use. We are also developing a graph clustering algorithm to enable automatic segmentation of either digital map or digital image data of urban areas on the basis of predefined land use classes.

4.7   Acknowledgements

This research is supported by the UK Natural Environment Research Council through the URGENT (Urban Regeneration and the Natural Environment) Thematic Program (Grant Ref. GST/02/2241). The authors would like to thank Prof. Paul Longley (UCL) and Dr Adrian Luckman (University of Wales Swansea) for constructive criticism in connection with this project. Thanks are also due to the Ordnance Survey (UK), particularly Dr David Holland, for advice and involvement in this project and for provision of digital map data, and to the UK Environment Agency for providing the LIDAR data.

4.8   References

Axelsson, P., 1999, Processing of laser scanner data – algorithms and applications. *ISPRS Journal of Photogrammetry and Remote Sensing,* **54,** 138–47.

Barnsley, M. J. and Barr, S. L., 1996, Inferring urban land use from satellite sensor images using kernel-based spatial reclassification. *Photogrammetric Engineering and Remote Sensing,* **62,** 949–58.

Barnsley, M. J. and Barr, S. L., 1997, A graph-based structural pattern recognition system to infer land use from fine spatial resolution land cover data. *Computers, Environment and Urban Systems,* **21,** 209–25.

Barnsley, M. J. and Barr, S. L., 2000, Monitoring urban land use by Earth observation. *Surveys in Geophysics,* **21,** 269–89.

Barr, S. L. and Barnsley, M. J., 1995, A spatial modelling system to process, analyse and interpret multi-class thematic maps derived from satellite sensor images. In P. F. Fisher (ed.) *Innovations in GIS 2* (London: Taylor and Francis), pp. 53–65.

Barr, S. L. and Barnsley, M. J., 1997, A region-based, graph-theoretic data model for the inference of second-order information from remotely sensed images. *International Journal of Geographical Information Science,* **11,** 555–76.

Barr, S. L. and Barnsley, M. J., 1998a, Application of structural pattern-recognition techniques to infer urban land use from Ordnance Survey digital map data. Paper given at Conference of GeoComputation '98, 3rd International Conference on GeoComputation (CD-ROM).

Barr, S. L. and Barnsley, M. J., 1998b, A syntactic pattern recognition paradigm for the derivation of second-order thematic information from remotely sensed images. In P. M. Atkinson and N. J. Tate (eds) *Advances in Remote Sensing and GIS Analysis* (Chichester, UK: John Wiley & Sons), pp. 167–84.

Barr, S. L. and Barnsley, M. J., 2000, Reducing structural clutter in land cover classifications of high spatial resolution remotely sensed images for urban land use mapping. *Computers and Geosciences,* **26,** 433–49.

Bauer, T. and Steinnocher, K., 2001, Per-parcel land use classification in urban areas applying a rule-based technique. *GeoBIT/GIS,* **6,** 24–7.

Biggs, N. L., 1989, *Discrete Mathematics* (Oxford: Clarendon Press).

Blair, J. B., Rabine, D. L. and Hofton, M. A., 1999, The laser vegetation imaging sensor: A medium-altitude, digitisation-only, airborne laser altimeter for mapping vegetation and topography. *ISPRS Journal of Photogrammetry and Remote Sensing,* **54,** 115–22.

DeLoach, S., 1998, Photogrammetry: A revolution in technology. *Professional Surveyor,* **18.**

Dubayah, R., Blair, J., Bufton, J., Clark, D., Ja Ja, J., Knox, R., Luthcke, S., Prince, S. and Weishampel, J., 1997, The vegetation canopy LiDAR mission. Paper given at Conference on Land Satellite Information in the Next Decade (Bethesda, MD: American Society for Photogrammetry and Remote Sensing), pp. 100–12.

Environment Agency, 1997, *Evaluation of the LiDAR technique to produce elevation data for use within the agency* (R&D summary). (Bristol: National Centre for Environmental Data and Surveillance, Environment Agency).

Environment Agency, 1998, *Airborne light detection and ranging feasibility study* (R&D Technical Summary ES41). (Bristol: National Centre for Environmental Data and Surveillance, Environment Agency).

Eyton, J. R., 1993, Urban land use classification and modeling using cover-type frequencies. *Applied Geography*, **13**, 111–21.

Flood, M. and Gutelius, B., 1997, Commercial implications of topographic terrain mapping using scanning airborne laser radar. *Photogrammetric Engineering and Remote Sensing*, **63**, 327–66.

Gong, P. and Howarth, P. J., 1992, Land use classification of SPOT HRV data using a cover-frequency method. *International Journal of Remote Sensing*, **13**, 1459–71.

Ichoku, C. and Karnieli, A., 1996, A review of mixture modelling techniques for sub-pixel land cover estimation. *Remote Sensing Reviews*, **13**, 161–86.

Jaromczyk, J. and Toussaint, G., 1992, Relative neighborhood graphs and their relatives. *P-IEEE*, **80**, 1502–17.

Krabill, W., Collins, J., Link, L., Swift, R. and Butler, M., 1994, Airborne laser topographic mapping results. *Photogrammetric Engineering and Remote Sensing*, **50**, 685–94.

Krabill, W., Thomas, R., Jezek, K., Kuivinen, K. and Manizade, S., 1995, Greenland ice sheet thickness changes measured by laser altimetry. *Geophysical Research Letters*, **22**, 2341–4.

Maas, H-G., 1999a, Fast determination of parametric house models from dense airborne laser scanner data. *International Archives of Photogrammetry and Remote Sensing*, **32**, 245–53.

Maas, H-G., 1999b, The potential of height texture measures for the segmentation of airborne laser scanner data. Paper given at Conference of the Fourth International Airborne Remote Sensing Conference and Exhibition/21st Canadian Symposium on Remote Sensing, Ottawa, Canada, p. 312.

Maas, H-G. and Vosselman, G., 1999, Two algorithms for extracting building models from raw laser altimetry data. *ISPRS Journal of Photogrammetry and Remote Sensing*, **54**, 153–63.

Mather, P. M., 1987, *Computer Processing of Remotely Sensed Images* (Chichester, UK: John Wiley & Sons).

Møller-Jensen, L., 1990, Knowledge-based classification of an urban area using texture and context information in Landsat-TM imagery. *Photogrammetric Engineering and Remote Sensing*, **56**, 899–904.

Mutula, D. and Sokal, R., 1980, Properties of Gabriel graphs relevant to geographic research and the clustering of points in the plane. *Geographic Analysis*, **12**, 205–22.

Norcliffe, G. B., 1982, *Inferential Statistics for Geographers* (London: Hutchinson).

Piff, M., 1992, *Discrete Mathematics: An Introduction for Software Engineers* (Cambridge: Cambridge University Press).

Ritchie, J., 1996, Remote sensing applications to hydrology: Airborne laser altimeters. *Hydrological Sciences Journal*, **41**, 625–36.

Sonka, M., Hlavac, V. and Boyle, R., 1993, *Image Processing, Analysis and Machine Vision* (London: Chapman & Hall).

Wharton, S. W., 1982, A contextual classification method for recognizing land use patterns in high-resolution remotely sensed data. *Pattern Recognition*, **15**, 317–24.

5 The use of wavelets for feature extraction of cities from satellite sensor images

Soe W. Myint

5.1 Introduction

Most image analysts would agree that, when extracting urban/suburban information from remotely sensed data, it is more important to have high spatial resolution (often finer than 5 m by 5 m) than high spectral resolution (i.e., a large number of spectral bands) (Jensen and Cowen 1999). However, some researchers (Latty and Hoffer 1981; Irons *et al.* 1985; Green *et al.* 1993; Muller 1997) have reported that finer spatial resolution image data do not necessarily improve traditional spectral-based image classification. Moreover, the spectral classification approach has been criticized when fine spatial resolution images are used, especially for urban features (Latty and Hoffer 1981; Markham and Townshend 1981; Woodcock and Strahler 1987; Cushnie 1987; Myint 2001). Traditional image classification methods, such as the maximum likelihood classifier, use spectral information (pixel values) as a basis to analyse and classify remote sensing images. They become less efficient when complex urban features are analysed (see Mesev, Chapter 9 in this book).

Urban land cover features are composed of various materials (e.g., plastic, glass, rubber, concrete, asphalt, metal, shingles, water, grass, shrubs, trees and soil). These materials have different spectral responses and are combined to form complex urban landscapes. As the spatial resolution becomes finer, smaller features or individual objects within an urban environment become more detectable. Therefore, the appearance of urban features becomes more complex, making traditional techniques more difficult for mapping urban land use and land cover types. Unfortunately, up to the present, traditional techniques have proven inadequate due to the lack of efficient tools to classify digitally the urban land use and land cover features in high spatial resolution image data. The problem is mainly due to their complex appearances. To extract the heterogeneous nature of urban features in high spatial resolution images, we need to consider the texture information contained in a group of neighbourhood pixels instead of individual spectral values. Traditional spectral classification algorithms use individual pixel values and ignore a huge amount of spatial information, which may be crucial in urban land use and land cover mapping. That is because most of the urban land cover classes generally contain a number of spectrally different pixels or objects. For example, roads, houses, grass, trees, bare soil, shrubs, swimming pools and footpaths, each of which may have a completely different spectral response, may need to be considered together as a residential class. Another problem for supervised classification is that it is extremely difficult to define suitable training sets for many categories within urban environments. This is due to variation in the spectral response of their component land cover types (Forster 1985; Gong and Howarth 1990; Barnsley *et al.* 1991). Therefore, the training statistics may exhibit a high standard deviation (Sadler *et al.* 1991) and violate one of the basic assumptions of the widely used maximum likelihood decision rule, namely that the pixel values follow a multivariate normal distribution (Barnsley *et al.* 1991; Sadler *et al.* 1991).

5.2   Image texture

The word "texture" comes from the Latin word *textura* meaning textile fabric, which is an example of a deterministic texture (Carstensen 1992). Texture is an elusive notion which mathematicians and scientists tend to avoid but engineers and artists fail to handle satisfactorily (Mandelbrot 1983). In any case, texture plays an important role in the human visual system for pattern recognition and interpretation. Pattern, in this respect, is defined as the overall spatial form of related features. The repetition of certain forms is a characteristic pattern found in many cultural objects and some natural features. Avery and Berlin (1992) defined texture as the visual impression of coarseness or smoothness caused by the variability or uniformity of image tone or colour. In fact, it is a crude description and does not reflect the complete characterization of a texture. Nevertheless, it is by far the most widely used explanation of texture in texture analysis and image processing. In general, researchers in image analysis and pattern recognition generally are not overly concerned with the precise definition of texture. All in all, there is no universally accepted mathematical definition of texture, although criteria include: (i) fine and coarse, (ii) regular and irregular, (iii) random and deterministic, (iv) smooth and rough, (v) orientation, (vi) high density and low density, (vii) phases, (viii) linear and non-linear and (ix) high and low frequency (Myint 2001).

Lark (1996) described a working definition of texture as two segments of an image that may be regarded as having the same texture if they do not differ significantly with respect to: (i) the variance of their Digital Numbers (DNs), (ii) the spatial dependence of DN variance at a characteristic scale (or scales), (iii) the directional dependence of DN variance and (iv) any spatial periodicity of the variation. This working definition is not comprehensive but is a useful basis for image analysis within remote sensing. It is clear from this definition that a simple texture transform such as the sample standard deviation within a local window will depend on the local image texture but will not characterize it with high precision. The same or similar value might be returned from image segments with rather different textures (Lark 1996).

Local variability in remotely sensed data could be characterized by computing statistics of a group of pixels (e.g., coefficient of variance or autocovariance), or by analysis of fractal relationships. There have been some attempts to improve the spectral analysis of remotely sensed data by using texture transforms in which some measure of variability in DNs is estimated within local windows. For example, contrast between neighbouring pixels (Edwards *et al.* 1988), the standard deviation (Arai 1993) or local variance (Woodcock and Harward 1992). The coefficient of variance gives a measure of the total relative variation of pixel values in an area and can be computed easily, but it gives no information about spatial patterns (De Jong and Burrough 1995). Snow and Mayer (1992), Klinkenberg (1992) and Burrough (1993) criticized many other neighbourhood operations such as diversity or variation filters. Their absolute outcome was easy to compare but they did not reveal any information on spatial irregularities. Lam and Quattrochi (1992) demonstrated that the fractal dimension of remotely sensed data could yield quantitative insight on the spatial complexity and information content contained within these data. Quattrochi *et al.* (1997) used a software package known as the Image Characterization And Modeling System (ICAMS) to explore how fractal dimension is related to surface texture. They also investigated how spatial resolution affects the computed fractal dimension of ideal fractal sets by using the isarithm method (Lam and De Cola 1993), the variogram (Mark and Aronson 1984) and the triangular prism methods (Clarke 1986). De Jong and Burrough (1995) analysed variograms of remotely sensed measurements to describe the spatial patterns quantitatively. Variogram interpretation of satellite sensor data was also carried out by Woodcock *et al.* (1988) and Webster *et al.* (1989). Emerson *et al.* (1999) analysed the fractal dimension using the isarithm method and the spatial autocorrelation of satellite imagery using Moran's I and Geary's C to observe the differing spatial structures of the smooth and rough surfaces in remotely sensed images. Details on the evaluation and limitations of different fractal approaches and spatial autocorrelation techniques are currently being prepared for publication by the author (see also Klinkenberg and Goodchild 1992; Xia 1993; Roach and Fung 1994; De Jong and Burrough 1995; Dong 2000; Myint 2001).

In general, the methods mentioned thus far have been successful to a certain extent. However, most of them focus primarily on the coupling between image pixels on a single scale and within a single band. They usually do not provide direct information on the orientation of texture features. These methods alone may not be able to provide satisfactory accuracy when they are applied to fine spatial resolution remotely sensed images with the use of relatively small local windows to differentiate between very closely related features or similar clusters. That is especially true when the above methods are applied to the original image data without transformation. Recent developments in the mathematical theory of wavelet transform approaches based on multi-channel or multi-resolution analysis have received overwhelming attention. There have been a number of developments in spatial frequency analysis of mathematical transforms that provide multi-resolution analysis. Of all transforms, wavelets play the most outstanding part in texture analysis of remotely sensed images. In this study, wavelet transforms are applied to the extraction of textural features from multispectral urban images.

5.3   Data and study area

Data from the Advanced Thermal and Land Applications Sensor (ATLAS)[1] (NASA – Stennis Space Center) at 10-m spatial resolution acquired with fifteen channels (0.45–12.2 µm) were used for this study. The data were collected by a NASA Lear Jet flying at 16,500 feet over Baton Rouge, Louisiana, on 11 May 1998, and the spectral characteristics are presented in Table 5.1. Figure 5.1 shows an ATLAS image.

Table 5.1 Spectral characteristics of the ATLAS data

Channel/Band	Wavelength (µm)
1	0.45–0.52
2	0.52–0.60
3	0.60–0.63
4	0.63–0.69
5	0.69–0.76
6	0.76–0.90
7	1.55–1.75
8	2.08–2.35
9	Removed by NASA
10	8.20–8.60
11	8.60–9.00
12	9.00–9.40
13	9.60–10.2
14	10.2–11.2
15	11.2–12.2

1   In 1999 ATLAS changed to Airborne Terrestrial Applications Sensor (resolution 2.5 m and 4 m). Some researchers still prefer to use "Advanced Thermal Land Application Sensor".

Figure 5.1 An ATLAS image of the Baton Rouge area (a subset) (reproduced by permission of the ATLAS Mission).

The performances of wavelet transforms were evaluated with the use of Shannon's entropy for the classification of seven different land use and land cover types derived from the ATLAS data. These classes include Residential 1 (single family homes with <30 per cent tree canopy), Residential 2 (single family homes with between 30 per cent and 60 per cent tree canopy), Residential 3 (single family homes with 60 per cent tree canopy), forest, commercial buildings and offices, water bodies and agricultural land. Samples of the classes are shown in Figure 5.2. The study used seventy samples for training each class, and for testing subsamples. The channels selected for this study include Channel 2 (0.52–0.60 µm: visible), Channel 6 (0.76–0.93 µm: reflected infrared) and Channel 13 (9.60–10.2 µm: thermal infrared). Figure 5.3 shows texture features of a sample (Residential 1) in each of the three selected channels. Therefore, the complete data set consists of 420 unique samples (20 training and testing samples, multiplied by 7 classes and multiplied by 3 channels).

5.4   Local window size

Windows are commonly used in digital image classification studies to define the local information content around a pixel. In general, the accuracy should increase with a larger local window size since it would contain more information and provide a fuller coverage of spatial variation, directionality and spatial periodicity of a particular texture than a smaller window size. As mentioned earlier, the texture of a residential area may consist of single family homes, lawns, shrubs, trees, tar roads, concrete roads, concrete footpaths, swimming pool, etc. If we were to use 2-m spatial resolution data it would be impossible for us to analyse the above residential texture features with the use of a local window size less than 25 m by 25 m since the window needs to cover at least one single family home with all the above mentioned objects or features. It is clear that we need to consider a minimum distance between two pixels that covers a particular texture or pattern of a land use or land cover type for a minimum window size.

Gong and Howarth (1990) generated edge density images with the use of sizes varying from 7 m by 7 m to 31 m by 31 m. The window size of 25 m by 25 m was selected and the resultant edge density image was used at the S1 band in their study. The identification of a method for determining optimal window size a priori classification is elusive (Gong and Howarth 1992). From a computational perspective, the ideal window size is the smallest size that also produces the highest accuracy. The most common approach for determining the appropriate window size is based on empirical results using automated classifications (Hodgson 1998). The seminal work on second-order texture statistics by Haralick *et al.* (1973) was based on windows of 64 rows by 64 columns in size or 20 rows by 50 columns. In this, the grey tone spatial dependency matrix was developed along with fourteen

(a) Agricultural land

(b) Commercial buildings and offices

(c) Forest

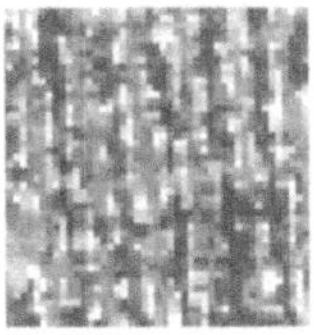

(d) Residential 1 (single family homes with <30 per cent tree canopy)

(e) Residential 2 (single family homes with between 30 per cent and 60 per cent tree canopy)

(f) Residential 3 (homes with >60 per cent tree canopy)

(g) Water bodies

Figure 5.2 Samples of seven classes (reproduced by permission of the ATLAS Mission).

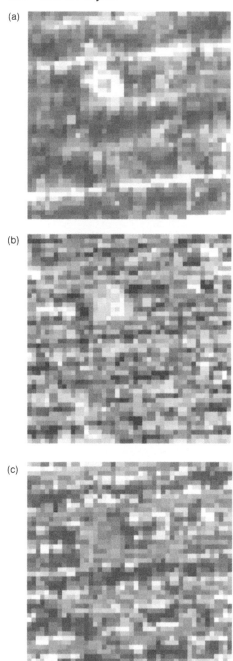

Figure 5.3 Different texture appearances of a sample (Residential 1) shown in each of the three selected channels: (a) Channel 2, (b) Channel 6 and (c) Channel 12 (reproduced by permission of the ATLAS Mission).

fundamental measures of texture from the spatial dependency matrix. Pesaresi (2000) experimented with forty-seven different square window sizes, ranging from 5 by 5 to 99 by 99 and showed an increase in the histogram separation index with the increase of window size. However, minimization of local window size is important in image texture and pattern recognition techniques since larger window sizes tend to cover more land covers and texture features and consequently create mixed boundary pixels or features problem. Another problem of using a larger window size is the fact that smaller land cover features will be lost in classification. In other words, the larger the window size the smaller the number of segmented regions or land use and land cover features identified on the image. It will also maximize missing pixels on the edges (e.g., fifteen missing pixels on the left, right, top and bottom of the image for a 32 by 32 window size) (Kershaw and Fuller 1992; Gong and Howarth 1992; Gong 1994; Hodgson 1998; Pesaresi 2000; Myint 2001).

In this chapter, different window sizes will not be evaluated. Instead, the objective is to examine the utility of wavelet transforms as an innovative classification algorithm for texture analysis and classification of urban features in relatively high spatial resolution image data. The local window size of 45 by 45, which is expected to be moderate, is used without prior testing. Ten training and ten testing samples with the size of 45 by 45 pixels are used in the analysis.

5.5   Wavelet transform and image analysis

In the past, one difficulty of texture analysis was the lack of adequate tools to characterize different scales of texture effectively. Recent developments in multi-resolution analysis such as Gabor (a windowed Fourier transform) and wavelet transforms have helped to overcome this difficulty (Zhu and Yang 1998). A key idea for wavelets is the concept of "scale". Sums and differences are at the finest scale, but we can move to a larger picture by taking sums and differences again. This is recursive – the same transform at a new scale. It leads to a multi-spatial resolution of the original signal (Strang and Nguyen 1997). The discrete wavelet transform proposed by Mallat (1989) initially decomposes an image into one "approximation" image and three "detail" images. It filters the original image with complementary low-pass and high-pass filters in each dimension. The filtered images are downsampled at every other pixel producing four images of half the resolution of the original (Tonsmann and Tyler 1999). This is the unique property of the wavelet transform technique, and our study is primarily aimed to examine whether wavelets could serve as an effective classification technique for urban land use and land cover mapping.

5.5.1   *Wavelet theory and analysis*

Wavelet analysis provides a windowing technique with variable-sized regions. The wavelet base function, unlike the sine waves used in Fourier analysis, has limited duration. These functions are not periodic and predictable and have irregular and asymmetric behaviour. Fourier analysis breaks up a signal into sine waves of various frequencies, however, wavelet analysis breaks up a signal into a combination of shifted and scaled versions of the mother wavelet. It allows the use of long time intervals for the cases where more precise low-frequency information is desired. Furthermore, it allows the use of short time intervals for high-frequency information, where wavelet analysis, unlike time frequency used in Fourier analysis, uses a timescale region. One of the advantages of using wavelets is the ability to perform local analysis, which is to analyse a localized area of a larger signal. Wavelet analysis is capable of revealing aspects of data that other signal analysis techniques miss, aspects like trends, breakdown points, discontinuities in higher derivatives, and self-similarity. Furthermore, because it affords a different view of data than those presented by traditional techniques, wavelet analysis can often compress or denoise a signal without appreciable degradation (Misiti *et al.* 1996). Wavelet theory is the mathematics associated with building a model for a signal, system or process with a set of "special signals". The special signals are just little waves or "wavelets". They must be oscillatory (waves) and have amplitudes that quickly decay to zero in both the positive and negative directions (Young 1998). Figure 5.4 shows an example of a set of wavelets including the one used in this study, called the "Haar mother wavelet" after its inventor.

Wavelet theory represents "things" by breaking them down into many interrelated component pieces, similar to the pieces of a jigsaw puzzle, This is when the pieces are scaled and translated, and this breaking down process is termed a wavelet decomposition or wavelet transform. Wavelet reconstruction or inverse wavelet transforms involve putting the wavelet pieces back together to retrieve the original object or process. Wavelet theory consists of the study of these pieces, their properties and interrelationships, and how to put them back together. Wavelet theory involves the scaling or warping operation. For example, watching a movie on TV with the VCR running on fast-forward is the scaling (time scaling function). A function (the movie) has been time scaled (fast-forwarded) and a time-compressed function (movie) is created. Although the time-compressed movie is still the same movie, its representation in terms of spatial–temporal parameters has changed. However, the information content of the movie does not change. It also involves the translation or shifting operation. Time delays are typical "translations" of the time axis or variable. For example, if the movie is shown at 5 p.m. instead of its original 2 p.m. airtime, then it has shifted or has been translated by 3 hours. If the translation operation is combined with the scaling action then the composite operation is referred to as an affined

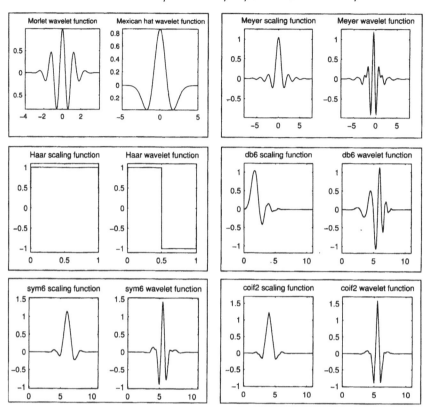

Figure 5.4 Examples of various wavelets (Misiti *et al.* 1996).

operation. The operation simultaneously scales and translates the independent variable (Young 1998).

In wavelet theory the scaling and translation operators act simultaneously on the mother wavelet function. The name "wavelet", literally meaning "little wave", originated from the study of Grossmann and Morlet (1984). The mother wavelet (the initial little wave) is the kernel of the wavelet transform. Performing affined operations on the mother wavelet creates a set of scaled and translated versions of this mother wavelet function. The general form of wavelet transform of one-dimension signal (Daubechies 1990) can be defined by the following expressions, and are translated and scaled versions of a function ψ:

$$Wf(a,b) = \int_{-\alpha}^{+\alpha} f(x)\psi_{a,b}(x)\,dx \tag{5.1}$$

$$\psi_{a,b}(x) = \frac{1}{\sqrt{|a|}}\psi\left(\frac{x-b}{a}\right), \quad (a,b \in R, a \neq 0) \tag{5.2}$$

where the wavelet function ψ is dilated by a factor a and shifted by b. The function ψ is called a mother wavelet. An integral transformation using the functions above as the basis is called a wavelet transformation. For digital signals, a discrete wavelet ψ, which is formed from ψ by sampling the parameters a and b following the Nyquist theorem, is used (Grossmann and Morlet 1984). Multiplying each wavelet coefficient by the appropriately scaled and shifted wavelet yields the constituent wavelets of the original signal. Wavelets are well located in both domains: space and scale (Daubechies 1990). The decomposition process can be iterated, with successive transformed signals being decomposed. In wavelet analysis, a signal is split into an approximation and a detail. The approximation is then itself split into a second-level approximation and detail, and the process is repeated. This is a standard procedure of wavelet decomposition and it is called the wavelet decomposition tree. In wavelet packet analysis, both details and approximation can be split. In theory the analysis process can be iterated indefinitely. In the real-world situation, the decomposition can proceed only until the individual details consist of a single sample or pixel. Selection of the number of levels depends on the nature of the signal or a particular subset of a feature and aim of the analysis.

5.5.2   *Wavelet decomposition example*

The following is an example to demonstrate briefly the process of a wavelet application. The image shown in Figure 5.5 is decomposed by a multi-resolution wavelet transform. The example below is similar to the research presented in Mallat (1989), Meyer (1990), Rioul and Vetterli (1991) and

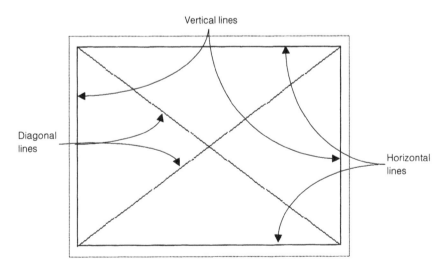

Figure 5.5 Image example.

Approximation

Horizontal details

Diagonal details

Vertical details

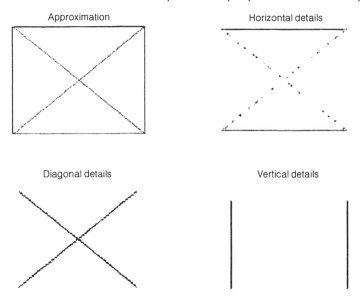

Figure 5.6 Wavelet representation of the example image on one resolution level.

Young (1998). Figure 5.5 is illustrated with the use of actual output subimages derived from a wavelet decomposition of the example image at two levels. Basically, the original image is formed by six lines, two horizontal, two vertical and two diagonal. The rest of the area is assigned zero. Only the image inside the dotted line is decomposed. The decomposition of this image is completed with the multi-resolution wavelet transform technique. The multi-resolution wavelet analysis is extended from one-dimensional signal processing to perform on two-dimensional images. The two-dimensional image is treated as two separate one-dimensional sequences (rows and columns). The outcome of the multi-resolution wavelet decomposition of the example image at two levels is illustrated in Figure 5.6. The output four subimages, approximation, horizontal details, vertical details and diagonal details are labelled. Approximation is also known as low frequency or trend subimage and the other three details, are called high-frequency or fluctuation subimages. A wavelet transform of an image consists of four subimages with a quarter area. This is one of the distinctive properties of wavelet transform. If an image is decomposed at one level, four subimages with different texture information will be obtained. The subimage composed of the low-frequency parts in both row and column direction is iteratively decomposed into four subimages level by level. This is the standard procedure of wavelet decomposition. However, decomposition can be performed on any subimage. If we perform wavelet decomposition of an image at two levels, we will generate eight different subimages of textures. In the example, the image decomposition is performed with the use of a Coiflet4 wavelet. However, Haar, the simplest wavelet, was used as an initial exploration of the performance of wavelet transforms for texture analysis and classification of remotely sensed images in the study.

5.5.3 Multi-resolution wavelet decomposition

Mallat (1989) developed the multi-resolution analysis theory using the orthonormal wavelet basis. A wavelet is orthogonal when all the pairs, formed from the basis functions $\psi_{j,k}$, are orthogonal to each other. An orthogonal wavelet where norm is normalized to one is called an ortho-normal wavelet (Fukuda and Hirosawa 1998). The multi-resolution wavelet transform decomposes a signal into low-frequency approximation and high-frequency detail information at a coarser spatial resolution. In satellite image analysis using the two-dimensional wavelet transform technique, rows and columns of image pixels are considered signals. The approximation and details of a two-dimensional image $f(x,y)$ at resolution 2^j can be defined by the coefficients computed by the following convolutions:

$$A_{2^j}^d f = ((f(x,y)^*\phi_{2^j}(-x)\phi_{2^j}(-y))(2^{-j}n, 2^{-j}m))_{(n,m)\in Z^2} \tag{5.3}$$

$$D_{2^j}^1 f = ((f(x,y)^*\phi_{2^j}(-x)\psi_{2^j}(-y))(2^{-j}n, 2^{-j}m))_{(n,m)\in Z^2} \tag{5.4}$$

$$D_{2^j}^2 f = ((f(x,y)^*\psi_{2^j}(-x)\phi_{2^j}(-y))(2^{-j}n, 2^{-j}m))_{(n,m)\in Z^2} \tag{5.5}$$

$$D_{2^j}^3 f = ((f(x,y)^*\psi_{2^j}(-x)\psi_{2^j}(-y))(2^{-j}n, 2^{-j}m))_{(n,m)\in Z^2} \tag{5.6}$$

where integer j is a decomposition level, m, n are integers, $\phi(x)$ is a one-dimensional scaling function and $\psi(x)$ is a one-dimensional wavelet function. In general, the $\phi(x)$ is a smoothing function that provides low-frequency information and $\psi(x)$ is a differencing function that provides high-frequency information. $A_{2^{j+1}}^d f$ can be perfectly reconstructed from $A_{2^j}^d f$, $D_{2^j}^1 f$, $D_{2^j}^2 f$, $D_{2^j}^3 f$. The expressions (5.3) through (5.5) show that in two dimensions, $A_{2^j}^d f$, $D_{2^j}^k f$ are computed with separable filtering of the signal along the abscissa and ordinate. The wavelet decomposition can thus be interpreted as a signal decomposition in a set of independent, spatially oriented frequency channels (Mallat 1989).

We need a filter bank (a set of filters) to perform multi-resolution wavelet decomposition. Filter banks used for signal analysis are commonly composed of low-pass and high-pass filters. They separate the input signal into frequency bands in a process referred to as sub band coding. Filters are linear time-invariant operators that act on input vectors x to generate output vectors y. In this case, y is the convoluted version of x with a fixed vector h. In discrete time $t = nT$, the input $x(n)$ and output $y(n)$ will appear at all times $t = -3, -2, -1, 0, 1, 2, \ldots$. There is a scaling function $\phi(x)$ related to the low-pass filter and a wavelet function related to the high-pass filter $\psi(x)$. They can be defined as:

Dilation equation $\phi(t) = \sqrt{2} \sum_{k} c(k)\phi(2t - k)$ $\qquad(5.7)$

Wavelet equation $\psi(t) = \sqrt{2} \sum_{k} d(k)\phi(2t - k)$ $\qquad(5.8)$

For example, Haar has coefficients: $c(0) = c(1) = 1/\sqrt{2}$, $d(0) = 1/\sqrt{2}$ and $d(1) = -1/\sqrt{2}$. Therefore, its dilation equation and wavelet equation can be expressed as (Strang and Nguyen 1997):

$\phi(t) = \phi(2t) + \phi(2t - 1)$ $\qquad(5.9)$

$\psi(t) = \phi(2t) - \phi(2t - 1)$ $\qquad(5.10)$

The approximation and detail coefficients can be computed with a pyramid algorithm based on convolutions with the two above mentioned one-dimensional parameter filters. Approximation of a signal $A_{2^j}^d f$, also known as trend, can be obtained by convolving the input signal $A_{2^{j+1}}^d f$ with the low-pass filter (Lo_F). First, the rows of an image are convolved with one-dimensional Lo_F. Next, filtered signals are downsampled. In the first step, downsampling is performed by keeping one column out of two. Then, the resulting signals are convolved with another one-dimensional low-pass filter and every other row is retained. For obtaining horizontal detail image, first the rows of the input image are convolved with a low-pass filter Lo_F and the filtered signals are downsampled by keeping one column out of two as in processing the approximation image. However, for the next stage, the columns of the signals are convolved with a high-pass filter Hi_F and again every other row is retained. For the vertical details, the original signals are convolved first with a high-pass filter Hi_F and then with a low-pass filter Lo_F following the above procedure. For the diagonal detail image, the same downsampling procedure is carried out using two high-pass filters consecutively. The algorithm for the application of the filters and downsampling procedure for computing approximation and details coefficients is illustrated in Figure 5.7. It shows how in the frequency domain the image $A_{2^{j+1}}^d f$ is decomposed into $A_{2^j}^d f$, $D_{2^j}^1 f$, $D_{2^j}^2 f$ and $D_{2^j}^3 f$ subimages. The subimage $A_{2^j}^d f$ corresponds to the lowest frequencies (approximation), $D_{2^j}^1 f$ gives the vertical high frequencies (horizontal edges), $D_{2^j}^2 f$ the horizontal high frequencies (vertical edges) and $D_{2^j}^3 f$ the high frequency in both directions (the diagonal edges).

Figure 5.8 represents standard orthonormal wavelet decomposition with two levels of an image. This is again illustrated by the decomposition of a randomly selected training sample of a residential area generated from ATLAS remotely sensed data. The pyramid decomposition can be

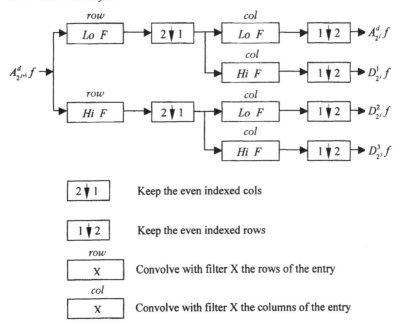

Figure 5.7 Decomposition of an image $A^d_{2^{j+1}}f$ into $A^d_{2^j}f$ (approximation), $D^1_{2^j}f$ (horizontal details), $D^2_{2^j}f$ (vertical details) and $D^3_{2^j}f$ (diagonal details).

continuously applied to the approximation image until the desired coarser resolution $2^{-j}(-1 \geq j \geq -J)$ is reached.

Let us suppose that initially we have an image $A_1 f$ measured at Resolution 1. For any $J > 0$, this discrete image can be decomposed between Resolutions 1 and 2^{-j}, and completely represented by the $3J + 1$ discrete images:

$$(A^d_{2^{-J}}f, (D^1_{2^j}f)_{-J \leq j \leq -1}, (D^2_{2^j}f)_{-J \leq j \leq -1}, (D^3_{2^j}f)_{-J \leq j \leq -1}) \qquad (5.11)$$

This set of images is called an orthogonal wavelet representation in two dimensions. If the original image has N pixels, each subimage $A^d_{2^j}f$, $D^1_{2^j}f$, $D^2_{2^j}f$, $D^3_{2^j}f$ will have $2^j N$ pixels ($j < 0$). The total number of pixels in this new representation is equal to the number of pixels in the original image. This is due to the orthogonality of the representation (Mallat 1989).

The Haar wavelet transform is the simplest orthonormal basis. More details can be observed in Strang and Nguyen (1997). Initially, Haar wavelets were used to build our prototypes. The usefulness of other wavelets such as Daubechies, Mallat and Symlet are currently being investigated. From the standard wavelet decomposition, it is understood that further decomposition is done using the low-frequency channels. However, the most important information for texture appears in the high-frequency channels

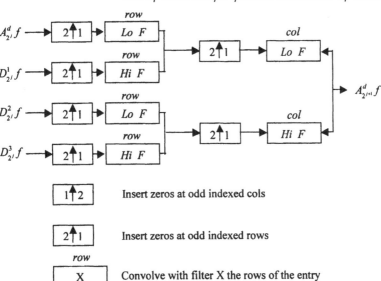

$1\uparrow 2$ Insert zeros at odd indexed cols

$2\uparrow 1$ Insert zeros at odd indexed rows

row
X   Convolve with filter X the rows of the entry

col
X   Convolve with filter X the cols of the entry

Figure 5.8 Reconstruction of an image $A^d_{2^{j+1}}f$ from $A^d_{2^j}f$ (approximation), $D^1_{2^j}f$ (horizontal details), $D^2_{2^j}f$ (vertical details) and $D^3_{2^j}f$ (diagonal details).

(the detail subbands). Therefore, upsampling is performed using the first Level 3 detail subimages. The approximation image was discarded in the upsampling process. Figure 5.9 shows the basic standard reconstruction steps for these images. However, it should be noted that reconstruction was done without using the approximation image in this study since the detail images are believed to contain more valuable textural information. The idea was to extract and fuse the edges (detail features) of an image in three different directions, horizontal, vertical and diagonal diagonals.

The details of subimages were reconstructed by adding a row of zeros, convolving the columns with a one-dimensional filter, adding a column of zeros between each column of the resulting image and convolving the rows with another one-dimensional filter. Further decomposition was later performed on the reconstructed image. In the study, further decomposition was also carried out on the horizontal edge since more textural information can be obtained in the mid-frequency channels. Zhu and Yang (1998) demonstrated that the decomposition of horizontal images was more efficient than the standard decomposition technique in their study. Therefore, three approaches were employed in our study: (i) the standard decomposition; (ii) decomposition with the horizontal details, and (iii) decomposition with the reconstruction of the first Level 3 detail subimages. Up to four levels of decomposition were performed with the local window

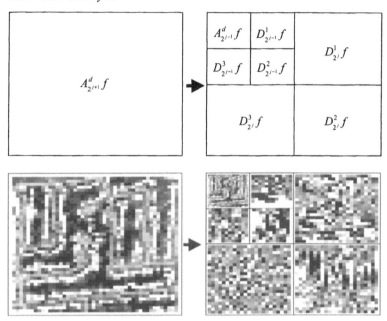

Figure 5.9 A standard orthonormal wavelet decomposition with two levels of a sample image (a sample of Residential 1).

size of 45 by 45. The multi-resolution approach of wavelet transforms provides textural information of images at different scales from coarse to fine. This is the unique property of wavelet transforms. Each level and each subimage yielded additional frequency and spatial properties. Different wavelet decomposition models, multi-channels and different mother bodies of the same features were examined in this study. Decomposition was carried out with the three channels and analysed the textural features separately for seven classes.

5.6   Classification procedure

Four levels for the first and second approaches and three levels for the third approach were performed for each training and testing sample. Sixteen subimages were obtained for every subsample with both standard and reconstructed details. Twelve subimages were obtained for the horizontal detail decomposition. Each level and each subband provided its spatial properties and characteristic frequency. Pesaresi (2000) tested contrast, angular second moment inverse difference moment and entropy with the use of original image data for urban pattern recognition. Albuz *et al.* (1999) used the sum of squares of the wavelet coefficients of each subband for their image retrieval system. Sheikholeslami *et al.* (1999) calculated the mean and variance of wavelet coefficients to represent the contrast of the image and counted the number of edge pixels in the horizontal, vertical and diagonal directions to have an approximation of directionality of the image. Zhu and Yang (1998) used the information entropy as the measure in their analysis. In our study, Shannon's entropy was used as the texture feature for image classification. Shannon's entropy was first computed for each subimage. Later its value was used as a distinct feature value in a vector. There are sixteen real values for each subimage for the first two approaches and twelve real values for the last model. The mean vector of 16 and 12 feature vectors of the 10 training samples of a class were used as a total feature vector of the samples. The total feature vector of the training samples was treated as representative of the classes and used for the classification. The total feature vector of the testing samples was also computed and used to test the accuracy of classes. The Shannon entropy measure (*ENT*) is defined as:

$$ENT = -\sum_{i-1}^{N}\sum_{j-1}^{N} c(i,j)^2 \log c(i,j)^2 \qquad (5.12)$$

where N is the number of rows and columns of an image and $c(i,j)$ is the wavelet coefficients of an image at (i,j).

The performance of a minimum distance classifier was evaluated for texture classification. Each pattern class C_k is represented by a prototype pattern P_k. In this study, P_k ($k = 1, 2, \ldots, 7$) were the total feature vectors of the training samples. The minimum distance classifier assigns an unknown class pattern Q to the class S_k, if the distance R_k between Q and P_k is minimum among all possible class prototypes. The Euclidean distance is defined as:

$$R_k = \|Q - P_k\| = \sqrt{\sum (q_j - p_{k,j})^2} \qquad (5.13)$$

Using the above wavelet decomposition approaches for individual channels, the classification of the seven texture images shown in Figure 5.5 was carried out. It was observed that textures of the same area in different bands (e.g., visible, reflected infrared and thermal infrared) were different in terms of contrast, smoothness/coarseness, spatial periodicity and spatial variation, etc. By visual observation, texture appearances in Channel 2 seemed to be weaker than the other two channels. Taking advantage of different textural information of the same windows or areas of the same class in different channels, multi-spectral texture analysis has been introduced in this study to improve the accuracy of the classification. Using this approach, the total feature vector of a sample with different channel combinations using the above minimum distance classifier was also examined. Channel combinations included: (i) Channels 13, 6 and 2, (ii) Channels 13 and 6, (iii) Channels 13 and 2 and (iv) Channels 6 and 2. The purpose of multispectral texture analysis and classification was to achieve complementary textural information benefits from different channels, which could be expected to improve the mapping accuracy.

5.7   Results and conclusions

With the use of wavelet transforms, seven types of urban image feature were classified using the minimum distance rule. The results of the classification with individual channels are shown in Table 5.2. Using the wavelet transform features of individual channel data, the accuracy was found to be low for all three approaches. The accuracy can be as high as 77.1 per cent when using Channel 6 data alone with the first approach (standard decomposition) of Level 1-4. It was originally expected that the overall accuracy would be low since similar texture features or land cover classes were included in the classification. One of the main objectives of this study

Table 5.2 Classification results of seven feature classes with Channel 2, Channel 6 and Channel 12 data separately

Level	Number of features	Approach 1		Approach 2		Approach 3	
		Error	Correct (%)	Error	Correct (%)	Error	Correct (%)
Channel 2							
1	4	27	61.4	28	60.0	33	52.9
2	4	32	54.3	40	42.9	36	48.6
3	4	35	50.0	48	31.4	43	38.6
4	4	42	40.0	51	27.1	–	–
1-2	8	23	67.1	30	57.1	30	57.1
1-3	12	23	67.1	30	57.1	30	57.1
1-4	16	23	67.1	30	57.1	–	–
Channel 6							
1	4	19	72.9	32	54.3	35	50.0
2	4	31	55.7	38	45.7	32	54.3
3	4	34	51.4	37	47.1	37	47.1
4	4	40	42.9	47	32.9	–	–
1-2	8	17	75.7	23	67.1	31	55.7
1-3	12	16	77.1	23	67.1	32	54.3
1-4	16	16	77.1	23	67.1	–	–
Channel 13							
1	4	31	55.7	28	60.0	30	57.1
2	4	24	65.7	45	35.7	44	37.1
3	4	35	50.0	42	40.0	38	45.7
4	4	37	47.1	42	40.0	–	–
1-2	8	22	68.6	21	70.0	27	61.4
1-3	12	22	68.6	26	62.9	27	61.4
1-4	16	22	68.6	20	71.4	–	–

was to examine how efficient the wavelet is to discriminate closely related features. The three decomposition models gave slightly different results. The second approach produced improved results for the combination of Level 1-2 and Level 1-4 in Channel 13. However, the first approach (standard decomposition procedure) consistently generated higher accuracy for all combinations of levels in Channels 6 and 2. The first approach also gave better results for almost all single-level classifications in all channels.

The feature vector for the combination of subimages at different levels proved to be better than any single-level subimages. In general, the higher decomposition level had lower accuracy for all approaches. The results may be due to the size of higher level subimages, which cover lesser spatial frequency information since they are just a quarter each of their mother bodies. In this analysis, Channel 6 alone was found to be more efficient than others for the texture analysis. It is obvious that the textures among three residential areas and the textures of forest and Residential 3 (single family homes with 60 per cent tree canopy) were similar to each other. As mentioned earlier it was one of the major limitations, which makes the classification more difficult than other land use and land cover texture analysis.

The results of the multi-spectral texture classification are shown in Table 5.3 for the channel combinations described in the classification procedure. As expected, the multi-channel approach largely improved the classification accuracy. In general, the accuracies obtained from the combination of different bands were higher than the single-band approaches. This accuracy was as high as 92.9 per cent when using the standard decomposition procedure with three channels or a combination of Channels 13 and 6. The accuracy in this type of analysis can be considered very high since closely related texture features were included and relatively small local window size was used in the classification. The standard decomposition procedure with the combination of Channels 13 and 2 achieved 80 per cent, 71 per cent, and 71 per cent for the combination of Level 1-2, 1-3 and 1-4, respectively. However, the same decomposition procedure with the combination of Channels 13, 6 and 2; Channels 13 and 6; and Channels 6 and 2 for all possible combinations of levels exceed the standard acceptable accuracy of 85 per cent (Townshend 1981) required for most resource management applications. It is interesting to note that the accuracy obtained in the standard technique is higher than the other approaches: the horizontal decomposition and the decomposition of reconstructed detail images. It is true for both single- and multi-band approaches. This finding is inconsistent with the remote sensing texture analysis reported by Zhu and Yang (1998), where they found the accuracy obtained by the horizontal decomposition was superior to the standard decomposition technique. However, there was no significant difference between the accuracy obtained by the standard procedure and by the horizontal decomposition technique in their analysis. The difference in this research

Table 5.3 Classification results of seven feature classes derived from the combination of different channels

Level	Number of features	Approach 1 Error	Approach 1 Correct (%)	Approach 2 Error	Approach 2 Correct (%)	Approach 3 Error	Approach 3 Correct (%)
Channel 13, 6 and 2							
1	4	7	90.0	9	87.1	19	72.9
2	4	14	80.0	32	54.3	20	71.4
3	4	25	64.3	32	54.3	28	60.0
4	4	26	62.9	35	50.0	–	–
1-2	8	5	92.9	10	85.7	16	77.1
1-3	12	5	92.9	14	80.0	17	75.7
1-4	16	5	92.9	11	84.3	–	–
Channel 13 and 6							
1	4	6	91.4	17	75.7	25	64.3
2	4	18	74.3	31	55.7	27	61.4
3	4	27	61.4	35	50.0	33	52.9
4	4	25	64.3	41	41.4	–	–
1-2	8	5	92.9	18	74.3	25	64.3
1-3	12	5	92.9	16	77.1	25	64.3
1-4	16	6	91.4	16	77.1	–	–
Channel 13 and 2							
1	4	19	72.9	20	71.4	15	78.6
2	4	22	68.6	22	68.6	26	62.9
3	4	25	64.3	25	64.3	34	51.4
4	4	31	55.7	31	55.7	–	–
1-2	8	14	80.0	14	80.0	15	78.6
1-3	12	16	77.1	16	77.1	15	78.6
1-4	16	16	77.1	16	77.1	–	–
Channel 6 and 2							
1	4	7	90.0	16	77.1	22	68.6
2	4	19	72.9	34	51.4	29	58.6
3	4	29	58.6	34	51.4	29	58.6
4	4	32	54.3	39	47.1	–	–
1-2	8	8	88.6	13	81.4	21	70.0
1-3	12	9	87.1	14	80.0	21	70.0
1-4	16	9	87.1	14	80.0	–	–

finding and their report might be due to a number of reasons such as classification specificity, spatial resolution of the data, local window size or image size, and nature of the study area.

Combinations of Channels 13 and 2 and Channels 6 and 2 gave low accuracy. This may be due to the weakness of texture features in Channel 2 as observed earlier by checking the texture appearances in different bands visually. In the future, it is recommended that more samples and feature classes be tested to yield better comparisons. Future work should investigate the optimum local window size that provides the satisfactory accuracy. There is also the question of another candidate, which could provide better accuracy than that using Shannon's entropy measure. The preliminary results of this research indicated that the accuracy of texture analysis in classifying fine spatial resolution image data could be drastically improved by using the wavelet transforms approach. The potential of the methods for wavelet analysis proposed in this study needs to be examined with higher spatial resolution image data and/or in other environments.

5.8 Acknowledgements

This research was supported by the Otis Paul Starkey Fund (The Association of American Geographers) under the AAG grants/awards scheme and a Robert C. West field research grant (Department of Geography and Anthropology, LSU). The author thanks John Tyler, Dept of Computer Science, LSU and Nina Lam, Dept of Geography and Anthropology, LSU for their suggestions during the analytical phases of this research. The author also expresses his appreciation to DeWitt Braud, Dept of Geography and Anthropology, LSU for providing the ATLAS data for this study.

5.9 References

Albuz, E., Kocalar, E. and Khokhar, A. A., 1999, Vector-wavelet based scalable indexing and retrieval system for large color image archives. *IEEE Transactions on Geoscience and Remote Sensing*, 44, 3021–4.

Arai, K., 1993. A classification method with a spatial-spectral variability. *International Journal of Remote Sensing*, 14, 699–709.

Avery, T. E. and Berlin, G. L., 1992, *Fundamentals of Remote Sensing and Airphoto Interpretation* (New York: Macmillan).

Barnsley, M. J., Barr, S. L. and Sadler, G. J., 1991, Spatial re-classification of remotely sensed images for urban land use monitoring. *Proceedings of Spatial Data 2000* (Nottingham, UK: Remote Sensing Society), pp. 106–17.

Burrough, P. A., 1993, Soil variability: A late 20th century view. *Soils and Fertilizers*, 56, 529–62.

Carstensen, J. M., 1992, Description and simulation of visual texture. Unpublished PhD thesis, Technical University of Denmark.

Clarke, K. C., 1986, Computation of the fractal dimension of topographic surfaces using the triangular prism surface area method. *Computers and Geosciences*, 12, 713–22.

Cushnie, J. L., 1987, The interactive effect of spatial resolution and degree of internal variability within land cover types on classification accuracies. *International Journal of Remote Sensing*, 8, 12–29.

Daubechies, I., 1990, The wavelet transform, time/frequency localization and signal analysis. *IEEE Trans. Inform. Theory*, 36, 961–1005.

De Jong, S. M. and Burrough, P. A., 1995, A fractal approach to the classification of Mediterranean vegetation types in remotely sensed images. *Photogrammetric Engineering and Remote Sensing*, 61, 1041–53.

Dong, P., 2000, Lacunarity for spatial heterogeneity measurement in GIS. *Geographic Information Sciences*, **6**, 20–6.

Edwards, G., Landary, R. and Thomson, K. P. B., 1988, Texture analysis of forest regeneration sites in high-resolution SAR imagery. Paper given at Conference of the International Geosciences and Remote Sensing Symposium (IGARSS 88), ESA SP-284 (Paris: European Space Agency), pp. 1355–60.

Emerson, C. W., Lam, N. S. N. and Quattrochi, D. A., 1999, Multi-scale fractal analysis of image texture and pattern. *Photogrammetric Engineering and Remote Sensing*, **65**, 51–61.

Forster, B. C., 1985, An examination of some problems and solutions in monitoring urban areas from satellite platforms. *International Journal of Remote Sensing*, **6**, 139–51.

Fukuda, S. and Hirosawa, H., 1998, Suppression of speckle in synthetic aperture radar images using wavelet. *International Journal of Remote Sensing*, **19**, 507–19.

Gong, P., 1994, Reducing boundary effects in a kernel-based classifier. *International Journal of Remote Sensing*, **15**, 1131–9.

Gong, P. and Howarth, P. J., 1990, The use of structural information for improving land cover classification accuracies at the rural urban fringe. *Photogrammetric Engineering and Remote Sensing*, **56**, 67–73.

Gong, P. and Howarth, P. J., 1992, Frequency based contextual classification and gray level vector reduction for land use identification. *Photogrammetric Engineering and Remote Sensing*, **58**, 423–37.

Green, D. R., Cummins, R., Wright, R. and Miles, J., 1993, A methodology for acquiring information on vegetation succession from remotely sensed imagery. In R. Haines-Young, D. R. Green and S. Cousins (eds) *Landscape Ecology and Geographic Information Systems* (London: Taylor & Francis), pp. 111–28.

Grossmann, A. and Morlet, J., 1984, Decomposition of Hardy functions into square integrable wavelets of constant shape. *SIAM Journal on Mathematical Analysis*, **15**, 723–36.

Haralick, R. M., Shanmugan, K. and Dinstein, J., 1973, Textural features for image classification. *IEEE Transaction on Systems, Man, and Cybernetics*, **SMC-3**, 610–21.

Hodgson, M. E., 1998, What size window for image classification? A cognitive perspective. *Photogrammetric Engineering and Remote Sensing*, **64**, 797–807.

Irons, J. R., Markham, B. L. and Nelson, R. F., 1985, The effect of spatial resolution on the classification of Thematic Mapper data. *International Journal of Remote Sensing*, **8**, 1385–403.

Kershaw, C. D. and Fuller, R. M., 1992, Statistical problems in the discrimination of land cover from satellite images: A case in lowland Britain. *International Journal of Remote Sensing*, **13**, 3085–104.

Klinkenberg, B., 1992, Fractals and morphometric measures: Is there a relationship? *Geomorphology*, **5**, 5–20.

Klinkenberg, B. and Goodchild, M. F., 1992, The fractal properties of topography: A comparison of methods. *Earth Surface Processes and Landforms*, **17**, 217–34.

Lam, N. S. N. and De Cola, L. (eds), 1993, Fractal simulation and interpolation. *Fractals in Geography* (Englewood Cliffs, NJ: Prentice Hall), pp. 56–74.

Lam, N. S. N. and Quattrochi, D. A., 1992, On the issues of scale, resolution, and fractal analysis in the mapping sciences. *Professional Geographer*, **44**, 88–97.

Lark, R. M., 1996, Geostatistical description of texture on an aerial photograph for discriminating classes of land cover. *International Journal of Remote Sensing*, 17, 2115–33.

Latty, R. S. and Hoffer, R. M., 1981, Computer based classification accuracy due to the spatial resolution using per point vs. per field classification techniques. *Proceedings of Symposium on Machine Processing of Remotely Sensed Data* (West Lafayette, IN), pp. 384–92.

Mallat, S. G., 1989, A theory for multi-resolution signal decomposition: The wavelet representation. *IEEE Transactions on Pattern Analysis and Machine Intelligence*, 11, 674–93.

Mandelbrot, B., 1983, *The Fractal Geometry of Nature* (New York: Freeman & Co.).

Mark, D. M. and Aronson, P. B., 1984, Scale dependent fractal dimensions of topographic surfaces: An empirical investigation with applications in geomorphology and computer mapping. *Mathematical Geology*, 16, 671–83.

Markham, B. L. and Townshend, J. R. G., 1981, Land cover classification accuracy as a function of sensor spatial resolution. *Proceedings of the 15th International Symposium on Remote Sensing of the Environment* (West Lafayette, IN), pp. 384–92.

Meyer, Y., 1990, *Ondelettes et Opérateurs* (Vol. 1: published by Hermann Ed.) [English translation: *Wavelets and Operators* (Cambridge: Cambridge University Press, 1993)].

Misiti, M., Misiti, Y., Oppenheim, G. and Poggi, J., 1996, *Wavelet Toolbox* (Natick, MA: The Math Works).

Muller, E., 1997, Mapping riparian vegetation along rivers: Old concepts and new methods. *Aquatic Botany*, 58, 411–37.

Myint, S. W., 2001, Wavelet analysis and classification of urban environment using high-resolution multispectral image data. Unpublished PhD dissertation, Louisiana State University.

Pesaresi, M., 2000, Texture analysis for urban pattern recognition using fine-resolution panchromatic satellite imagery. *Geographical & Environmental Modelling*, 4, 43–63.

Quattrochi, D. A., Lam, N. S. N., Qiu, H. and Zhao, W. E. I., 1997, Image Characterization and Modeling System (ICAMS): A geographic information system for the characterization and modeling of multiscale remotely sensed data. In Quattrochi and Goodchild (eds) *Scale in Remote Sensing and GIS* (Boca Raton, FL: CRC Press), pp. 295–308.

Rioul, O. and Vetterli, M., 1991, Wavelets and signal processing. *Signal Processing Magazine*, 8, 14–38.

Roach, D. and Fung, K. B., 1994, Fractal-based textural descriptors for remotely sensed forestry data. *Canadian Journal of Remote Sensing*, 20, 59–70.

Sadler, G. J., Barnsley, M. J. and Barr, S. L., 1991, Information extraction from remotely sensed images for urban land analysis. Paper given at Conference of the Second European Conference on Geographical Information Systems (EGIS '91) (Utrecht, The Netherlands: EGIS Foundation), pp. 955–64.

Sheikholeslami, G., Zhang, A. and Bian, L., 1999, A multi-resolution content-based retrieval approach for geographic images. *Geoinformatica*, 3, 109–39.

Snow, R. S. and Mayer, L. 1992, Fractals in geomorphology. *Geomorphology*, 5, 194.

Strang, G. and Nguyen, T., 1997, *Wavelets and Filter Banks* (Wellesley, MA: Wellesley-Cambridge Press).

Tonsmann, G. and Tyler, J. M., 1999, Estimation of oceanic surface velocity fields using wavelets. *Proceedings of SPIE Conference on Wavelet Applications VI, Orlando, Florida,* **3723,** 122–9.

Townshend, J. R. G., 1981, *Terrain Analysis and Remote Sensing* (London: George Allen and Unwin).

Webster, R., Curran, P. J. and Munden, J. W., 1989, Spatial correlation in reflected radiation from the ground and its implication for sampling and mapping by ground-based radiometry. *Remote Sensing of Environment,* **29,** 67–78.

Woodcock, C. and Harward, V. J., 1992, Nested-hierarchical scene models and image segmentation. *International Journal of Remote Sensing,* **13,** 3167–87.

Woodcock, C. E. and Strahler, A. H., 1987, The factor of scale in remote sensing. *Remote Sensing of Environment,* **21,** 311–32.

Woodcock, C. E., Strahler, A. H. and Jupp, D. L. B., 1988, The use of variograms in remote sensing. *Remote Sensing of Environment,* **25,** 323–48.

Xia, Z., 1993, The uses and limitations of fractal geometry in digital terrain modelling. Unpublished PhD dissertation, City University of New York.

Young, R. K., 1998, *Wavelet Theory and Its Applications* (6th edn) (Norwell, MA: Kluwer Academic).

Zhu C. and Yang, X., 1998, Study of remote sensing image texture analysis and classification using wavelet. *International Journal of Remote Sensing,* **13,** 3167–87.

Part II

Cities by day

6 Refining methods for dasymetric mapping using satellite remote sensing

Mitchel Langford

6.1 Introduction

It is technically very demanding for remote sensing techniques to decipher and provide accurate and meaningful information about the cityscape, since it is comprised of a highly complex physical surface. For remote sensing "purists" this may be reason enough to wish to study urban areas, and they are likely to be more than satisfied by the challenges faced when so doing. However, at risk of stating the obvious, what makes cities particularly important for most practitioners interested in applied remote sensing is not that they present a complex physical surface, but simply that they contain people. It is a fact that in most countries today the majority of the population resides within city boundaries, and the trend toward ever-increasing urbanization across the globe shows few signs of slowing down. It is within the confines of cities, therefore, that most people conduct their daily lives. In so doing, they create numerous problems and issues that can potentially be addressed, or at least partially addressed, by information extracted from remotely sensed imagery.

Examples (and more can be found in Longley and Clarke 1995, and throughout this book) include those relating to transportation problems, housing and industrial development, the control of urban sprawl, the regeneration and recycling of derelict urban land, urban planning in general, the allocation of resources such as health care and policing, the development of emergency contingency plans, controlling and optimizing the develop-ment of retail outlets and so on. Central to all these issues are the people themselves and characteristics such as their age, their ethnic composition, their social, financial and medical conditions, but most importantly of all *where they are located*. Evidently, for any effective planning of a city's transport system you need to know, among other things, how many people who are present in one place will want to get to some other place. Likewise, a key factor when deciding where to place a new retail outlet is the number of people who reside within a given travel time of any potential building sites. So, we can see from these examples that the distribution of population within a city is of critical value when attempting to confront topics of this nature.

From this standpoint it is perhaps somewhat ironic that remote sensing is offered as a partial solution since it is seldom able to record directly, and thus place, individual people. Indeed, there are relatively few sources of information about population characteristics that also provide a geographical reference, and of those that do the richest source is typically a national population census. A census is, by its nature, about as far removed from a remote sensing system as it is possible to get – being based on individually completed questionnaire returns collected by hand or by post and conducted on a decennial cycle, perhaps with some form of mid-cycle estimates and updates. A census will provide highly detailed and highly accurate information about the population, typically broken down by age, sex, ethnicity and with numerous indicators of social standing, wealth and health. However, invaluable as this is, the geographical component of the census (i.e., the "where" factor) is always somewhat compromised since census statistics are published as counts that have been aggregated across administrative zones. This is done primarily to protect the confidentiality of the individual, but is also a mechanism to reduce the volume of data involved. So, for example, in England the smallest spatial unit for which a population count can be obtained is the Enumeration District (ED) containing approximately 150 households. The actual area of ED needed to achieve this degree of aggregation is likely to be quite small in densely populated inner city areas, but will by necessity become much larger in wealthy suburbs and especially so in rural areas. This variability in size creates problems of its own, but it is the process of spatial aggregation that imposes the greatest limitations on our ability to map population distribution using census statistics accurately. More importantly, the aggregated nature of census-derived population data restricts the accuracy with which any subsequent spatial analyses can be performed. Finally, it should also be noted that census information can be further devalued by the decennial cycle, which may mean that data are effectively obsolete at the time of analysis, particularly in parts of the world where cities are expanding rapidly, population growth rates are high, the census is frequently inaccurate and may take years to become available. How then can remote sensing technology assist?

As stated earlier, remote sensing is seldom able to map individual people directly but instead relies on recording urban physical properties (land cover types) that are indirectly related to the distribution of the population (land use). Viewed in the simplest terms, the population contained within a city will not be distributed uniformly within its boundaries but will reside in houses, flats, tower blocks and suchlike. These structures may have properties recorded by a remotely sensed image (e.g., their spectral reflectance, shape, texture or pattern) that allow their accurate identification and placement within the city limits. Having identified these areas or objects we can utilize the data integration capabilities of a Geographical Information System (GIS) to populate them with information obtained from census sources.

The aim of this opening chapter of Part II is to demonstrate how remote sensing can be used to improve our ability to map the distribution of population both within and outside of city limits. It will illustrate how one of the most appropriate methods for integrating census statistics with satellite imagery is through a process known as "dasymetric mapping". Despite its simplicity this technique can create population density surfaces that facilitate better visualization of population patterns, and can also create population distribution maps that enhance our ability to utilize population information in spatial analytical modelling tasks. The discussion starts off by reviewing the purposes for population mapping and currently adopted practices. It then outlines the principles behind dasymetric mapping and shows how remotely sensed images of cities may be used to operationalize this technique. The relative strengths and weaknesses of dasymetric mapping are compared with a number of alternative solutions, and, finally, the prospects for further development are considered especially in relation to the latest sources of remotely sensed imagery. Essentially this involves populating the census tract (the ED). Subsequent chapters in Part II of this book will further develop zone-based techniques (Lo, Chapter 7), examine techniques for populating the actual pixel (Harvey, Chapter 8), review the Bayesian method for incorporating land use information into conventional classification (Mesev, Chapter 9), introduce geodemographic characteristics into remote sensing (Harris, Chapter 10) and, finally, verify measurements of urban heat islands in the developing world (Nichol, Chapter 11). All will make cases for the importance of urban monitoring by establishing conceptual and technical methodologies for greater interaction between remotely sensed data and socio-economic data.

6.2 Purposes and practices of population mapping

Turning our attentions to zone-based population mapping it is important to appreciate fully the benefits that can be gained from utilizing remote sensing in the dasymetric mapping of a city's population. However, it is first necessary to consider both the purpose of creating a population map and the traditional methodologies used to perform this task. In fact, two quite distinct reasons for mapping population distributions can be readily identified. The first of these is simply to create a cartographic expression of the phenomena, wherein our primary purpose is to communicate to an observer an effective visual representation of the spatial distribution and pattern of population across the study region. The second reason for mapping population is to create a utilitarian object (i.e., a model of population distribution) from which further information can be extracted using spatial analysis techniques. It turns out that, despite these very differing objectives, much the same methodology is traditionally used to accomplish either task.

The traditional approach adopted for the cartographic representation of census-derived population counts is to use the choropleth map (sometimes called a graduated colour or a shaded area map). To create this we compute a density value for each reporting unit (e.g., a census ED), then impose a classification scheme onto these values and, finally, adopt a suitable gradational shading scheme to indicate the class membership of each reporting unit and provide some indication of its relative value. The aggregated nature of census data lends itself well to this approach and the almost ubiquitous availability of functionality in GIS software to facilitate this style of mapping has further encouraged its widespread use. This is despite the fact that many problems with this methodology have been identified (Langford *et al.* 1990; Langford and Unwin 1994). In particular, it is known to be vulnerable to the effects of the Modifiable Areal Unit Problem (MAUP) and ecological fallacy (Openshaw 1984), sensitive to the classification scheme adopted (Evans 1976) and, perhaps most importantly of all, it suggests there is homogeneity within zones and instant transition of values across zonal boundaries, both of which often bear little relationship to reality. In fact, within most census zones there are likely to be areas in which population is absent altogether (e.g., land covered by managed woodlands, water bodies, industrial or commercial complexes, quarries and so on). Even in those areas where population is present it is still unlikely to be uniformly distributed since people tend to live in clusters of homes, such as in villages in rural areas or on housing estates in cities. These are interspersed by less densely populated land covers such as agricultural land in rural areas, or spaces occupied by parks, local amenities, factories and institutional buildings within cities. Given these inherent characteristics of population distribution, dasymetric mapping is an appropriate tool for creating cartographic expressions of the phenomena since it is able to accommodate and communicate these features accurately.

Turning now to the second purpose for mapping population. It is found that the traditional approach used to create a distribution model for subsequent exploitation in a GIS is based on very similar principles to choropleth mapping. The primary objective in this situation is not to communicate an understanding of spatial pattern, but rather to create a model that allows population estimates to be derived for arbitrary areal units (or zones) that do not match those for which population counts have been reported. The task of generating estimated values across incompatible spatial units is called "areal interpolation", and the need for this is frequently encountered in GIS-based applications. Typically this is because it is found that other sources of data needed in an analysis have been aggregated to a different set of spatial units. It therefore becomes necessary to generate population estimates for these so-called "target units" to

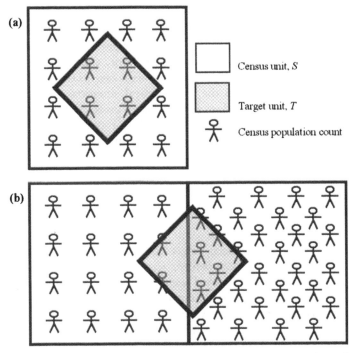

Figure 6.1 The choropleth population distribution model and simple areal weighted interpolation.

facilitate any further modelling and integration. Alternatively, a population estimate may be required for an entirely new spatial unit that has been created by the GIS as a direct result of spatial analytical operations such as buffering and overlay. The most widely used areal interpolation method is "simple areal weighting" in which the properties that are implied on a choropleth map are explicitly modelled.

The process is illustrated in Figure 6.1a, where the reported population of a census unit is assumed to be uniformly distributed within the zone's boundary. A population estimate for any spatially incompatible target unit that is wholly contained within this census zone can be obtained from the following formulae:

$$\hat{P}_T = A_T \frac{P_S}{A_S} \tag{6.1}$$

or

$$\hat{P}_T = P_S \frac{A_T}{A_S} \tag{6.2}$$

where $\hat{P}_T$ = estimated population for target unit T, P_S = known population of census unit S, A_T = area of target unit T and A_S = area of census unit S.

This is simply the area of the target unit multiplied by the mean population density in the source unit (Equation 6.1) or, equivalently, the population of the source unit scaled by the relative areal extents of the target and source units (Equation 6.2). In the more typical situation (shown in Figure 6.1b) where a target unit overlaps several source units, the same basic computation is performed but is applied separately for each overlapping component part of the target unit, and the results are summed to provide a final estimated total, as represented by Equation 6.3:

$$\hat{P}_T = \sum_{S=1}^{n} A_{TS} \frac{P_S}{A_S} \tag{6.3}$$

where A_{TS} = area of overlap between target zone T and census unit S, n = number of census units, *others as before*.

Like the choropleth map, simple areal weighted interpolation is easily implemented using only the most basic functionality of GIS software, and this encourages its widespread use. However, it is an appropriate methodology only when there is no additional information available that can be used better to inform the process. This is where remotely sensed images and dasymetric mapping techniques can begin to play a part, but before we see exactly how this is done it is necessary to understand the characteristics and methodology of dasymetric mapping.

6.3   Dasymetric population mapping

Dasymetric mapping is, like choropleth mapping, essentially an area-based cartographic method. Although it has not been as widely practised, and is not as well known as its choropleth counterpart, it has nevertheless had a long tradition in cartography (McCreary 1984). Wright (1936) was the first to introduce many of the basic concepts of dasymetric mapping in a seminal paper in which he explains how the term itself is of Russian origin and can be interpreted as "density measuring".

The basic principle of dasymetric mapping is a very simple one: *to subdivide source zones into smaller spatial units that possess greater internal consistency in the density of the variable being mapped.* Within this rather loose definition there are no specific rules for its implementation, and therefore any available sources of spatial information that can provide further insight into the probable internal structure of the source zones may be utilized, together with any appropriate methodology that can make use of these data. McCreary (1984) has explored numerous methodologies for dasymetric mapping as, more recently, have Eicher and Brewer

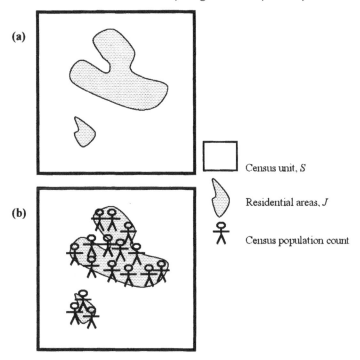

(a)

(b)

☐ Census unit, *S*

Residential areas, *J*

⚇ Census population count

Figure 6.2 The dasymetric principle for population distribution modelling.

(2001). While seeking to uncover the internal distribution of a source zone, dasymetric mapping methods will also ensure that any marked discontinuities in the level of the variable being mapped are better reflections of the true underlying geography and that, unlike a choropleth map, they are not merely artefacts thrown up by the relatively arbitrary location of the areal unit's boundaries.

The dasymetric principle is illustrated in Figure 6.2a, where the additional source of spatial information takes the form of a set of polygons that depict the location of residential areas within the study area. Given the premise that the people recorded by a census as present within this zone will reside in the houses that make up the residential areas, it immediately becomes apparent that a uniform population distribution model, such as depicted in Figure 6.1a, is no longer reasonable. In the light of this additional information it is rational to assume that at least the majority, and possibly the totality, of the reported population should be distributed to lie within the boundaries of the residential polygons. It is at this point that the lack of any specific rules for performing dasymetric mapping becomes very apparent – for instance, should all the population be redistributed into the residential polygons, or just some proportion of it? If only a proportion of the population is to be distributed in the residential polygons how should the relative densities in the residential and non-residential areas be determined? Perhaps it is this lack of procedural clarity that has generated a degree of unwillingness by some to adopt the dasymetric mapping approach.

The simplest scheme for implementing dasymetric population mapping is to utilize an ancillary data source that creates a binary mask. This has been referred to as the "binary dasymetric method" by a number of workers (e.g., Fisher and Langford 1995; Eicher and Brewer 2001). In other words, a map is employed that has just two classes which identify the locations of "occupied" and "unoccupied" areas. The use of a binary mask is actually just a special case of the more general "limiting variable" technique first proposed by Wright (1936), in which one class has been limited to the value of zero. Within the confines of each source zone (i.e., census unit) the recorded population is evenly distributed into just those internal parts that carry the "occupied" label, as is demonstrated in Figure 6.2b.

Looking at Figure 6.2b it is immediately apparent that distributing a census zone's population dasymetrically is likely to have significant effects on any subsequent areal interpolation estimates. The areal interpolation process when using the binary dasymetric method can be represented algebraically by the following formula:

$$\hat{P}_T = \sum_{S=1}^{n} A_{TS_j} \frac{P_S}{A_{S_j}} \tag{6.4}$$

where $\hat{P}_T$ = estimated population of target zone T, P_S = population of census unit S, A_{TS_j} = area of land cover j within overlap of target zone T and census unit S, A_{S_j} = area of census unit S having *land cover j* (identified as "occupied") and n = number of census units.

Figure 6.3 illustrates how remotely sensed images can play a significant role in facilitating dasymetric mapping of city populations. It shows a number of census EDs (in this case census wards) at the urban fringe of the city of Leicester, UK. Within two of these wards has been superimposed corresponding remotely sensed imagery obtained from a Landsat ETM scene. It is immediately apparent that satellite sensor images of cities allow us to "see inside" the rather arbitrary aggregation units adopted by the census, and that they can therefore be used to disaggregate associated population statistics using dasymetric principles. The residential areas can be clearly identified as the mid-grey dark-grey tones, and inter-city green spaces as bright or light-grey areas.

All that is required to implement the binary dasymetric mapping method is for a suitable classification technique to be applied to the satellite image that will identify to a high degree of spatial detail, and typically thematic accuracy, the location of residential land cover within the confines of the source zones. Typically, to minimize classification error this requires that a relatively large number of land cover types must first be identified to account

Figure 6.3 The internal structure of census enumeration districts revealed by satellite imagery.

for the wide variety of natural surfaces present across the urban landscape and beyond. However, once this initial classification is completed the classes may be collapsed to provide the simple binary map required. After over-laying the boundary of census reporting units with the binary map, the associated census population count is distributed evenly among only those internal pixels that are labelled as "occupied" (see Chapter 9 by Mesev for more on urban classification).

Langford and Unwin (1994) first investigated binary dasymetric mapping using classified satellite imagery as a potential method for creating better cartographic representations of population density. They concluded that the initial dasymetric product was not in fact particularly suitable for carto-graphic purposes due to the fine degree of spatial disaggregation achieved and the consequent visual complexity of the resultant map. Fortunately, the subsequent application of simple low-pass filtering operations to this output did allow highly effective isarithmic and pseudo-three-dimensional products to be created. Therefore, remotely sensed images of cities can help to provide improved visualizations of their internal patterns of population density, particularly compared with the principal alternative of choropleth mapping.

However, undoubtedly the biggest benefits to be gained from dasymet-rically mapping the population are when the resulting output is used as a distribution model for subsequent areal interpolation tasks. A number of studies have now shown that modelling population distribution dasymet-rically prior to areal interpolation yields very substantial improvements in the accuracy of the estimates obtained. For example, Fisher and Langford

(1995) devised a means of testing the relative performances of dasymetric and choropleth distribution models when using a binary residential mask created from a classified Landsat Thematic Mapper (TM) image of the city of Leicester, UK. They applied their distribution models at the ward level of UK census enumeration units and then constructed overlapping "target zones" that were created by randomly agglomerating EDs into new contiguous spatial units. Since the target zones were created from EDs that have a known true population, the target zones also had a known population, despite the fact that they intersected the source zones in a complex manner similar to that experienced in real world applications. True population values could thus be compared with estimates derived from the dasymetric- and choropleth-based interpolations. These experiments showed that by adding intelligence to the areal interpolation process through the addition of a classified satellite image, substantial improvements in areal interpolation accuracy could be gained. Accuracy is always dependent to some degree on the relative sizes of the target units to the source units, with better results achievable under conditions of spatial aggregation (i.e., where target units are larger than source units) compared with spatial disaggregation. Using a data set consisting of forty-nine source units and repeatedly predicting population totals for 100 randomized target units (i.e., spatial disaggregation) yielded errors of (± 59 per cent when using simple areal weighting, but only ± 27 per cent using dasymetric mapping. These figures declined to ± 13 per cent and ± 6 per cent, respectively, when the number of target units was reduced to ten (i.e., spatial aggregation). The strong relative performance of dasymetric mapping has also been confirmed in more recent studies conducted by Martin *et al.* (2000).

6.4   Comparison with alternatives

It is appropriate to compare and contrast the binary dasymetric method utilizing satellite imagery with alternative methods for redistributing and disaggregating census counts. Perhaps one of the most widely known and publicized methods is that developed by Bracken and Martin (Bracken and Martin 1989; Martin 1989; Martin and Bracken 1991; Bracken 1994). This makes use of a mathematical distance decay function together with the population-weighted centroids that are supplied in the UK for each census ED. The modelling procedure is illustrated in Figure 6.4. First, a regular grid is superimposed across the study area (a resolution of 200 m is the most frequently adopted) and the entire population of the enumeration district is initially placed in the grid cell in which the centroid location falls (Figure 6.4a). Population is then redistributed out of the centroid's cell and into neighbouring cells, guided by their degree of proximity and a mathematical distance decay function (Figure 6.4b–c). The population received by each

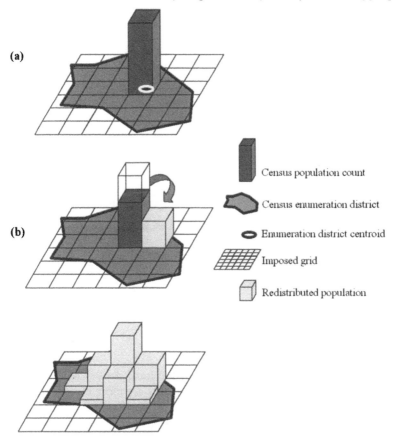

Figure 6.4 The centroid method of generating a population distribution model.

cell is estimated as:

$$\hat{P}_i = \sum_{j=1}^{n} P_j w_{ij} \tag{6.5}$$

where the weighting function w_{ij} is determined by:

$$w_{ij} = \left(\frac{k^2 - d_{ij}^2}{k^2 + d_{ij}^2} \right)^{\alpha} \tag{6.6}$$

and P_i = the total population received by cell i, P_j = the population at centroid j, w_{ij} = weighting of cell i with respect to centroid j, k = initial kernel width, d_{ij} = the distance between cell i and centroid j, α = a parameter to control the distance decay rate.

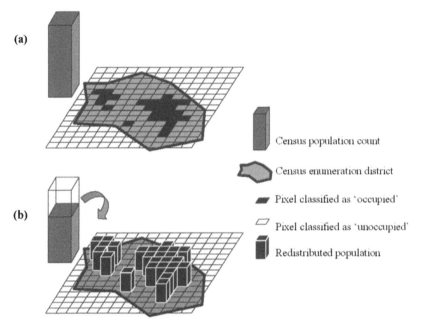

(a)

(b)

Census population count

Census enumeration district

Pixel classified as 'occupied'

Pixel classified as 'unoccupied'

Redistributed population

Figure 6.5 The binary dasymetric method of generating a population distribution model.

The distance decay function is adapted to the local density of centroids, resulting in a faster rate of decline where centroids are more closely spaced. Using this original algorithm it is possible for a proportion of the population count for a source zone to be redistributed into grid cells lying outside the unit's boundary, and conversely for areas within a source unit to receive a proportion of population originating from neighbouring source zones.

If we compare this approach with the dasymetric modelling method based on satellite imagery, as illustrated in Figure 6.5, a number of distinct differences can be identified. First, as implied in Figure 6.5b, the dasymetric method will typically utilize a grid with a much finer spatial resolution than the centroid method. The grid resolution of the dasymetric model is of course fundamentally controlled by the source of satellite imagery used. To date this has largely been implemented using Landsat TM imagery and consequently a grid size of around 30 m has been used. However, SPOT XS data and a corresponding grid resolution of 20 m would most likely prove to be equally successful. In the centroid method the grid spacing is a parameter that is set by the user, and, although it is theoretically possible to match or exceed these resolutions, Martin (1996) has indicated that this is unlikely to prove beneficial. Experiments indicated that a grid resolution of 50 m tended to cause over-smoothing of the population density surface resulting in the generation of clearly visible, concentric decay patterns around each centroid location. It was therefore concluded that grid resolutions much finer than 200 m cannot be supported by centroid data supplied at the ED level.

Given this contrast it could be argued that the binary dasymetric method based on satellite remotely sensed imagery offers the possibility for much finer spatial discrimination of population distribution. And in future the very high spatial resolution imagery derived from systems discussed earlier in this book could be utilized, possibly leading to even more spatial precision in the placement of population, provided of course that sufficient classification accuracy can be maintained.

Another contrast between these two methods is that the binary dasymetric approach does not redistribute population according to a mathematical rule (i.e., a distance decay function) but instead is guided by empirically derived physiographic evidence. Therefore, situations where more than one population cluster exists within a source zone (e.g., two housing estates situated at either end of an elongated enumeration zone and separated by a city park) should not pose any particular problem, unlike in the centroid method. Likewise, a situation where population distribution within the source zone is essentially linear, such as ribbon development occurring along arterial roads at a city's periphery, is more likely to be correctly modelled by a classified satellite image than by a mathematical distance decay function. Martin (1996) has claimed that the centroid method creates "an intuitively correct residential geography" but it is debatable whether this is simply a plausible simulation, somewhat akin to the very convincing appearance of an artificial fractal landscape, rather than a factual representation of reality. It could certainly be argued that the amount of geographical information obtained from a high spatial resolution (pixel size <50 m) remotely sensed image greatly exceeds that of the population-weighted centroids, and for this reason alone should bestow a more accurate population distribution model.

It will be noted when comparing Figure 6.4c with Figure 6.5b that as no distance decay function is involved in the binary dasymetric method each residential grid cell within a source zone receives the same proportion of population. Furthermore, the population count reported by the census is always redistributed only within the boundaries of the source zone, so preservation of the original volume at the source zone level is an inherent feature of this technique. This is in contrast to the centroid model originally proposed by Bracken and Martin (1989) and as used to generate a national population distribution model for the UK (Bracken and Martin 1995), in which the total population count (i.e., across the study area as a whole) is preserved, but not at the source zone level. In fact the correlation coefficient between original ED census counts and those obtained from the centroid-derived surface when retotalled over the same spatial units was only 0.6586 (Martin 1996). Of course, it is possible to apply a suitable scaling factor to the grid cells falling within each source zone to correct this discrepancy or, as introduced by Martin (1996), to modify the centroid algorithm so that it is constrained to redistribute population only to those grid cells lying within the confines of the original source zones.

Overall, the centroid method is quite closely tuned to the characteristics and properties of the UK Census. This is not to say that it can be only applied in the UK, but that to apply it elsewhere requires the local census to display similar features. First, it needs the provision of population-weighted centroids, which may not always be available from a census conducted in other countries. Second, it also requires the high level of spatial resolution provided by the UK's ED polygons, since it is known to perform best when source zones are relatively small and when their internal homogeneity is relatively high (Martin *et al.* 2000). Again this level of spatial detail may be something that is not available in a census conducted in other parts of the world. In contrast, the binary dasymetric method only requires the availability of census zone boundaries and a suitable satellite image from which urban or residential land cover can be identified. It is in effect scale-independent too, since it is possible to match the resolution of satellite imagery to whatever level of spatial zonation is locally available. So, for instance, it is quite possible to work with more coarsely scaled census units if these are the only available data (something approximating the local government districts in the UK Census hierarchy, for example) and match them with lower spatial resolution satellite imagery such as that supplied by Moderate Resolution Imaging Spectrometer (MODIS) or Advanced Very High Resolution Radiometry (AVHRR). Obviously the final results would be less accurate than when working with higher resolution data sets, but at any given scale the gains to be made over choropleth mapping or simple areal weighting can still be expected to be equally effective and valuable.

Having stressed some of the largely favourable contrasts between these two approaches it is only fair to concede that there are also some clear disadvantages with using satellite imagery and dasymetric mapping. Perhaps the most obvious of these is the volume of data associated with using satellite imagery, which has significant impacts both in terms of storage requirements and processing speed. In comparison with satellite imagery, an ED-level centroid dataset and a mathematical function to specify the distance decay properties are very compact. To actually implement dasymetric mapping from satellite remote sensing also requires a much larger skills set. Image processing skills are needed to register the image to the census boundary files and to perform the classification routine that identifies residential land cover. The latter may also necessitate a considerable degree of local knowledge. In contrast, practitioners of the centroid method need do little more than supply file names for the boundary and centroid data sets and specify a few optional parameters such as the grid resolution and study area limits.

A potential problem encountered with the binary dasymetric method when working with fine scale enumeration districts is the situation where no internal portion is identified as "occupied" in the remotely sensed image.

This implies that the ancillary data source is unable to provide any insight into the internal organization of the source zone and a sensible solution to this situation is simply to trap any such occurrences and revert to the choropleth model in these instances.

A final issue that must be considered in the dasymetric approach is error in the classification of the satellite image. It is well known that land cover classification from satellite imagery is imperfect and there must therefore be errors in the dasymetric distribution map. Due to the complexity of the surface, relatively high levels of spectral confusion and resulting misclassifications are likely to be present when working in urban environments. Problems encountered in the urban environment that are of particular significance to this application are misclassifications between residential homes and other buildings such as factories, offices and commercial outlets. The important question is to what extent any misclassification of residential areas will corrupt the dasymetric mapping process. Fisher and Langford (1996) address this issue directly and have shown that the areal interpolation of population using dasymetric mapping is actually quite robust to classification error. They found that a simulated error of up to 40 per cent was needed before population estimates based on dasymetric mapping deteriorated to the level of the next best interpolation method tested (based on a regression model). Since accuracy figures of around 80 per cent or more are typically quoted for land cover maps derived from satellite imagery it seems reasonable that we should not become overly concerned with this issue. So, although classification error is undoubtedly present and will degrade overall performance, dasymetric mapping remains a powerful tool for the areal interpolation of population and still outperforms alternative methods.

One of the key reasons identified by Fisher and Langford (1996) for this apparent robustness is the fact that we are normally concerned with population estimates for areas formed by relatively large aggregations of image pixels. Research has shown that classification error at the pixel level can be quite large without this causing the accuracy of estimates of regional amounts, or inventory accuracy, to be unduly affected (Franklin *et al.* 1986; Strahler 1981). Fisher and Langford (1996) have noted that, provided the error is randomly distributed over space, a dasymetric model with 100 per cent error (i.e., white noise) still performs as well as simple areal weighting! This highlights the fact that the spatial properties of error are also important, and of course classification error is typically not randomly distributed. Overall, therefore, classification error should not be ignored and it should be minimized whenever possible, but equally it should not detract from the gains to be made by dasymetric mapping over its alternatives.

Before leaving the topic of alternative methods for redistributing and disaggregating census counts a brief mention should also be made of the set of techniques based on regression modelling. The basic premise here is that a relationship will exist between the population count of an enumeration zone and its composition as measured by the pixel counts of its component land covers derived from a classified satellite image. Given a sufficiently large number of cases a regression equation of the following form can be calibrated to model this relationship:

$$P = \left(\sum_{j=1}^{k} \beta_j n_j \right) + \varepsilon \tag{6.7}$$

where P = population of a zone, β_j = coefficient for land cover j, n_j = area of land cover j within the zone, k = number of classified land covers and ε = a random error term.

The absence of an intercept term indicates that the best fit line is forced to pass through the origin, thereby ensuring that a population cannot be generated without reference to an associated area. The coefficients in this model are considered to provide estimates of the mean population density to be associated with each land cover class. Langford *et al.* (1991) created three such regression models based on one, two and five land cover classes and applied over a study area consisting of three Leicestershire districts. One of the attractions of this approach is that it provides a way to take advantage of the multiple land cover classes typically derived from remotely sensed imagery. However, calibration by ordinary least squares can be problematic as it has a tendency to give rise to negative correlation coefficients. In the areal interpolation experiments conducted by Fisher and Langford (1995) the binary dasymetric method was found to outperform these regression models consistently. This is probably due to the fact that the dasymetric model is fitted individually to each source zone, while regression models of the type specified above are calibrated at the global level (i.e., across the data set as a whole). They are also inferior to the binary dasymetric method in the sense that they do not ensure correct population totals are maintained when pixels are aggregated across original source zones. A number of developments have taken place in recent years to address some of these short-comings (e.g., Yuan *et al.* 1997; Harvey 2000) and the reader is urged to maintain interest and read Chapter 8 in this book by Harvey, where a methodology to allow census population statistics to be directly related to the spectral vectors of unclassified image pixels is examined.

6.5 Future prospects

Given that binary dasymetric mapping has shown itself to be a robust and effective method for enhancing areal interpolation and population mapping, what lies in the future in terms of methodological development? Obviously the aim must be to enhance its performance further, and to find ways to achieve this we need to consider where the major areas of current difficulty lie.

When using the binary dasymetric method to provide an estimate of a target area's population any resultant error can arise from one of two sources. The first of these is through the incorrect identification of residential areas in the satellite image. If areas that are not residential are classified as residential (i.e., commission errors) the effects will be twofold. It will lower the mean dasymetric density of all residential areas contained within the reporting unit (potentially causing a calibration error), and it will distribute population into places where it should not be present (thus causing a location error). Conversely, failing to identify any areas that are truly residential (i.e., omission errors) will have the effects of raising the mean dasymetric density of all residential areas lying within the reporting unit and failing to distribute population into places where it should be present. So, higher classification accuracy is clearly a desirable goal.

The second source of error is in the calibration of the dasymetric densities themselves. In a binary dasymetric map the assumption is made that all residential areas in any specific reporting unit will have the same population density (see Figure 6.4b). As reported above, this simple model has been shown to be surprisingly effective at enhancing the accuracy of areal interpolation when compared with a choropleth or uniform density model. Despite this it still seems logical to assume that, in most situations, there is likely to be some degree of variation in housing type, occupancy and thus population density, within the boundaries of the reporting units. The differentiation of more than one class of occupied land cover, each with an associated population density, is therefore another area of potential development.

It seems logical, then, that these are the two directions in which the dasymetric methodology can be further advanced. First, we can concentrate efforts on improving the classification accuracy of the satellite image source. Of course, this is one of the themes of this book and the reader is urged to examine accompanying chapters for evidence of what the future may hold in terms of better urban mapping using the latest sources of satellite imagery. It has long been argued that very high spatial resolution satellite imagery is likely to prove more of a hindrance than a help in the accurate identification of urban land cover, due to the complexity and heterogeneity of the cityscape (Cushnie 1987). However, this point of view is generally founded on the assumption that a traditional per-pixel classification approach has to be adopted (see Chapter 2 in this book by Aplin on *per-parcel* classification). Very high spatial resolution imagery may result in better mapping of cities provided alternative classification methods are adopted that can fully utilize texture measurements, image segmentation, fuzzy classification and object-based approaches (see Tso and Mather 2001). Adaptations to the classification algorithms to make use of ancillary data resources available in a GIS (Harris and Ventura 1995), or even census-derived statistics such as housing density measures (Mesev 1998), also offer an opportunity to enhance urban image classification.

Second, we can develop methods for the calibration of population densities associated with multiple dasymetric classes. It is already plausible to distinguish between several residential categories when classifying urban areas from satellite imagery, but the difficulty lies in determining how appropriate density values should be assigned. In the simple binary model this does not present a problem since, with one of the classes set to a value of zero, the density for the remaining occupied class is effectively "self-calibrating". However, once more residential classes are introduced we lose this convenient property and a key issue is how to calibrate the relative densities of each category. The solution most often adopted in previous cartographic applications of dasymetric mapping has been to allocate a fixed proportion of the source zone population to each class (e.g., Eicher and Brewer 2001). Other solutions have been to use subjective estimates based on expert knowledge of the study area or empirical evidence obtained from limited field survey (Wright 1936). Possible future solutions include the use of global estimates derived from regression models, or some form of regionally derived regression estimates that allow for spatial non-stationarity in these parameters. The relative merits of these and other schemes, and the degree to which they enhance areal interpolation accuracy, are areas of current research.

6.6   Conclusions

Dasymetric mapping is an area-based cartographic tool that has shown itself to be useful in developing better visualizations of population density, and has also proven to be an extremely powerful tool for enhancing the areal interpolation of population-related statistics. It is conceptually simple and is readily implemented in a GIS where ancillary spatial data that can shed light on the internal structure of census zones are likely to reside. Remotely sensed images are a particularly relevant source of ancillary data since they are widely available at a variety of spatial scales and effectively allow us to "see inside" the arbitrary spatial zones used to aggregate census statistics. Dasymetric mapping with satellite imagery only requires that the boundaries of census EDs are available and can provide an accurate reconstruction of the underlying population geography based on empirical observation.

The simplest dasymetric approach is to classify the satellite image into land cover classes that are subsequently collapsed to, from a binary mask of occupied and unoccupied areas. Despite its rudimentary nature, this binary dasymetric method enables a high degree of spatial disaggregation to be achieved, and experiments have shown that the resultant population distribution model performs very well in areal interpolation tasks. Some of the key advantages of this approach are that the population distribution model is locally fitted, retaining volume within the original reporting zones, and is equally applicable at a variety of spatial scales.

The key issues that need to be addressed by future research, in order to refine dasymetric population mapping, are improving classification accuracy and further methodological development to allow multiple residential classes to be effectively incorporated into the process.

6.7 References

Bracken, I., 1994, A surface model approach to the representation of population-related social indicators. In A. S. Fotheringham and P. A. Rogerson (eds) *Spatial Analysis and GIS* (London: Taylor & Francis), pp. 247–60.

Bracken, I. and Martin, D., 1989, The generation of spatial population distributions from census centroid data. *Environment and Planning A*, **21**, 537–43.

Bracken, I. and Martin, D., 1995, Linkage of the 1981 and 1991 UK censuses using surface modelling concepts. *Environment and Planning A*, **27**, 379–90.

Cushnie, J. L., 1987, The interactive effect of spatial resolution and degree of internal variability within landcover types on classification accuracies. *International Journal of Remote Sensing*, **8**, 15–29.

Eicher, C. L. and Brewer, C. A., 2001, Dasymetric mapping and areal interpolation: Implementation and evaluation. *Cartography and Geographic Information Science*, **28**, 125–38.

Evans, I., 1976, The selection of class intervals. *Transactions of the Institute of British Geographers*, **2**, 98–124.

Fisher, P. F. and Langford, M., 1995, Modelling the errors in areal interpolation between zonal systems by Monte Carlo simulation. *Environment and Planning A*, **27**, 211–24.

Fisher, P. F. and Langford, M., 1996, Modelling sensitivity to accuracy in classified imagery: A study of areal interpolation. *The Professional Geographer*, **48**, 299–309.

Franklin, J., Logan, T. L., Woodcock, C. E. and Strahler, A. H., 1986, Coniferous forest classification and inventory using Landsat and digital terrain data. *IEEE Transactions on Geoscience and Remote Sensing*, **GE-24**, 139–49.

Harris, P. M. and Ventura, S. J., 1995, The integration of geographic data with remotely sensed imagery to improve classification in an urban area. *Photogrammetric Engineering and Remote Sensing*, **61**, 993–8.

Harvey, J. F., 2000, Small area population estimation using satellite imagery. *Statistics in Transition*, **4**, 611–33.

Langford, M. and Unwin, D. J., 1994, Generating and mapping population density surfaces within a geographical information system. *The Cartographic Journal*, **31**, 21–6.

Langford, M., Unwin, D. J. and Maguire, D. J., 1990, Generating improved population density maps in an integrated GIS. Paper given at Conference of the European Conference on Geographical Information Systems (EGIS90) (Utrecht, The Netherlands: EGIS Foundation), pp. 651–60.

Langford, M., Maguire, D. J. and Unwin, D. J., 1991, The area transform problem: Estimating population using remote sensing in a GIS framework. In I. Masser and

R. Blakemore (eds) *Handling Geographical Information: Methodology and Potential Applications* (London: Longman), pp. 55–77.

Langford, M., Fisher, P. F. and Troughear, D., 1993, Comparative accuracy measurements of the cross-areal interpolation of population. Paper given at Conference of the European Conference on Geographical Information Systems (EGIS93) (Utrecht, The Netherlands: EGIS Foundation), pp. 663–74.

Longley, P. A. and Clarke, G. (eds), 1995, *GIS for Business and Service Planning.* (New York: John Wiley & Sons), 316 pp.

McCreary Jr, G. F., 1984, Cartography, geography and the dasymetric method. Paper given at Proceedings 12th Conference of International Cartographic Association, Perth, Australia (Elsevier: Marrickville, Australia), pp. 599–610.

Martin, D., 1989, Mapping population data from zone centroid locations. *Transactions of the Institute of British Geographers*, 14, 90–7.

Martin, D., 1996, An assessment of surface and zonal models of population. *International Journal of Geographical Information Systems*, 10, 973–89.

Martin, D. and Bracken, I., 1991, Techniques for modelling population-related raster databases. *Environment and Planning A*, 23, 1065–79.

Martin, D., Tate, N. J. and Langford, M., 2000, Refining population surface models; experiments with Northern Ireland Census data. *Transactions in Geographical Information Systems*, 4, 342–60.

Mesev, V. 1998, The use of census data in urban image classification. *Photogrammetric Engineering and Remote Sensing*, 64, 431–8.

Openshaw, S., 1984, *The Modifiable Areal Unit Problem*, CATMOG 38 (Norwich, UK: Geo Books).

Strahler, A. H., 1981, Stratification of natural vegetation for forest and rangeland inventory using Landsat digital imagery and collateral data. *International Journal of Remote Sensing*, 2, 15–41.

Tso, B. and Mather, P. M., 2001, *Classification Methods for Remotely Sensed Data* (London: Taylor & Francis).

Wright, J. K., 1936, A method of mapping densities of population: With Cape Cod as an example. *Geographical Review*, 26, 103–10.

Yuan, Y., Smith, R. M. and Limp, W. F., 1997, Remodelling census population with spatial information from Landsat TM imagery. *Computers Environment and Urban Systems*, 21, 245–58.

7 Zone-based estimation of population and housing units from satellite-generated land use/land cover maps

Chor Pang Lo

7.1 Introduction

Although it was announced optimistically that population censuses could be completed more accurately with space-age technology as early as 1983 (Brugioni 1983), an operational approach to extract population data accurately from remotely sensed imagery has yet to be developed. Previous research has established the usefulness of large-scale aerial photographs (1:40,000 and larger) for accurately estimating population in a small area by manually typing and counting the dwelling units, to which the appropriate household size per dwelling unit is applied (Watkins and Morrow-Jones 1985). This approach can produce population estimates with relative errors of less than ±10 per cent. However, such an approach is not satisfactory in a computer age when satellite sensor data in digital form are readily available. Also, population censuses should not only be limited to small areas, but should be applied to large regional estimates as well.

Early attempts to estimate population from satellite sensor images (notably Landsat Multispectral Scanner Systems [MSS] data) were carried out by Hsu (1973) and Iisaka and Hegedus (1982), who linked the spectral radiance characteristics of individual pixels with population densities using a polynomial equation. The accuracy of such an approach varies according to the level of areal unit to which the population is aggregated. This is generally known as the *pixel-based approach*. This approach is handicapped by the fact that pixel-level population density is not readily available and has to be estimated through a process of interpolation based on the distance decay principle (Martin 1996). A new development of this approach for automated population estimation as applied to Ballarat, Australia from Landsat Thematic Mapper (TM) images was advocated by Harvey (1999). He classified the pixels into residential and non-residential. For the residential pixels, he applied a statistical approach known as the expectation maximization technique for iterative re-estimation of the ground truth pixel population within a census district. The pixel-based approach minimizes the effects of the Modifiable Areal Unit Problem

(MAUP) caused by differences in the size of the census tracts or enumeration units (see Chapter 8 in this book by Harvey).

Another approach of population estimation is to delineate different types of residential land use area from aerial photographs. Initially developed by Kraus *et al.* (1974), population densities associated with a type of residential land use were applied to produce estimates of population. Such an approach is now quite straightforward and can be automated with the use of satellite sensor images, producing a land use/land cover map using an appropriate digital image classification method. Regression models, linking population with spatially aggregated pixels of residential land use/land cover matched to the census tract or enumeration unit, are developed. Known as the *zone-based approach* a basic assumption of this approach is that population growth will result in a spatial expansion of residential land area, which can be detected by comparing two satellite sensor images acquired at different times. In the instances where existing structures are demolished and replaced by denser ones, no change of land area is involved, and, as a result, population increase cannot be detected. Fortunately, such cases occur in limited parts of the city (such as the central area) and can be compensated by overestimation caused by vacant or abandoned housing units found in the same area. The general trend of development of cities, in North America at least, is to expand outward, a process known as suburbanization, thus making the zone-based approach to population estimation applicable.

The zone-based approach has been successfully employed by Weber (1994) using high-spatial-resolution multi-spectral and panchromatic SPOT images to estimate the population of Strasbourg, France. First, all the residential areas were extracted from the satellite sensor images using a zone-based, contextual, digital image classification approach so that different housing types could be differentiated. The population of the city was then estimated by means of a linear regression equation in which the constant *a* was suppressed, using the different housing types as independent variables. Another variant of the zone-based approach is to extract all the residential areas from the satellite sensor images. Then, the population density of each residential area was determined either by fieldwork or from census data. Population can be estimated simply as the product of the residential area and its associated population density within the limits of the census tract or enumeration unit. Such an approach has been successfully applied to Hong Kong (Lo 1995) and South Africa (De Klerk 1998). See also Chapter 6 in this book by Langford.

From a practical point of view, the zone-based approach with image classification is particularly appealing as a cost-effective operational approach to extract population estimates from satellite sensor images automatically. This is because population estimation can be generated as a by-product of land use/land cover mapping – the most important application of satellite sensor images today. As a result, land use/land cover statistics will be available to develop the residential land use population regression model. However, further evaluation of this zone-based approach to population estimation and housing units is needed to understand the advantages and disadvantages of this approach further.

Since 1996, the author has been involved in Project ATLANTA (ATLANta land use analysis: Temperature and Air quality) funded by NASA, which generated a land use/land cover database for the metropolitan area of Atlanta from 1973 to 1997, using Landsat MSS and TM images (Quattrochi and Luvall 1999, Lo and Yang 2000). This database was used to evaluate the zone-based population and housing unit estimation approach in an urban environment at the census tract level. This chapter reports the findings from such an evaluation.

7.2 Study area

The study area consists of ten urban counties in the service area of the Atlanta Regional Commission (ARC) (Figure 7.1). On 1 April 1997, the population of the ten-county Atlanta region had just passed the three million mark – 3,033,400 (ARC 1997). Atlanta is a rapidly growing city in the US South, and has expanded in size through the process of suburbanization (Research Atlanta, Inc. 1993). Since 1970, much of the population growth occurred in northern counties. This trend continued in the 1990s, although population growth in southern counties (counties to the south of Interstate Highway I-20) began to pick up in recent years (Figure 7.1). Atlanta is also noted for having four downtowns. Apart from the traditional Central Business District (CBD) at the centre of the city, three other downtowns sprang up in Buckhead/Lenox, Cumberland/I-75 and Perimeter/GA-400, all to the north of the city at highway intersections with the I-285 (the perimeter loop) (Fuji and Hartshorne 1995). More and more Atlanta residents, particularly those who live in the suburbs, need to commute to work by car every day. Traffic congestion is a daily occurrence, and urban sprawl has brought about loss of forest land and worsening air pollution. Because of the north–south differences in the rate of growth of population, Atlanta provides an ideal urban environment to test the methodology of zone-based population estimation.

7.3 Land use/land cover classification

Project ATLANTA created a historic database of land use/cover maps for the Atlanta region extracted from Landsat MSS and TM sensor data using digital image classification. Eleven cloud-free Landsat scenes between 1973 and 1998 in the spring or summer season were employed. Eight of these are Landsat MSS images obtained in 1973, 1979, 1983, 1987, 1988 and 1992, while the remaining three are Landsat TM images acquired in 1987, 1997

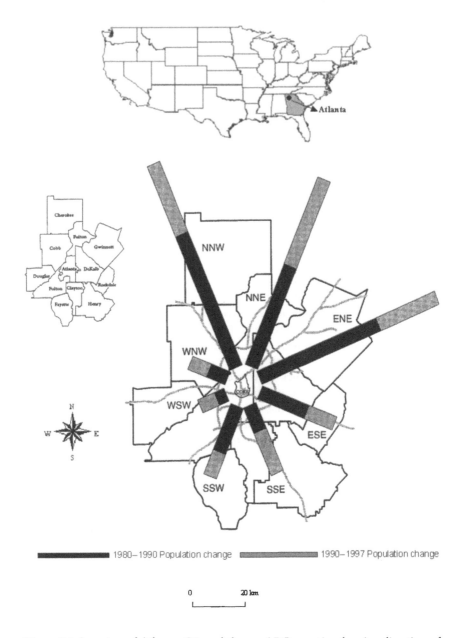

Figure 7.1 Location of Atlanta, GA and the ten ARC counties showing direction of growth in population during the period 1980–1997. The black bars show 1980–1990 population changes, while the grey bars show 1990–1997 population changes. The major interstate system is also shown as grey lines. (Data from Atlanta Regional Commission, 1997, used with kind permission.)

and 1998 (Table 7.1). For 1973 and 1979 first-generation Landsat MSS images, two scenes were required to cover the whole of Atlanta. Before land use/land cover classification, these two scenes were mosaicked together to form one scene centring over Atlanta. The remaining images belong to second-generation Landsat data, which allow the scene centre to be shifted along the orbiting path, so that the whole Atlanta metropolitan area can be covered by one scene, thus saving money and processing time.

Image preprocessing involving geometric rectification and radiometric normalization was then performed. Geometric rectification made use of ground control points extracted from 1:24,000 US Geological Survey topographic maps. The total number of control points used varied between twelve and fifteen as indicated in Table 7.1. All the Landsat images were registered to the 1997 Landsat TM image supplied by Space Imaging EOSAT, which has been rectified and georeferenced to the Universal Transverse Mercator (UTM) coordinate system (Zone 16) with North American Datum (NAD) 83 and ellipsoid Geodetic Reference System (1980) (GRS 80). The accuracy of registration for all the Landsat images was very high, as shown in Table 7.1, the most accurate for a Landsat TM image being root mean square error (RMSE) = 0.22 pixel and the least accurate for a Landsat MSS image being RMSE = 0.61 pixel.

Because the Landsat MSS data were acquired by different sensors at different times, radiometric normalization was necessary to eliminate differences in radiometry caused by variations in sensor properties, atmospheric conditions and sensor target illumination geometry. This was carried out using the method developed by Hall *et al.* (1991). The choice of method took into consideration the nature of terrain and land–water distribution pattern in the Atlanta environment (Yang and Lo 2000). Basically, the method selected a reference scene from which Radiometric Control Sets (RCSs) for dark and bright targets were selected. The corresponding RCS in a subject image was also selected. Corrected pixel values in the subject image were computed using the following linear regression equation band by band:

$$S'_k = m_k S_k + b_k \qquad (7.1)$$

where S_k is the digital number (DN) of band k in subject image, S'_k is the normalized (corrected) DN of band k, m_k is the slope or gain and b_k is the intercept or offset. Both m_k and b_k were computed through a linear least-squares regression performed on the two RCSs from the reference–subject image pair. The 1988 Landsat MSS scene was selected as the reference image because of its superior quality and the fact that large-scale aerial photographs of the study area are available for the same year, thus making it possible to verify the RCS. All the Landsat MSS images underwent this radiometric normalization. No radiometric normalization was required for the Landsat TM data (1987 and 1997) because both were obtained from

Table 7.1 Characteristics of the satellite data used for land use/cover mapping in the Atlanta Metropolitan area

Date	Type of imagery	Landsat No.	Nominal IFOV (m) [†]	Sun elevation (degree)	Sun azimuth (degree)	Scene location	Rectification RMSE in pixel unit (Control Point No.)	Radiometric normalization
4-13-73	MSS	1	57 × 79	54.16	129.37	North Atlanta	0.58 (13)	Yes
4-13-73	MSS	1	57 × 79	54.83	127.39	South Atlanta		
6-11-79	MSS	3	57 × 79	61.65	105.65	North Atlanta	0.46 (13)	Yes
6-11-79	MSS	3	57 × 79	61.74	102.77	South Atlanta		
5-9-83	MSS	4	57 × 79	60.50	117.64	Centre-shifted*	0.61 (13)	Yes
6-29-87	MSS	5	57 × 79	61.84	103.71	Centre-shifted*	0.44 (13)	No
6-29-87	TM	5	28.5 × 28.5	61.84	103.71	Centre-shifted*	0.22 (13)	No
5-14-88	MSS	5	57 × 79	61.61	115.51	Centre-shifted*	0.51 (12)	Reference
4-23-92	MSS	5	57 × 79	56.00	121.80	Centre-shifted*	0.47 (15)	Yes
7-29-97	TM	5	28.5 × 28.5	61.00	106.00	Centre-shifted*	Reference	No
1-2-98	TM	5	28.5 × 28.5	27.00	150.00	Centre-shifted*	0.27 (14)	No

* The centre of the north scene has been shifted 50% southward along the orbit so that the whole Atlanta area is covered in one scene, thus saving on data and processing costs. This type of shifting is only possible for the second- and third-generation Landsat images. The area of one scene is approximately 185 × 185 km².
[†] IFOV (m) = Instantaneous Field of View converted to metres.

Table 7.2 Land use/cover classes and definitions

No.	Classes	Definitions
1	High-density urban use	Approximately 80% to 100% construction materials (e.g. asphalt, concrete, etc.); typically commercial and industrial buildings with large open roofs as well as large open transportation facilities (e.g., large airports, parking lots, and multilane interstate/state highways); with low percentage of residential development residing in the city cores.
2	Low-density urban use	Approximately 50% to 80% construction materials; often residential development including most single/multiple family houses and public rental housing estate as well as local roads and small open (transitional) space as can always be found in a residential area; with certain amount of vegetation cover (up to 20%).
3	Cultivated/exposed land	Areas of sparse vegetation cover (less than 20%) that are likely to change or be converted to other uses in the near future; including clear-cuts, all quarry area, cultivated land without crops, and barren rock or sand along river/stream beaches.
4	Cropland or grassland	Characterized by high percentages of grass, other herbaceous vegetation and crops; including lands that are regularly mowed for hay and/or grazed by livestock, golf courses and city parks, and regularly tilled and planted cropland.
5	Forest	Including coniferous, deciduous and mixed forests (90% to 100%).
6	Water	All areas of open water, generally with greater than 95% cover of water, including streams, rivers, lakes and reservoirs

the same sensor and their Sun elevation and azimuth are very similar (Table 7.1). The radiometric quality of the Landsat TM images is very high.

A land use/land cover classification scheme using hybrid Levels I and II categories of the US Geological Survey land use/land cover classification scheme for remote sensor data (Anderson *et al.* 1976) was adopted. To ensure compatibility of the land use/land cover data extracted over time using Landsat images of different spatial resolutions, the following six land use/land cover classes were adopted: (i) high-density urban use, (ii) low-density urban use, (iii) cultivated/exposed land, (iv) cropland or grassland, (v) forest and (vi) water. The image criteria for these six land use/land cover classes are shown in Table 7.2.

In order to ensure high accuracy in the land use/land cover mapping, an unsupervised classification approach followed by cluster labelling and spatial reclassification was adopted. Such an approach is possible because of the abundance of ground truth data in the form of old large-scale aerial photographs, specially flown, new, colour, infrared aerial photographs by NASA and field data. The unsupervised classification made use of the ISODATA (Iterative Self-Organizing DATA Analysis) algorithm in the ERDAS Imagine program to identify spectral clusters from the Landsat images. Too many clusters will break up homogeneous pixels, while too few clusters will cause heterogeneous pixels to combine. Through empirical evaluation, starting with thirty clusters and ending with ninety clusters, a total of sixty clusters were determined to be optimum. Each cluster was then assigned to one of the six land use/land cover classes with reference to the ground truth data obtained from large-scale, black-and-white and colour, infrared aerial photographs, supplemented by data collected in the field. If a cluster contained more than one type of land use/land cover class, it was split into smaller clusters using spatial and contextual properties. The correct land use/cover classes were then assigned to these smaller clusters. The whole process is known as spatial reclassification. The same number of clusters and the same approach were used for Landsat images of each year. The accuracy of the resultant land use/land cover map for the most current year (1997) (Figure 7.2, see colour plates) was evaluated using a stratified random sample of 488 points so that each land use/cover class has at least 50 points. An overall classification accuracy of about 90 per cent and a kappa index of 0.878 were achieved. Each land use/land cover category exhibited a producer's accuracy of over 80 per cent. In particular, the category of "Low-Density Urban Use" showed a producer's accuracy of over 93 per cent. The land use/land cover map is therefore of sufficient accuracy for population estimation. The same conclusion was considered to be true for land use/land cover maps produced for the Atlanta region for 1973, 1979, 1983, 1987 and 1992.

7.4   Population estimation

The zone-based approach to population estimation was to establish the link between residential use and population at the census tract level in the Atlanta region. For the ten counties under study, there are 418 census tracts based on the 1990 Census from the US Census Bureau. These census tracts were delineated in order to maintain a uniform population density. As a result, the areas of the census tracts vary from a minimum of 21.8 acres (8.82 hectares) to a maximum of 54,466.2 acres (22,042.17 hectares). As expected, the large census tracts are found in the outer periphery of the city (Figure 7.4, see p. 168). Despite this problem of variable areal units, census tracts are the most appropriate areal units to use for this population estimation evaluation because accurate population census and housing data are available at the census tract level for 1990. In addition, the ARC produced population estimates for each year by census tracts.

Residential land use (low-density urban use) areas at the census tract level were extracted from the satellite-generated land use/land cover maps for 1987 and 1997 of the Project ATLANTA database, the two most accurate land use/cover maps produced from Landsat TM data. These two years matched up approximately with the population census data for 1990 and the population estimates for 1997 at the census tract level, respectively, to calibrate and evaluate the accuracy of the results. In order to see the strength of relationship between population and residential use, Pearson correlation coefficients were calculated between "low-density urban use area" for 1987 and 1997 and "population" for 1990 and 1997, respectively. It was found that for 418 census tracts the correlation coefficient (R) between "low-density urban use area" for 1987 and population for 1990 is 0.57 while that between "low-density urban use area" for 1997 and population for 1997 is 0.66, all at over the 99.9 per cent level of confidence. While significant, the relationships are still not too strong. This means that a linear regression equation is probably not good enough to provide the link between residential use and population. There is also a problem known as heteroscedasticity in the population and residential area data set, which contains a large number of observations of small values relative to large values (Johnston 1980).

An enhanced equation is the allometric growth model, which has been successfully used in population estimation (Tobler 1969; Lo and Welch 1977). The model takes the following form:

$$P = aA^b \tag{7.2}$$

where P is population and A is residential land area. Its computation form is a logarithmic transformation as follows:

$$\log P = \log a + b \log A \tag{7.3}$$

The logarithmic transformation of both the dependent and independent variables minimizes the problems caused by heteroscedascity. When the function is plotted on a double logarithmic coordinate system, it describes a straight line. This is known as the allometric growth law in biology but Pareto's law in economics (Lee 1989). This equation expresses that the relative growth rate of the population (P) is proportional to the relative growth rate of the residential land area (A), with b as the scaling factor. It is interesting to note that when the value of b is larger than 1, positive allometry occurs, implying that P increases at a faster rate than A. If the value of b is smaller than 1, negative allometry results, suggesting that P increases at a slower rate than A. However, if the value of b equals 1, isometry results, indicating that both P and A increase at the same rate. Finally, the coefficient a is known as the *proportionality coefficient*. Clearly,

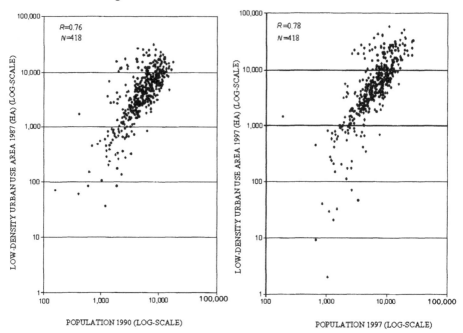

Figure 7.3 Correlation between population and low-density urban use area in logarithmic scale for 1990 and 1997, indicating a strong correlation.

by rearranging Equation (7.1), $a = P/A^b$, showing some form of residential population density or space standard.

Figure 7.3 shows that by applying the logarithmic transformation to the population data for 1990 and 1997 and to the residential land use area data for 1987 and 1997, a linear relationship appeared, thus suggesting the suitability of the allometric growth model as the linking model between population and residential land use area. Indeed, the correlation coefficients computed for these two sets of variables are 0.76 and 0.78, respectively, all at over the 99.9 per cent level of confidence. Before the model can be used, it needs to be calibrated with known population and residential land use area data. A random sample of 45 census tracts were used to provide the data for the calibration. The remaining 373 census tracts were withheld for use to evaluate the accuracy of the population estimation. The following two models resulted for the two different periods:

$$\log POP90 = 1.899\,795\,5 + 0.509\,005\,8 * \log RESID87 \quad (R^2 = 0.76)$$

$$(7.4)$$

$$\log POP97 = 2.356\,577\,5 + 0.387900\,5 * \log RESID97 \quad (R^2 = 0.68)$$

$$(7.5)$$

Table 7.3 Accuracy of population and housing unit estimation of Atlanta at the census tract level

Year	No. of tracts	Relative error (%)			Absolute relative error (%)		
		Mean	Standard deviation	Max	Mean	Standard deviation	Max
Population							
1990	373	18.13	77.72	725.77	40.78	68.57	725.77
1997	373	14.80	110.55	1,886.35	40.03	104.09	1,886.35
Housing units							
1990	373	38.32	245.21	3,226.23	63.82	239.83	3,226.23
1997	373	36.43	288.70	4,227.11	63.94	283.86	4,227.11

In both models the value of b is less than 1, implying that population increases at a slower rate than the residential area at the census tract level. This is more so for the 1997 model because of its even smaller b value, 0.388 compared with 0.509. This is not surprising because, in the process of suburbanization, people move further outward to build larger houses, thus occupying more land per capita, and this tendency has become much stronger in recent years. The proportionality coefficient a has also shown a great increase in value in 1997 (2.357 compared with 1.9) suggesting an increase in population density.

The two models were applied to the 373 census tracts withheld to test the accuracy of the models. The low-density urban use area for the appropriate year was input as the independent variable (as residential area) in the model. The antilogarithms of $\log POP90$ and $\log POP97$ from the models would give the estimated population. Table 7.3 provides a summary of the various measures of accuracy for the estimation of population of the ten counties of Atlanta at the census tract level. The mean relative error of the population estimate for 1990 is 18.13 per cent. But if the sign of the difference between the estimated and actual population is used, the mean absolute relative error is 40.78 per cent. This error varies from a low of 0.02 per cent to a high of 725.77 per cent, according to the accuracy of the residential use extracted from the Landsat images. However, the estimated total population of the 373 census tracts (2,262,558) is very close to the actual total population of 2,257,351, or a mere 0.23 per cent overestimation. In comparison, the mean relative error of the population estimate for 1997 is 14.80 per cent and the mean absolute relative error is 40.03 per cent, showing a slight improvement over the 1990 results. However, its total estimated population of 2,511,137 for the 373 census tracts is quite different from the actual population of 2,731,839 in 1997, or an 8.07 per cent underestimation, which is worse than the 1990 estimation. As indicated before, the much smaller b value of the 1997 model compared with the 1990 model predicted a much slower

1990   **1997**

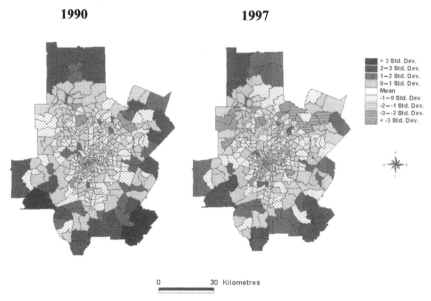

Figure 7.4 Spatial patterns of residual errors of population estimation for 1990 and 1997.

rate of population increase and rate of increase for residential area for 1997 than for 1990, hence explaining the underestimation. The error in population estimation in both cases could be caused by the classification error of the low-density urban use extracted from the Landsat TM images. The boundary between low-density urban use and forest is not always very clear, because the new suburb houses always have extensive forested yards and lawns.

In order to better understand the causes of population estimation errors, the residual errors by census tracts were mapped for both years (Figure 7.4). It was clearly shown that the peripheral area of the city (which also has larger census tracts) tended to exhibit overestimation while the central area (with smaller census tracts) exhibited underestimation. This is particularly distinct for the 1990 pattern. For 1997 some peripheral census tracts began to show underestimation, although the overall periphery–centre contrast remained intact. These spatial patterns of residual errors can be explained by population change that occurred between the two years (Figure 7.5). Clearly, the peripheral census tracts have shown large increases in population while the central census tracts have exhibited a decline.

7.5   Improvements in population estimation

The analysis given above suggested some possibilities to improve the accuracy of population estimation. An obvious approach is to divide the

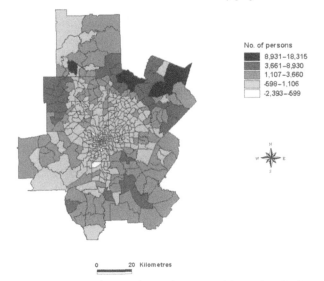

No. of persons
- 8,931–18,315
- 3,661–8,930
- 1,107–3,660
- -598–1,106
- -2,393–-599

Figure 7.5 Population change between 1990 and 1997 by census tracts.

census tracts into two groups: the periphery and the centre. For each group, an allometric growth model was developed. By examining the population density maps (Figure 7.6), it was decided that a density of 5.0 persons/ hectare (2.0 persons/acre) was a good cut-off point between the periphery (with a population density fewer than 5.0 persons/hectare) and the centre (with a population density equal to and larger than 5.0 persons/hectare). This results in: (i) 132 census tracts with a population density of fewer than 5.0 persons/hectare and 286 census tracts with a population density of 5.0 persons/hectare or larger for the 1990 data and (ii) 113 census tracts with a population density of fewer than 5.0 persons/hectare and 305 census tracts with a population density of 5.0 persons/hectare or larger for the 1997 data. Clearly, population density increased in some previously low-density census tracts in 1997.

For each set of data, a random sample of 10 per cent of the census tracts in each category was drawn, based on which the following allometric growth models were developed:

$$\log POP90 = 1.639\,575\,2 + 0.509\,925\,1 * \log RESID87 \quad (R^2 = 0.14)$$
$$\text{(for the periphery)} \tag{7.6}$$

$$\log POP90 = 2.189\,056\,4 + 0.441\,896\,7 * \log RESID87 \quad (R^2 = 0.67)$$
$$\text{(for the centre)} \tag{7.7}$$

$$\log POP97 = 2.718\,073\,1 + 0.270\,042\,5 * RESID97 \quad (R^2 = 0.20)$$
$$\text{(for the periphery)} \tag{7.8}$$

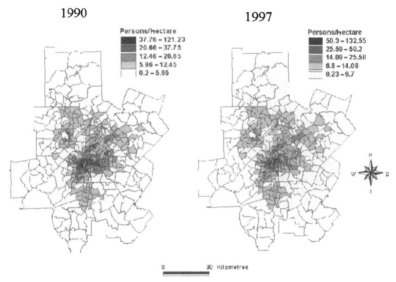

Figure 7.6 Population density patterns of Atlanta by census tracts, 1990 and 1997.

$$\log POP97 = 2.250\,122\,1 + 0.428\,848\,6 * RESID97 \quad (R^2 = 0.50)$$
$$\text{(for the centre)} \tag{7.9}$$

The models for the peripheral census tracts (Equations 7.6 and 7.8) revealed a lower R^2 than that for the central census tracts, which is also inferior to that for the overall 1990 model in Equation (7.4). The b value for Equation (7.6) is almost identical to that for the overall 1990 estimation model in Equation (7.4). The models for the central census tracts for both years were quite similar. However, the 1997 model for the periphery is very different from its 1990 counterpart. These four models (Equations 7.6 to 7.9) were applied to estimate the population of the withheld census tracts in each category using the low-density urban land use area extracted from the land use/cover map of the appropriate year as the independent variable (residential land area). The accuracy of population estimation for each group of census tracts is shown in Table 7.4.

The model used to estimate population at the peripheral census tracts (with a population density of fewer than 5.0 persons/hectare) in 1990 produced a somewhat better accuracy than that from the overall 1990 allometric growth model as evidenced by a smaller standard deviation and a lower maximum relative error. There is a clear underestimation of the population by the model, so much so that the total population for this set of census tracts is underestimated by almost 29 per cent. The improvement is much greater for the estimation of population in the central census tracts (with a population density larger than or equal to 5.0 persons/hectare). The mean absolute relative error is 29.16 per cent, and both the standard

Table 7.4 Accuracy of population estimation of Atlanta at the census tract level for the peripheral and central census tracts

Year	No. of tracts	Relative error (%)			Absolute relative error (%)		
		Mean	*Standard deviation*	*Max*	*Mean*	*Standard deviation*	*Max*
(1) Peripheral census tracts							
(with a population density of fewer than 5 persons/ha)							
1990	118	−9.69	59.77	356.68	42.54	42.92	356.68
1997	101	45.73	406.87	4,032.89	86.05	400.22	4,032.89
(2) Central census tracts							
(with a population density of larger than or equal to 5 persons/ha)							
1990	257	9.18	39.84	220.96	29.16	28.61	220.96
1997	274	10.71	54.35	427.80	34.83	43.03	427.80

deviation and the maximum relative error are small. This certainly represents an improvement over the overall model of population estimation for 1990. On the other hand, the total population for this set of census tracts is still slightly underestimated by about 7 per cent. In combining the population of the central and peripheral census tracts together, there is an underestimation of about 15 per cent. Overall, there has been improvement in the accuracy of the population estimates at the census tract level in 1990 by using two separate models, one for the peripheral and one for the central census tracts (Equations 7.6 and 7.7, respectively).

The results of population estimation for the peripheral census tracts for 1997 are disappointing. All the measures (mean relative error, mean absolute relative error, standard deviation, and maximum), which show extreme overestimations, were poorer than those obtained from the overall model for 1997 (Equation 7.5) and those of the 1990 counterpart (Table 7.4). The very small *b* value in the model (Equation 7.8), which implied a slower growth rate of population in relation to residential land area, does not match with reality. On the other hand, estimating the 1997 population of the central census tracts (with a population density larger than or equal to 5.0 persons/hectare) using Equation 7.9 gave a higher accuracy than the overall model (Equation 7.5). The mean absolute relative error is 34.83 per cent, and both the standard deviation and maximum relative error are small. However, compared with its 1990 counterpart, its accuracy is lower. There is still the same tendency of underestimation. The combined population of the peripheral and central census tracts for 1997 is underestimated by 7.27 per cent, very similar to that for 1990.

Why is there a difference in the performance of the allometric growth models between the peripheral and central census tracts for 1990 and 1997? Why is the accuracy of the population estimation for 1997 worse than that for 1990? One of the reasons is clearly the very rapid increase in population in the peripheral census tracts that violates the assumption of the allometric growth model (i.e., the relative growth rate of population is proportional to the relative growth rate of the residential land area). The rate of urban sprawl in Atlanta has increased rapidly since 1990. It has already been stated that, despite efforts made to improve the accuracy of land use/cover classification, residential area tended to be confused with forests in the peripheral census tracts, where suburban houses are usually built on densely wooded lots.

7.6   Estimation of housing units

Land use/land cover data can also be employed to estimate housing units because there is also a strong correlation between housing units and low-density urban use area for both 1990 and 1997 at the census tract level ($R = 0.40$ and $R = 0.53$, respectively, both at the over 99.9 per cent level of confidence). However, their relationship is not as strong as that between population and low-density urban use area. By taking the logarithmic transformation of both the housing units and the low-density urban use area, the correlation coefficients (R) have improved to 0.64 and 0.70 for 1990 and 1997, respectively. As a result, allometric growth models were also adopted for use to estimate housing units.

A random sample of forty-five census tracts was employed to calibrate the allometric growth models for estimating housing units at the census tract level of the ten urban counties of Atlanta in 1990 and 1997. The following models resulted:

$$\log \mathrm{HU90} = 1.733\,158\,9 + 0.449\,941\,3 * \mathrm{RESID87} \quad (R^2 = 0.64)$$

$$\tag{7.10}$$

$$\log \mathrm{HU97} = 2.053\,298\,5 + 0.365\,132\,7 * \mathrm{RESID97} \quad (R^2 = 0.63)$$

$$\tag{7.11}$$

Both models are statistically significant. They were applied to estimate the housing units of the 373 withheld census tracts for 1990 and 1997. The antilogarithms of $\log \mathrm{HU90}$ and $\log \mathrm{HU97}$ would give the housing units estimates. Table 7.3 gives all the measures of accuracy of the housing unit estimation, which shows no great difference between the two years. The mean absolute relative errors are 63.82 per cent and 63.94 per cent for 1990 and 1997, respectively, and the standard deviations and maximum errors are also large. In other words, the housing unit estimates are less accurate than the corresponding population estimates. However, the total number of housing units for 1990 is underestimated only by 2.36 per cent while that for 1997 is underestimated by 9.13 per cent. They are acceptable results.

1990   **1997**

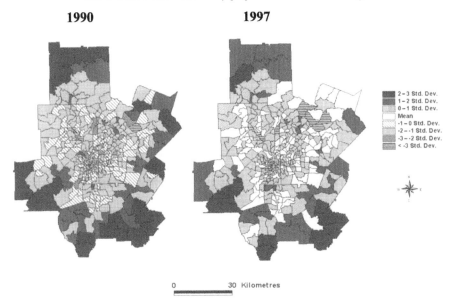

Figure 7.7 Spatial patterns of residual errors of housing units estimation for 1990 and 1997.

The poorer housing unit estimation accuracy for 1997 than for 1990 echoes that for population estimation. The residual error maps of housing unit estimation for 1990 and 1997 are very similar to those for population estimation (Figure 7.7), thus indicating the same periphery–centre differences.

Housing units can be used to estimate population if the household size is known. In an attempt to evaluate the usefulness of the estimated housing unit data for population estimation, household size data for each county in which the census tracts belong are used. These household size data for 1990 and 1997 for the ten counties are available from ARC (1997). Population estimates for 1990 and 1997 were computed as follows:

$$ESPOP90 = ESHU90 * Hhsize90 \tag{7.12}$$

$$ESPOP97 = ESHU97 * Hsize97 \tag{7.13}$$

where ESPOP90 and ESPOP97 are estimated population for 1990 and 1997, ESHU90 and ESHU97 are estimated housing units for 1990 and 1997 from Equations (7.10) and (7.11), and Hsize90 and Hsize97 are average household sizes of the county in which the census tracts fall for 1990 and 1997. These were applied to estimate the population of 373 census tracts withheld for accuracy evaluation. The accuracy measures are shown in Table 7.5. It is interesting to note that the results are comparable with those

Table 7.5 Accuracy of population estimation using estimated number of housing units and average household sizes from the counties of Atlanta at the census tract level

Year	No. of tracts	Relative error (%)			Absolute relative error (%)		
		Mean	Standard deviation	Max	Mean	Standard deviation	Max
1990	373	29.54	90.22	897.37	46.71	82.63	897.37
1997	373	22.95	126.89	2194.24	44.14	121.15	2,194.24

obtained using the population estimation models (Equations 7.4 and 7.5). The mean absolute relative errors for 1990 and 1997 are 45.22 per cent and 43.82 per cent, slightly higher than the corresponding values from the population estimation (Table 7.5). The accuracy of the estimated total population is also quite high, being +8.04 per cent and +1.79 per cent, respectively, for 1990 and 1997.

7.7   Estimation of population and housing units at the county level

In an attempt to understand the impact of areal units on the accuracy of population and housing unit estimation, the census tract data were aggregated into the ten urban counties of Atlanta. This aggregation produced highly significant correlation between population and housing units with low-density urban use area, the surrogate for residential land area, for both years. The correlation coefficients (R) between population and low-density urban use for 1990 and 1997 are, respectively, 0.95 and 0.94, all at the over 99.9 per cent level of confidence. After the logarithmic transformation of the two variables, the correlation coefficients (R) are still highly significant (being 0.93 and 0.92, all at the over 99.9 per cent level of confidence for 1990 and 1997, respectively). The same is true between housing units and low-density urban use. The correlation coefficients for 1990 and 1997 are 0.94 and 0.93, respectively, before (and 0.93 and 0.91, respectively, after) logarithmic transformations. For comparison purposes, the same allometric growth model was adopted for population and housing unit estimation at the county level.

Because only ten counties are involved, the whole data set was used to calibrate the models for population and housing unit estimation, the results of which are shown below. For population estimation:

$$\log POP90 = -2.835\,174\,6 + 1.512\,969\,2 * RESID87 \tag{7.14}$$

$$\log POP97 = -3.939\,527\,9 + 1.683\,535\,9 * RESID97 \tag{7.15}$$

Table 7.6 Accuracy of population estimation of Atlanta at the county level

County	Population 1990			Population 1997		
	Actual	Estimated	Relative error (%)	Actual	Estimated	Relative error (%)
Fulton	648,951	598,966	−7.70	760,100	690,396	−9.17
DeKalb	545,837	382,965	−29.84	594,400	386,839	−34.92
Cobb	447,745	464,725	3.79	535,000	517,270	−3.31
Clayton	182,052	96,007	−47.26	209,500	110,093	47.45
Gwinnett	352,910	408,209	15.67	478,900	540,433	12.85
Rockdale	54,091	49,345	−8.77	4,800	63,505	2.00
Henry	58,741	127,136	116.44	95,900	203,296	111.99
Douglas	71,120	58,139	−18.25	88,400	77,650	−12.16
Cherokee	90,196	112,334	24.54	122,300	144,017	17.76
Fayette	62,415	75,730	21.33	84,100	115,468	32.30
Total	2,514,058	2,373,556	−5.59	3,033,400	2,848,967	−6.08

Relative error (%)			Relative error (%)		
Mean	Standard deviation	Max	Mean	Standard deviation	Max
6.99	44.70	116.43	7.09	44.26	111.99

Absolute relative error (%)			Absolute relative error (%)		
Mean	Standard deviation	Max	Mean	Standard deviation	Max
29.36	33.08	116.43	28.89	32.97	111.99

For housing unit estimation:

$$\log HU90 = -3.824\,609\,5 + 1.622\,478\,5 * RESID87 \tag{7.16}$$

$$\log HU97 = -4.957\,664\,4 + 1.794\,135\,4 * RESID97 \tag{7.17}$$

The negative a coefficient and the large b value (exceeding 1.5) in these equations should be noted, suggesting larger space and positive allometry (i.e., population increases at a faster rate than the low-density urban use area), which differs from the negative allometry (i.e., population increases at a slower rate than the low-density urban use area) of the models at the census tract level (cf. Equations 7.4, 7.5, 7.10 and 7.11). In other words, the models better reflect the real situation in Atlanta.

These models were applied to estimate the population and housing units for 1990 and 1997 of the ten counties. The results are shown in Tables 7.6 and 7.7. Overall, the accuracy of population and housing unit estimation at the county level is higher than that at the census tract level. The mean absolute relative errors for population estimation are 29.36 per cent for 1990 and 28.89 per cent for 1997. Both of their relative errors are low,

Table 7.7 Accuracy of housing unit estimation of Atlanta at the county level

County	Housing unit 1990			Housing unit 1997		
	Actual	Estimated	Relative error (%)	Actual	Estimated	Relative error (%)
Fulton	297,503	25,783	−13.33	334,984	290,647	−13.24
DeKalb	231,416	159,605	−31.03	251,399	156,773	−37.64
Cobb	189,872	196,411	3.44	233,887	213,672	−4.56
Clayton	71,926	36,199	49.20	80,769	41,082	−49.14
Gwinnett	137,607	170,914	24.67	185,561	223,884	20.65
Rockdale	19,963	17,730	−11.18	23,780	22,856	−3.89
Henry	21,275	48,921	129.94	35,383	78,980	123.21
Douglas	26,440	21,140	−20.05	32,720	28,318	−13.45
Cherokee	33,836	42,840	26.61	46,127	54,697	18.58
Fayette	22,428	28,068	25.15	30,517	43,222	43.32
Total	1,052,266	979,666	−6.90	1,244,767	1,154,132	−7.28

	Relative error (%)			Relative error (%)		
	Mean	Standard deviation	Max	Mean	Standard deviation	Max
	8.41	49.72	129.94	8.39	48.71	123.21

	Absolute relative error (%)			Absolute relative error (%)		
	Mean	Standard deviation	Max	Mean	Standard deviation	Max
	33.46	36.15	129.94	32.77	35.46	123.21

being 6.99 per cent for 1990 and 7.09 per cent for 1997. The standard deviation and maximum relative errors are much smaller at the county level than at the census tract level. The total population of the ten counties combined is underestimated at 5.59 per cent for 1990 and 6.08 per cent for 1997, all acceptable results. As in census tract level estimation, housing unit estimation at the county level performed slightly worse than population estimation, but the accuracy is much higher than that achieved at the census tract level. The mean relative errors are 7.41 per cent for 1990 and 8.39 per cent for 1997, and the mean absolute relative errors are 33.46 per cent for 1990 and 32.77 per cent for 1997. Standard deviations and the maximum relative errors are low. In examining Tables 7.6 and 7.7 it is clear that the greatest error (overestimation) in population and housing unit estimation occurred in Henry county, which had a rather low population density of 0.29 persons/acre in 1990 and 0.46 persons/acre in 1997.

Clearly, because of the use of larger areal units, variations in population densities found at the smaller census tracts have been concealed as a result of spatial averaging. Population and housing unit estimation using the allometric growth model with the zone-based approach is more accurate at the county level than at the census tract level. However, census tracts are probably the better areal units to use than counties for in-depth population mapping and analysis tasks conducted by geographers, demographers and planners.

7.8 Conclusions

This research has shown that reasonably accurate estimates of population and housing units of a city at both the census tract and county levels can be obtained from land use/land cover maps produced from automated image classification of satellite sensor images in a zone-based approach. Residential land use area, as extracted from the category of low-density urban map, is a good surrogate of population size and the number of housing units. Using the land use/cover maps of Atlanta extracted from Landsat TM images for 1987 and 1997, the low-density urban use area extracted was used as the independent variable to estimate the population of 418 census tracts for 1990 and 1997. In order to minimize the problem caused by heteroscedasticity, logarithmic transformations of the variables were used. The allometric growth model was found to be most suitable for use in both population and housing unit estimation. An examination of the residual errors of population estimation indicated that the city of Atlanta has been split into a peripheral and a central zone as a result of its rapid suburbanization. The peripheral zone experienced rapid population growth while the central zone showed population decline. These gave rise to overestimation of population for census tracts in the peripheral zone but underestimation of population for census tracts in the central zone when the allometric growth model was applied. Dividing census tracts into the peripheral and central groups, and then developing an allometric growth model for each achieved a slight improvement in the accuracy of population estimation, especially for 1990. Housing unit estimation at the census tract level also made use of the allometric growth model, but the accuracy is slightly poorer than that for population estimation. However, by using the average household size of the county in which a particular census tract falls, reasonably accurate population estimates can be obtained from the estimated number of housing units. In all cases, the total number of population and housing units for Atlanta as a whole were quite accurately estimated.

As expected, the MAUP affected the accuracy of population and housing unit estimation. At the county level (only ten counties for Atlanta in this study), the relationship among population, housing units and low-density urban use is very much stronger than that at the census tract level. More accurate population and housing unit estimation results are obtained at the county level, but the larger areal units dilute the population density differences within a county, which only the smaller census tracts can capture.

To conclude, the zone-based population and housing unit estimation is operationally a better approach than the pixel-based approach because of its simplicity and seamless integration with land use/land cover classification. The population and housing unit estimation is a by-product of the land use/land cover map. In other words, it is cost-effective. With the future availability of high-spatial-resolution satellite sensor images (such as the 1-metre IKONOS data from Space Imaging), improved image classification software, such as the zone-based contextual classification (Barnsley *et al.* 2001), subpixel processing (Huguenin *et al.* 1997), knowledge-based image segmentation (Bähr 2001), modified maximum likelihood classification (Mesev *et al.* 2001; Mesev, Chapter 9 in this book), all enabled by the power of geographical information systems, will substantially improve the accuracy of land use/land cover mapping in the urban area. This should in turn improve the accuracy of population and housing unit estimation particularly at the census tract level, using the allometric growth model.

7.9   Acknowledgements

The author gratefully acknowledges funding from the NASA EOS IDS program, which made possible the land use/cover mapping of the Atlanta region. Research assistance provided by Mr Xiaojun Yang in land use/cover mapping is also acknowledged.

7.10   References

Anderson, J. R., Hardy, E. E., Roach, J. T. and Witmer, R. E., 1976, *A Land Use and Land Cover Classification System for Use with Remote Sensor Data*, Geological Survey Professional Paper 964 (Washington, DC: United States Government Printing Office).

ARC, 1997, *1997 Population and Housing* (Atlanta, GA: Atlanta Regional Commission).

Bähr, H-P., 2001, Image segmentation for change detection in urban environments. In J-P. Donnay, M. J. Barnsley and P. A. Longley (eds) *Remote Sensing and Urban Analysis* (London: Taylor & Francis), pp. 95–113.

Barnsley, M. J., Møller-Jensen, L. and Barr, S. L., 2001, Inferring urban land use by spatial and structural pattern recognition. In J-P. Donnay, M. J. Barnsley and P. A. Longley (eds) *Remote Sensing and Urban Analysis* (London: Taylor & Francis), pp. 115–44.

Brugioni, D. A., 1983, The census: It can be done more accurately with space-age technology. *Photogrammetric Engineering and Remote Sensing*, **49**, 1337–9.

De Klerk, T. C., 1998, *The efficacy of satellite remote sensing technology for rural population estimation.* Unpublished MA thesis in Geography and Environmental Studies, Potchefstroomse Universiteit vir Christelike Hoer Onderwys, Potchefstroom, South Africa.

Fuji, T. and Hartshorne, T. A., 1995, The changing metropolitan structure of Atlanta, Georgia: Locations of functions and regional structure in a multinucleated urban area. *Urban Geography*, 16, 680–707.

Hall, F. G., Strebel, D. E., Nickeson, J. E. and Goetz, S. J., 1991, Radiometric rectification: Toward a common radiometric response among multidate, multi-sensor images. *Remote Sensing of Environment*, 35, 11–27.

Harvey, J., 1999, Modelling population associated with individual TM pixels. Paper given at Conference of ASPRS 1999 Annual Conference (Bethesda, MD: American Society for Photogrammetry and Remote Sensing), pp. 827–39.

Hsu, S. Y., 1973, Population estimation from ERTS imagery: Methodology and evaluation. Paper given at Conference of the American Society of Photogrammetry 39th Annual Meeting (Bethesda, MD: ASPRS), pp. 583–91.

Huguenin, R. L., Karaska, M. A., Van Blaricom, D. and Jensen, J. R., 1997, Subpixel classification of bald Cypress and Tupelo gum trees in Thematic Mapper imagery. *Photogrammetric Engineering and Remote Sensing*, 63, 717–25.

Iisaka, J. and Hegedus, E., 1982, Population estimation from Landsat imagery. *Remote Sensing of Environment*, 12, 259–72.

Johnston, R. J., 1980, *Multivariate Statistical Analysis in Geography* (London: Longman).

Kraus, S. P., Senger, L. W. and Ryerson, J. M., 1974, Estimating population from photographically determined residential land use types. *Remote Sensing of Environment*, 3, 35–42.

Lee, Y., 1989, An allometric analysis of the US urban system: 1960–80. *Environment and Planning A*, 21, 463–76.

Lo, C. P., 1995, Automated population and dwelling unit estimation from high-resolution satellite images: A GIS approach. *International Journal of Remote Sensing*, 16, 17–34.

Lo, C. P. and Welch, R., 1977, Chinese urban population estimates. *Annals of the Association of American Geographers*, 67, 246–53.

Lo, C. P. and Yang, X., 2000, Mapping the dynamics of land use and land cover change in the Atlanta metropolitan area using time sequential Landsat images. *ASPRS 2000 Proceedings* (Bethesda, MD: American Society for Photogrammetry and Remote Sensing) (CD-ROM).

Martin, D., 1996, An assessment of surface and zonal models of population. *International Journal of Geographical Information Systems*, 10, 973–89.

Mesev, V., Gorte, B. and Longley, P. A., 2001, Modified maximum-likelihood classification algorithms and their application to urban remote sensing. In J-P. Donnay, M. J. Barnsley and P. A. Longley (eds) *Remote Sensing and Urban Analysis* (London: Taylor & Francis), pp. 71–94.

Quattrochi, D. A. and Luvall, J. C., 1999, Assessing the impacts of Atlanta's growth on meteorology and air quality using remote sensing and GIS. *GeoInfo Systems*, 9, 26–33.

Research Atlanta, Inc., 1993, *The Dynamics of Change: An Analysis of Growth in Metropolitan Atlanta over the Past Two Decades* (Atlanta, GA: Policy Research Center, Georgia State University).

Siegel, S. and Castellan, Jr, N. J., 1988, *Nonparametric Statistics for the Behavioral Sciences* (New York: McGraw-Hill).

Sutton, P., Roberts, D., Elvidge, C. and Meij, H., 1997, A comparison of night-time satellite imagery and population density for the continental United States. *Photogrammetric Engineering and Remote Sensing*, 63, 1303–13.

Tobler, W., 1969, Satellite confirmation of settlement size coefficients. *Area*, 1, 30–4.

Watkins, J. F. and Morrow-Jones, H. A., 1985, Small area population estimates using aerial photography. *Photogrammetric Engineering and Remote Sensing*, 51, 1933–5.

Weber, C., 1994, Per-zone classification of urban land cover for urban population estimation. In G. M. Foody and P. J. Curran (eds) *Environmental Remote Sensing from Regional to Global Scales* (Chichester, UK: John Wiley & Sons), pp. 142–8.

Yang, X. and Lo, C. P., 2000, Relative radiometric normalization performance for change detection from multi-date satellite images. *Photogrammetric Engineering and Remote Sensing*, 66, 967–80.

8 Population estimation at the pixel level

Developing the expectation maximization technique

Jack T. Harvey

8.1 Introduction

As Lo pointed out in Chapter 7, there was considerable optimism a generation ago over the potential of using remote sensing for population estimation. Although Brugioni (1983) went so far as to predict that the US Census would be rendered obsolete, Lo (1986), in a more considered and pragmatic assessment, distinguished four approaches to population estimation from remotely sensed imagery: counts of dwelling units, measurement of areas of urbanization, measurement of areas of different land use and automated digital image analysis.

The first three methods have long been applied to the visual interpretation of analogue images from aerial photography at various scales. The fourth, particularly applied to orbitally acquired imagery, was seen as an emergent and radically different methodology with considerable potential. However, in the 15 years since, population-related research has not generally involved new approaches that are specific to digital image analysis (Harvey 2002b). Rather, it has generally consisted of digital implementations of Lo's second and third categories, often at medium to large scales and often in an ancillary role in combination with other data from ground-based sources.

Only a few studies have realized Lo's fourth category. Digital image analysis techniques have been applied to estimate directly population or dwelling counts at the small-area scale (in the order of $1\,km^2$ or less) from satellite sensor imagery alone. For at least three reasons this is a daunting task. First, satellite sensor data relate to land surface and land cover characteristics, and, while it is true that in many applications the variables of interest are characteristics of land use, which are only indirectly linked to the measured characteristics, in the case of population the link is particularly tenuous and conjectural. For example, reflectances alone capture little if any information about the height and type of urban structures, which presents calibration problems at high densities. Second, estimating a quantitative variable like population across the spatial dimensions of an image is inherently more ambitious than the more usual qualitative objectives such as segmentation or classification. If the methodology is also required to be broadly applicable and robust to differences in season, geographical location, culture, etc., then the undertaking is doubly difficult. Third, and most importantly, while the resolution of remotely sensed imagery is quite adequate for most demographic purposes (nominally 30-m by 30-m pixels for Landsat Thematic Mapper [TM], for instance), ground reference population data for model development and training are generally only available at a much lower resolution – either for census-related zones such as UK Census enumeration districts (Langford *et al.* 1991), Hong Kong tertiary planning units (Lo 1995), Australian Census collection districts (Harvey 2002b) or for regions defined by some reference grid which is much coarser than the individual image pixels (Iisaka and Hegedus 1982; Webster 1996).

To date, the response to this problem of spatial incompatibility of data has been to aggregate the finer resolution spectral data to the zonal level of the available ground reference populations, and then build regression models for estimating the population density of these larger zones. Aggregated remote sensing predictor variables have included mean reflectances of individual spectral bands (Iisaka and Hegedus 1982; Lo 1995; Harvey 2002b), numbers of pixels in various land use categories (see Chapter 6 in this book by Langford, and Langford *et al.* 1991; Lo 1995), measures of variability and image texture (Webster 1996; Harvey 2002b) and various band-to-band ratios and other mathematical transformations of the multi-spectral data (Harvey 2002b). These studies have demonstrated that there is a substantial level of correlation between population density of small zones and various aggregated remote sensing indicators, with coefficients of determination (R^2) in the range 80–90 per cent (corresponding to correlations of 0.90 to 0.95) typically being reported for the training set data. However, if such models are to be seriously considered as predictive tools for population estimation, it is necessary to move beyond exploratory analysis of relationships in a particular image to independent predictive validation on data sets other than those used for model development. Validation on an enlarged data set was first reported by Lo (1995), though Langford *et al.* (1991) carried out an interpolatory form of cross-validation by deriving estimates for a second set of zones overlaid on the same geographical area as the training set. Harvey (2002b) undertook a more thorough validation study, by using one image to identify and train models and then applying these models to a second image of a nearby culturally and demographically similar area obtained on the same date. He concluded that the potential of zone-based models is limited by heterogeneity of both land cover and population density within each zone, and that further improvements are in principle unlikely using this approach (see also Langford, Chapter 6, and Lo, Chapter 7, both in this book for discussions on zone-based models based on dasymetric models and linear relationships respectively).

8.2   Advantages of a pixel-based modelling approach

There are a number of potential advantages to be gained from modelling at the level of single pixels rather than larger spatial zones. Prima facie, the relationship between human habitation and spectral reflectances should be improved and expressed at the pixel level. Of course, this assumes pixels are commensurate in size with individual residences, and also assumes that both spectral characteristics and population densities can vary greatly within a more extended area. Nevertheless, the pixel-based approach has many advantages in all phases of model building, validation and application. Table 8.1 summarizes these advantages.

With regard to modelling, simple linear models fitted at the level of individual pixels have obtained population estimates at least as accurate as those generated from zone-based models. To obtain comparable accuracy with zone-based models requires more complex functional forms and more exploratory model development (Harvey 2002b). Such finely tuned models are potentially less robust. However, some common problems are reported in almost all the research cited. These are underestimation of high population densities, overestimation of low population densities and occasional negative population estimates. Refinements to deal with such problems are much more easily implemented in pixel-based models.

With regard to outcomes, pixel-based population estimates provide more flexibility. They are more amenable to integration with other spatial data in Geographical Information Systems (GISs), and they enable population density maps to be produced at a sufficient resolution for most demographic and planning purposes, rather than less informative choropleth maps. Why then the reliance on zone-based models? Essentially because of the lack of reference population data for model building at the spatial resolution of individual pixels. Disaggregation of the population of a zone into populations associated with each constituent pixel is less straightforward conceptually and computationally than the inverse process of spatially aggregating the spectral data. A statistical approach to this problem is at the nub of the methodology of this chapter.

8.3   Methodology

8.3.1   Form and conceptual basis of models

Population density is both a time-dependent and a scale-dependent concept. With regard to time dependence, the instantaneous spatial population density distributions of a city at midday and at midnight are very different. For most demographic and planning purposes, population is assigned by place of residence. In spatial terms, the concept of residential population density changes in character as one reaches the quantum scale of the

Table 8.1 Comparison of zone-based and pixel-based estimation models. (Data from Harvey 1999b, used with kind permission of ASPRS.)

Aspect	Zone-based models	Pixel-based models
Model building and training		
Information about the relationship between population and spectral response	Loss of detailed information about spectral response of individual pixels	Pixel-level population not known, but can be estimated
Mathematical form of model	Complex – variable selection required – risk of data dependence and lack of robustness	Simple, no selection, robust
Availability of texture measures	Wider range of measures available for larger areas (e.g., pattern-based)	Local neighbourhood measures only
Available sample size and degrees of freedom	Small relative to pixel-based models	Large
Area (and population) required for training	Large relative to pixel-based models	Small
Classification and stratification	Difficult, though possible via areal interpolation models for some forms of model (e.g., Langford et al. 1991)	Straightforward
Morphological approaches to classification	Difficult – zonal incompatibility	Straightforward
Suppression of anomalous spurious features	Difficult – zonal incompatibility	Straightforward
Addition of anomalous concentrations of population	Feasible	Straightforward
Incorporation of ancillary information (e.g., differential weighting by building heights or occupancy ratios)	Difficult if available on an incompatible zonal basis	Straightforward
External validity: estimation beyond the training set		
Estimates for incompatible zones of comparable size with the training zones, defined for the same region	Feasible but non-trivial: predictors must be calculated for the new zones	Straightforward aggregation
Estimates for zones of arbitrary size and shape	Feasible, but models may not be robust to changes in scale	Straightforward aggregation
Estimates for other regions	Normalization probably not feasible	May be feasible through normalization of models
General		
Resolution, GIS compatibility and mapping	Resolution very coarse. Not directly GIS-compatible. Without further analysis, limited to choropleth mapping of irregular areas	Pixel resolution is finer than the scale of administrative units and adequate for most demographic applications. GIS-compatible and mappable. However, at finer resolutions than the scale of the areas for which reference population training data is available, the estimates can only be qualitatively validated – the accuracy is not quantifiable

individual residence. On the scale of census zones, an individual residence is unambiguously and discretely located within a particular zone. The same cannot be said at the smaller scale and on the regular grid of the pixels of a satellite image, whose boundaries intersect property lines, structures and even rooms. We can conceptualize either an instantaneous, time-dependent, discrete pixel population, or a notional residential pixel population that is time-invariant for most pixels on a scale of days or weeks, but which is not discrete. Since a place of residence is not a single point but an extended area that might contribute fractionally to a number of pixels, in the latter case the residential population associated with a single pixel can be conversely conceptualized as a continuous variable, whose value is the sum of the fractional contributions of the persons whose residences lie partly within the area of the pixel.

Ordinary Least-Squares (OLS) models with Gaussian or normal errors, used by the author and in all of the studies cited above, are congruent with this conceptualization, though Langford *et al.* (1991) obtained very similar results using Poisson regression models, which are more congruent with the discrete count conceptualization. The simplest OLS regression models for population estimation are additive linear models of the form:

$$p_i = \beta_0 + \sum_{j=1}^{v} \beta_j r_{ij} + \varepsilon_i \qquad (8.1)$$

where p_i = population or population density of entity or case i (pixel, zone, grid square, etc.), r_{ij} = remote sensing indicator j for case i, n = number of explanatory variables or indicators and ε_i = random error in population or population density of entity or case i. The random errors represent the variation in population unexplained by the remote sensing indicators, and are assumed to be independent and identically normally distributed with constant variance. In the simplest models, r_{ij} is the reflectance of entity i in band j, but the basic form can be extended to include derived variables such as band-to-band ratios, spatial texture measures and logarithmic and other mathematical transformations of the dependent population variable. The incorporation of neighbourhood transformations of the independent variables allows for the possibility that population is related to spatial or textural characteristics of the remotely sensed imagery. In principle, explicit provision could also be made for spatial autocorrelation in the random error term (i.e., in any other unobserved correlate of population), but this would come at considerable computational complexity and cost.

8.3.2 Multilevel regression analysis with incompletely determined data

In the absence of ground reference data for the dependent variable – the population of each pixel – an initial estimate is made for each pixel and then iteratively estimated.

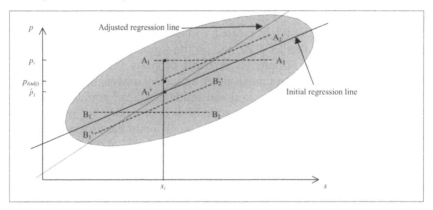

Figure 8.1 Regression with incompletely determined data: schematic representation. (Data from Harvey 1999b, used with kind permission of ASPRS.)

The first step is to perform a comprehensive land cover/land use classification of the image, using some standard technique such as the maximum likelihood classification. Pixels in non-residential classes are assigned zero population. Modelling proceeds with reference to pixels in the residential class(es).

Figure 8.1 portrays the relationship between pixel population p and a multivariate vector of remote sensing predictors s, represented schematically by a single dimension s. The ellipse schematically represents the outline of a set of data points, one for each residential pixel. Suppose a particular zone has a ground reference population P, and includes n residential pixels. Initially the simplest assumption of constant population density is assumed, and then each residential pixel is assigned an equal share of the zone population (i.e., all residential pixels in the zone are assigned equal populations) such that:

$$p_i = \frac{P}{n} \qquad i = 1, \ldots, n \tag{8.2}$$

The residential pixels of one particular zone are represented in Figure 8.1 by the line $A_1 A_2$. Since these pixels have the same imputed population but different spectral characteristics (i.e., different s values), they fall along a line parallel to the s axis. Similarly the line $B_1 B_2$ represents the initial imputed data for the residential pixels of another zone with a lower average population density.

The regression coefficients are calculated and the pixel populations estimated. For points near A_1, the regression estimate $\hat{p}_i$ will be less than the assigned population p_i (as illustrated in Figure 8.1), and the converse will be true for points near A_2. Assuming that there is in fact a linear relationship between p and s, the regression estimates provide additional information about the population of individual pixels. This now refines the initial imputed pixel populations by redistributing population within each zone away from underestimated pixels and toward overestimated pixels, in such a way that the known zone totals are maintained. Intuition suggests, and it can be shown using the method of Lagrange multipliers (Harvey 1999a), that the optimal such redistribution in a least-squares sense – which minimizes the sum of squared residuals about the regression line while holding the sum of the p values constant – is to make all the residuals equal by adjusting within each zone as follows:

$$p_{i(\text{adj})} = \hat{p}_i + \bar{r} \tag{8.3}$$

where $\hat{p}_i$ is the regression estimate, and

$$\bar{r} = \sum_{i=1}^{n} (p_i - \hat{p}_i) \bigg/ n \tag{8.4}$$

This has the effect of mapping $A_1 A_2$ onto $A_1' A_2'$, parallel to the regression line. Similarly, the data for the pixels from another zone might be reassigned from the line $B_1 B_2$ onto $B_1' B_2'$, and so on. Figure 8.1 schematically illustrates the first such adjustment for the pixels of two zones. In practice there are many zones. These adjusted reference populations lie closer to a line than the initially assigned populations (i.e., the ellipse is made narrower and more elongated). Consequently, when the regression line is re-estimated (the dotted regression line in Figure 8.1) the new line fits the adjusted points more closely, and so R^2 is increased. When the process is iterated R^2 continues to increase monotonically but at a decreasing rate toward some limiting value $R_L^2 < 1$. A value of $R_L^2 = 1$ would imply that it was possible to distribute the fixed zone populations among their component pixels in such a way that some linear combination of the s variables would reproduce them exactly. While this could happen if only one zone was involved, with more than one any non-linearity precludes the relationship between the zone means and the s variables.

Table 8.2 shows regression coefficients and coefficients of determination (R^2) for ten iterations of this procedure on a model utilizing the six reflective TM bands as predictors. The limiting value of R^2 gives an indication of the upper limit to the accuracy of prediction that could be attained were the individual pixel populations known. While intuition suggests that the iteratively re-estimated pixel populations should be more realistic and hence the regression coefficients more accurate than the initial estimates, there is no way of knowing *ipso facto* whether this is the case. The apparent increase in accuracy as indicated by the increase in R^2 means nothing in itself and may be quite spurious. However, the procedure can be both justified theoretically and validated empirically. On a theoretical level, this iterated regression procedure has been shown (Harvey 1999a) to be a least-squares approximation to an EM (Expectation Maximization) algorithm

Table 8.2 Iterative refinement of a regression model for pixel population: based on six TM bands. (Reproduced by permission of ASPRS.)

TM band coefficient	Iteration										
	0	1	2	3	4	5	6	7	8	9	10
Constant	1.033	1.161	1.357	1.556	1.736	1.889	2.019	2.126	2.215	2.289	2.350
b_1	0.087	0.119	0.128	0.129	0.127	0.124	0.120	0.117	0.115	0.112	0.110
b_2	0.114	0.164	0.187	0.199	0.206	0.211	0.214	0.217	0.219	0.221	0.222
b_3	−0.136	−0.192	−0.215	−0.225	−0.229	−0.230	−0.231	−0.231	−0.232	−0.232	−0.232
b_4	−0.003	−0.005	−0.008	−0.009	−0.011	−0.013	−0.014	−0.015	−0.016	−0.017	−0.017
b_5	−0.039	−0.058	−0.068	−0.074	−0.079	−0.082	−0.084	−0.086	−0.088	−0.089	−0.090
b_7	0.064	0.096	0.113	0.124	0.131	0.137	0.142	0.146	0.149	0.152	0.154
R^2	0.444	0.751	0.819	0.838	0.845	0.849	0.851	0.852	0.854	0.854	0.855
% increase		69.1	9.1	2.3	0.8	0.5	0.2	0.1	0.2	0.0	0.1

(Dempster *et al.* 1977). This is an established generic statistical procedure for estimating parameters from incompletely specified data. While this procedure was devised independently by the author, the EM approach was also suggested but not implemented by Flowerdew and Green (1989) in the conceptually related context of areal interpolation for combining data from two incompatible sets of spatial zones.

Both empirical results and simulations have shown that there are convergence problems associated with multi-collinearity between the TM bands (Harvey 1999a). There is also the logical problem of negative population estimates, which can occur because of the unconstrained form of the linear model. This cannot occur in Poisson models, and it can be overcome less elegantly in Gaussian models by *ad hoc* adjustments after each iteration. In practice, the number of residential pixels in an image can be very large. The iterative re-estimation procedure can also be applied to a random sample of residential pixels. The only modification being that the sampling fraction in each zone is determined, and the same fraction of the zone population is then distributed among the pixels in the sample. The implicit assumption is that this proportion of the zone population is in fact contained in the sampled pixels. This is a much weaker assumption than the initial supposition of uniform population in all pixels. The validity and utility of iterated regression applied to a random sample of pixels has been established empirically by Harvey (1999a, 1999b, 2002a, 2001) in two ways. First, by examining the accuracy of the estimates of total zone population obtained for the set of zones from which the training sample was drawn, and, second, by applying the estimation equation to the pixels in a second set of zones beyond the training set, whose populations were also known.

8.4 Some applications of pixel-based regression models

This methodology can be applied in three ways: (i) to estimate unknown populations directly, (ii) to enhance spatial flexibility of population estimates by disaggregation of the published census counts for fixed zones and (iii) to enhance temporal flexibility of population estimates by estimation of inter-censal and post-censal population change. Each of these is now discussed and illustrated.

8.4.1 *Direct estimation of unknown populations*

The modelling of population at the pixel level has great potential for estimating population and mapping population distributions in countries where the demographic infrastructure is not well developed. The training requirements for direct population estimation across a region include a land cover/land use classification for the whole region and population counts for a moderate number of small geographical zones within the region. It is to be expected, and it has been demonstrated by Harvey (1999a), that particular regression relationships, even when normalized, do not apply robustly across all geographic locations, seasons or cultural variations, even within a single nation. A population estimation exercise will always require an initial investment in ground referencing and calibration for both the classification and regression phases. Nevertheless, this is a small investment in comparison with the cost of a full census.

Tables 8.3 and 8.4 and Figures 8.2 and 8.3 show some results from an Australian study designed to test the feasibility of such a methodology (various aspects of which are described in more detail in Harvey 1999a, 1999b, 2000, 2002a, 2002b). In the initial phase, census Collection District (CD) populations were estimated for a primary image centred on the provincial city of Ballarat, Victoria, using a number of models based on both zones (the CDs themselves) and on the individual pixels of a TM image.

The first two models in Table 8.3 were the most basic and the best fitting of the zone-based models, the latter involving transformations of both dependent and explanatory variables. The last three models were pixel-based. In each case, the procedure consisted of an initial, supervised, maximum likelihood classification phase, followed by regression analysis on a training set of pixels sampled from all pixels classified as residential. The resulting population estimation equation was then applied to all residential pixels in the image. Negative estimates were reset to zero, and positive estimates were aggregated to produce CD estimates, which were then compared with the ground reference figures to provide the results summarized in Table 8.3. The statistical criteria in Table 8.3 were chosen to enable the evaluation of both overall accuracy and the constituent components of variability and bias, while being resistant to the inflationary effects of a small number of extreme outliers (see Harvey 2002a, 2002b). In Model 3, just the six reflective TM bands were used for both classification and regression. In Model 4, twenty-four classification variables were selected by stepwise linear discriminant analysis from a set of sixty spectral and spatial transformations, but only the six TM bands were used at the regression stage. In Model 5, stepwise regression was performed on eighty variables (the same sixty plus twenty multiplicative interaction terms), of which thirteen were chosen. In each case, the regression fitting was iterated six times.

Table 8.3 shows that the pixel-based Model 3, utilizing just the six TM bands for both classification and regression, performed as well as any of the more complex models, whether zone-based or pixel-based. The regression equation for pixel population was:

$$\text{Pixel pop} = 2.019 + 0.120b_1 + 0.214b_2 - 0.231b_3$$
$$- 0.014b_4 - 0.084b_5 + 0.142b_7 \tag{8.5}$$

Table 8.3 Comparison of estimates of census collection district population densities and populations: derived from CD-based and pixel-based models. (Data from Harvey 1999b, used with kind permission of ASPRS.)

Model	Basis of model	Form of model[1]	R^2	Region n = 138			Urban area n = 122		
				Mean (% error)	Median (% error)	Total[3] (% error)	Mean (% error)	Median (% error)	Total[3] (% error)
1	CD	Basic TM bands	0.54	553.6	31.0	−142.3	83.2	28.5	+25.6
2	CD	Complex[2]	0.84	39.4	17.4	+13.5	28.2	13.6	+0.6
3	Pixel	Basic TM bands (6/6)	0.82	36.6	17.0	+7.8	24.9	14.3	−4.1
4	Pixel	Complex (25/6)	0.81	38.8	17.8	+9.9	26.3	16.2	−3.4
5	Pixel	Complex (25/13)	0.79	26.0	16.2	−10.7	21.7	15.4	−15.6

[1] Values in parens indicate number of classifiers/number of predictors.
[2] Square root transformation of dependent variable; six explanatory variables: means and standard deviations of various spectral transformations.
[3] Ground reference populations: Region 79,179; Urban Area 70,222.

Table 8.4 Comparison of census CD population density estimates for training samples and full images. (Data from Harvey 1999b, used with kind permission of ASPRS.)

Study area	Scope	Population ('000s)	n (CDs)	Sample fraction (% of CDs)	R^2	Region			Urban area		
						Mean (% error)	Median (% error)	Total (% error)	Mean (% error)	Median (% error)	Total (% error)
Ballarat 1988	Sample	79	14	10	0.92	15.5	7.1	−7.8	12.6	6.2	−1.2
	Region		137		0.81	29.3	17.0	+2.2	19.6	15.8	+0.8
Ballarat 1994	Sample	36	14	20	0.86	26.8	12.7	+22.3	11.0	9.5	+5.6
	Region		72		0.86	22.1	12.2	+7.5	16.8	9.7	+4.2
Geelong	Sample	148	11	5	0.90	60.7	8.2	+50.8	11.4	7.4	+4.3
	Region		224		0.75	75.4	18.4	+28.6	20.6	15.2	−0.8
Adelaide	Sample	1,159	118	5	0.83	31.9	16.2	+10.2	20.4	14.9	−0.2
	Region		2,412		0.74	51.4	15.5	+16.3	18.5	12.9	−0.2
Sydney	Sample	3,284	112	2	0.09	43.7	22.3	+12.3	29.2	21.8	+0.7
	Region		5,628		0.03	161.0	26.5	+7.1	32.4	25.1	−2.2

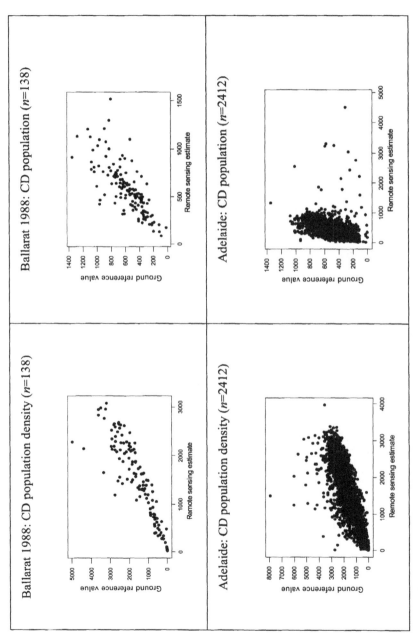

Figure 8.2 Population densities and populations of census collection districts: ground reference estimates *vs.* remote sensing estimates for two study areas. (Data from Harvey 1999b, used with kind permission of ASPRS.)

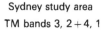

Sydney study area
TM bands 3, 2 + 4, 1

Sydney study area
Estimated population density
(light-grey = high, dark = low)

Figure 8.3 Sydney study area: Landsat TM and population density images.

The specific details of this equation were subsequently found to be quite data-dependent, but, even so, consistent levels of estimation accuracy were obtained in validation sets. There were a number of consistent character-istics in all variants of the equation, which can be speculatively related to the reflectance properties of various manufactured and natural materials.

In the second phase of the study, this simple model was fitted to several further images, including some representing Australia's largest cities. In each case, an image classification was performed, CD boundaries were overlaid, the regression model was fitted to a training set consisting of a random sample of residential pixels from a random sample of CDs and the fitted equation was applied to all pixels in the image. To address a consistent tendency toward the overestimation of population at low densities – incidentally, also reported with zone-based models by Langford *et al.* (1991); Webster (1996); Harvey (2002a) – a contextual reclassification step was incorporated at this point. Using a mean-based averaging filter, pixels in areas of low average population density with population below a chosen threshold value were reassigned a zero population, effectively reclassifying them as non-residential (see Harvey 2002a for details). The resulting estimates were validated by aggregating them to produce an estimate of the population of each CD in the image. Some typical results are shown in Table 8.4 and Figure 8.2.

The first three images were much less extensive than the last two; hence the sampling fractions were relatively large though the sample sizes were relatively small. In the case of Sydney and Adelaide, macro- and micro-level estimates were produced for urban populations of around one and three millions, respectively, on the basis of ground reference population training sets of around 5 per cent and 2 per cent of the total, respectively, without substantial loss of accuracy at the small-area level beyond the training sets. With the exception of the Sydney results, which are dominated by a relatively small number of "high-rise" CDs with extremely high densities, R^2 values based on the results for the training set CDs are comparable with reported training set results in the previous zone-based studies cited. Total populations were overestimated for both training samples and images in all but one case, due to misclassification as residential of many natural and constructed features in rural areas (a problem also reported in previous studies). However, urban totals were estimated quite accurately in most cases, both for the urban sections of the training sample and for the whole urban area. The accuracy of individual CD estimates, indicated by mean and median relative errors, was similar for all images other than Sydney. For the larger images and samples, there was a close correspondence between the results for the sample and the image, a fact that was used to reduce bias by selectively deleting outliers from the training sample. Overall, the best results were obtained for the Adelaide image, which was large enough for training samples to be of reasonable size, but which does not have the same extent of high-density urbanization as Sydney. Average relative errors for individual CDs in the Adelaide urban area were typically in the range 15–20 per cent. In Sydney, the corresponding errors were higher (25–35 per cent), but much lower than the 64–99 per cent reported by Lo (1995) using zone-based methods for tertiary planning units in Hong Kong. Figure 8.3 shows an Landsat TM image and an estimated population density image of the Sydney study area.

8.4.2 *Spatial disaggregation of known populations*

Even in nations where regular comprehensive censuses are taken, publication of population data is limited by confidentiality restrictions to aggregated figures for fixed, predefined spatial units. Many spatial analyses are based by default on available census zones, though the boundaries of these units are seldom chosen with the needs of the end-users of the data in mind (Pollé 1996). Though census figures can be linked to other spatially incompatible data in a geographical information system (GIS), the precision of the linkages is limited by the zonal nature of the census data. When the zones of interest are "non-census" areas such as grid squares, or circular or irregular areas which might represent the "catchments" of businesses, services or installations, or areas exposed to environmental hazards, it is common practice to assign the known populations of census zones to the centroid of each zone, or to assume a uniform spatial distribution throughout the zone (the "choropleth" approach). These approaches lead to biased estimates of the relevant or exposed populations (Arnold *et al.* 2000). Considering these limitations, the use of remotely sensed data has the potential to enable "finer grained studies than are possible with typical social science data" (Rindfuss and Stern 1998).

An improvement over a uniform distribution is the "dasymetric" approach of Langford *et al.* and others (Langford and Unwin 1994; Fisher and Langford 1995; Fisher and Langford 1996; Yuan *et al.* 1997) in which ancillary information from remote sensing imagery is used to classify pixels as residential or non-residential. The population of each census zone is then distributed uniformly, but only among those pixels classified as residential (see also Langford, Chapter 6 and Dobson *et al.*, Chapter 12, both in this book).

The pixel regression methodology, which essentially uses the dasymetric distribution as its starting point, can be used to refine the dasymetric estimates further. Rather than assigning equal populations to all residential pixels in a census zone, each pixel is assigned a different population determined by its individual spectral characteristics. Joint work is currently under way (Langford and Harvey 2001) to test the validity and effectiveness of this approach. Models based on TM imagery are first calibrated using larger spatial zones, and then used to estimate the populations of smaller constituent zones. Initial work has calibrated models using the published population counts for the 187 census wards (the second smallest spatial unit in the UK census hierarchy) in the county of Leicestershire, and then validated by calculating population estimates for the 1907 census Enumeration Districts (EDs – the smallest spatial unit in the UK census hierarchy) and comparing the estimates with the published ED census counts.

Some early results are shown in Table 8.5. The three disaggregation methods have been compared using nine criteria: three types of statistical average are reported for each of three measures of error in the ED estimates. Each outcome from the dasymetric method and the pixel regression method has been expressed as a percentage of the corresponding uniform choropleth outcome. The results from the pixel regression method have also been compared with those from the dasymetric method. On the whole, the error distributions are very positively skewed, with some extreme outliers. As a consequence in every case root mean square error (RMSE) > Mean absolute error > Median absolute error. Broadly, the error in the dasymetric estimates is some 50–60 per cent of the error in the corresponding measure for the choropleth method (i.e., a reduction of 40–50 per cent in the estimation error). The pixel-based regression refinement results in a further reduction of the order of 10 per cent in the choropleth error (or around 20 per cent in the dasymetric error). The level of accuracy attained (e.g., a median absolute relative error of 25 per cent) is comparable with the

Table 8.5 Comparison of three methods for spatial disaggregation of census ward populations

Performance criterion		Disaggregation method					
		(1) Uniform (choropleth)	(2) Classification only (dasymetric)		(3) Classification + Regression		
Measure	Statistic	Value	Value	As a % of (1)	Value	As a % of (1)	As a % of (2)
ED population	Root mean square error	578	310	54	253	44	82
	Mean absolute error	362	202	56	154	42	76
	Median absolute error	254	146	58	108	43	74
ED population density	Root mean square error	3,608	2,592	72	2,197	61	90
	Mean absolute error	2,262	1,384	61	1,086	48	79
	Median absolute error	1,347	601	46	449	33	73
Relative error (%)	Root mean square error	222	132	60	119	54	90
	Mean absolute error	97	56	58	45	46	79
	Median absolute error	61	34	55	25	40	73

26.5 per cent figure reported by Harvey (2000) for Sydney CDs (Table 8.3) but is not as good as the other CD results in that study, which ranged from 12.2 per cent to 18.4 per cent. In addition to the obvious difference in location, the two studies differed in another important aspect. In the Australian study, models were calibrated at a particular spatial level (CDs) and tested on a larger set of CDs (i.e., at the same spatial scale). In the Leicestershire study, models were calibrated at a coarser spatial scale (wards) and tested on the smaller EDs (similar in scale to Australian CDs). The latter is a more spatially stringent validation test for two reasons – the calibration zones were larger and the target zones were at a finer spatial resolution than the calibration zones – so it is not surprising that the accuracy was somewhat lower.

8.4.3   Estimates of population change

National census agencies produce inter-censal and post-censal population estimates using models based on ancillary data such as residential building approvals and school enrolments. The resulting estimates are usually limited both spatially and temporally, because they are only published annually and then only for larger zones than the smallest census zones. The provision of timely and spatially detailed inter-censal and post-censal population estimates is of interest not only to users of census data such as infrastructure planners and service providers (see, e.g., Arnold *et al.* 2000; Elliot *et al.* 2000) but also potentially to census agencies themselves, for the purposes of planning field operations for subsequent censuses and maintenance of sampling frames for national sample surveys (e.g., Booz Allen Hamilton 2000).

There is substantial remote sensing literature on change detection in many contexts. This usually involves co-registration and comparison of images of the same place at different times, with allowance being made for such effects as seasonal change. In the context of population change, pixel-based regression techniques provide an alternative strategy. Rather than comparing two images, a simpler and more pertinent comparison is between the ground reference census figures from one date and the population estimates derived from a remotely sensed image on another date, with the model being calibrated using areas known to be demographically stable. Longitudinal population-related studies are often further complicated by the fact that even consecutive censuses lack a common geographic basis, particularly at small area level (Elliot *et al.* 2000; Staines and Jarup 2000). This problem lies within both the spatial and temporal domains, and both aspects can be addressed by modelling at the pixel level. The essentials of the technique were demonstrated by Harvey (1999a). In that study, population estimates for the Australian city of Brisbane were generated from a remotely sensed image, with census data from 7 years later being used for calibration.

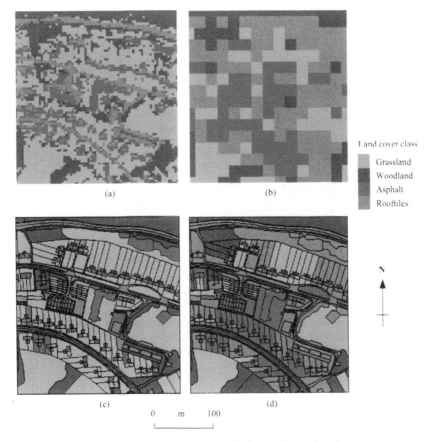

(a)

(b)

Land cover class

Grassland
Woodland
Asphalt
Rooftiles

N

(c)

(d)

0   m   100

Figure 2.2 Arundel study area (a) per-pixel classified simulated IKONOS image, (b) per-pixel classified simulated SPOT HRV image, (c) per-parcel classified simulated IKONOS image and (d) per-parcel classified simulated SPOT HRV image.

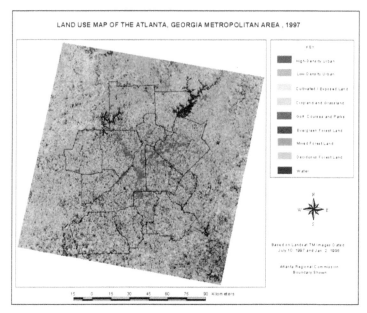

LAND USE MAP OF THE ATLANTA, GEORGIA METROPOLITAN AREA , 1997

Figure 7.2 The land use/cover map of the Atlanta Metropolitan Region for 1997 with county boundaries superimposed. (Data from Landsat TM images dated 10 July 1997 and 2 January 1998, with kind permission of Atlanta Regional Commission.)

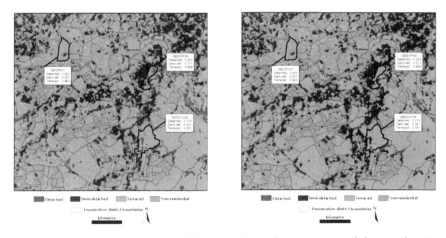

Figure 9.4 ML classification of the Bristol study area using (left) equal prior probabilities and (right) unequal prior probabilities. Notice the (slight) differences in the distribution of the three housing types. In all three highlighted EDs, detached housing (low density) is overestimated and terraced housing (high density) underestimated.

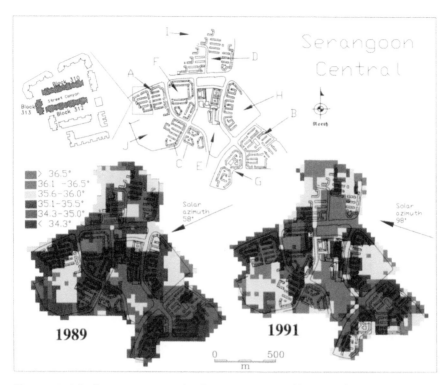

Figure 11.6 Surface temperature for Serangoon Central housing for May 1989 and March 1991

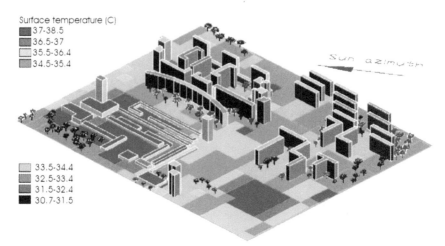

Figure 11.9 Rendered model of Toa Payoh housing estate viewed from the south‑east.

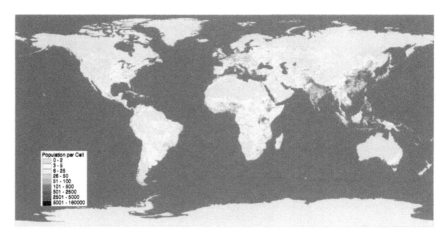

Figure 12.1 LandScan2000 global population.

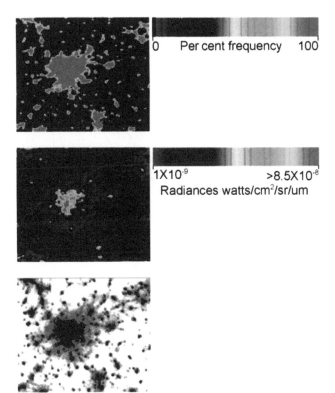

0   Per cent frequency   100

$1X10^{-9}$ $>8.5X10^{-8}$
Radiances watts/cm^2/sr/um

Figure 13.8 Comparison of the three types of night-time lights products for the city of Atlanta, GA. The per cent frequency of light detection from the 1994-95 stable lights product are shown in (a). The radiance-calibrated lights from 1999-2000 are shown in (b). The night-time lights change image from 1992-93 to 2000 are shown in (c): black = lights saturated in both time periods, red = lights substantially brighter in 2000, grey = faint lighting detected in both time periods, but little change in brightness.

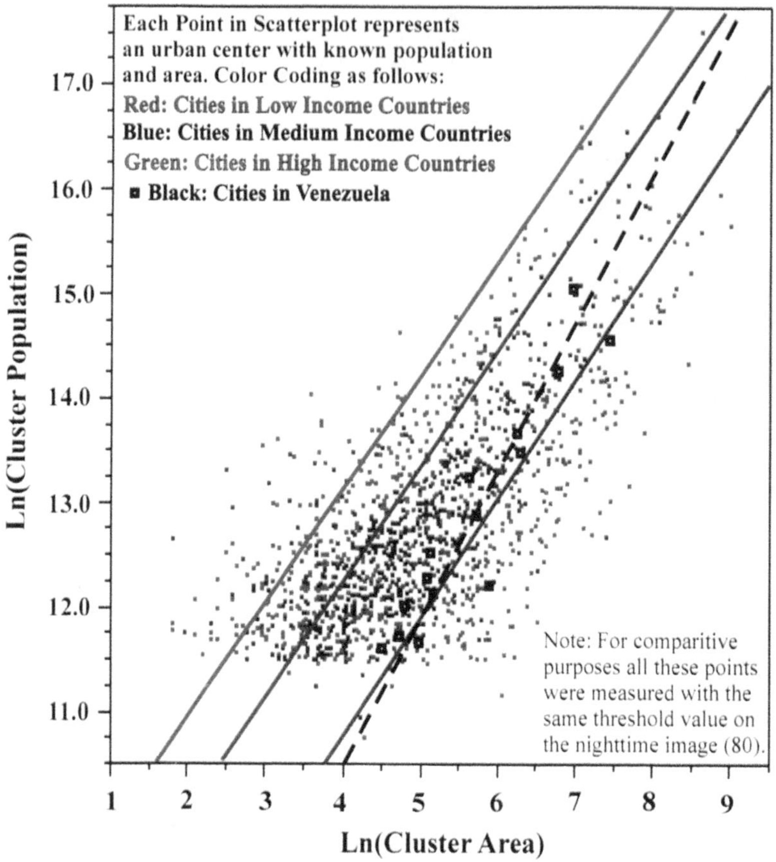
Each Point in Scatterplot represents
an urban center with known population
and area. Color Coding as follows:
Red: Cities in Low Income Countries
Blue: Cities in Medium Income Countries
Green: Cities in High Income Countries
Black: Cities in Venezuela
Note: For comparitive
purposes all these points
were measured with the
same threshold value on
the nighttime image (80).
17.0
16.0
15.0
14.0
13.0
12.0
11.0
Ln(Cluster Population)
1
2
3
4
5
6
7
8
9
Ln(Cluster Area)

Regression Statistics	(Linear Regression w/points weighted by population)
All Cities (N= 1,404):	Ln(pop) = .850* Ln(Area) + 9.107 $R^2 = 0.45$
High Income Cities (N=471):	Ln(pop) = 1.065*Ln(Area) + 7.064 $R^2 = 0.77$
Medium Income Cities (N=575):	Ln(pop) = 1.011*Ln(Area) + 8.174 $R^2 = 0.78$
Low Income Cities (N=358):	Ln(pop) = 0.989*Ln(pop) + 8.889 R^2=0.48
Venezuelan Cities (N=15):	Ln(pop) = 1.164*Ln(pop) + 6.475 R^2=0.84

Figure 14.3 Scatterplot of ln(cluster area) vs. ln(cluster population) for all cities of the world with known populations whose areal extent could be measured with the night-time satellite image.

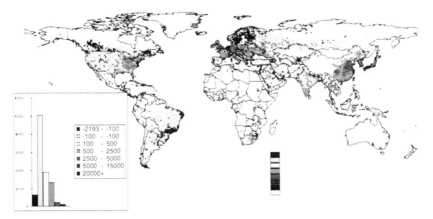

Figure 15.1 Difference map and histogram from CDIAC OLS (CO_2 in kilotonnes of carbon [ktC]).

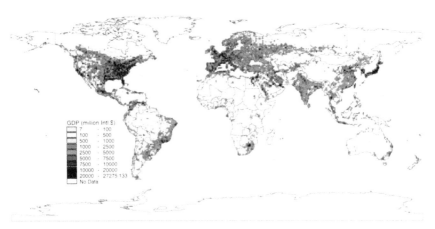

Figure 15.2 Global economics activity map at 1° x 1° resolution created from the relationship between country level GDP and lit area from the DMSP OLS sensor.

Regions with substantial population change were identifiable as statistical outliers with respect to the model.

8.5 Directions for future research

The issues of overestimation at low density and underestimation at high density, reported in most studies, remain problematical. The pixel-based procedure as presently implemented has produced reasonably accurate estimates of CD populations within the range of typical Australian suburban densities. But further research is needed if acceptably accurate estimates of scattered rural populations or of high-density inner city populations are to be attained. Directions for further research will be considered in three parts:

- improving estimation at low population densities;
- improving estimation at high population densities; and
- other aspects.

8.5.1 Improving estimation at low population densities

8.5.1.1 Specific approaches to known anomalies

The essential problem at low population densities is misclassification. Some types of large non-residential feature – built (airports, oil refineries), agricultural (orchards, vineyards) and natural (swamps, vegetated dunes) – can result in many pixels being wrongly classified as residential and hence assigned spurious population. In a practical population estimation exercise, at least some such features should be clearly identifiable. It is a routine matter to construct a binary masking GIS layer that sets population to zero in these areas.

Another common source of misclassification of residential is lineal features such as country roads and shorelines. This aspect might be addressed by incorporating specific morphological approaches for road detection (Karathanassi *et al.* 1999; Couloigner and Ranchin 2000).

8.5.1.2 Overall improvements to classification

The results reported in this chapter have been based on established rather than innovative classification methodologies. Refined and improved classification methods may enable a clearer separation of the faint, genuine, human habitation signal from the noise of either sparsely inhabited landscapes or areas of more intense agricultural activity. The range of available approaches reported in the recent literature includes: Bayesian adjustments to maximum likelihood classification (see Mesev, Chapter 9 in this book), analysis of fractal dimensions, fuzzy set theory, mixed pixel or end member analysis, knowledge-based systems, artificial neural networks, genetic programming and various contextual reclassification techniques.

8.5.2   *Improving estimation at high population densities*

At high densities, a major problem is one of hidden population. Most orbital sensors essentially operate in two dimensions, and, to the extent that they can detect population, really do so on a "population per floor level" basis. From this perspective, deleting high-density outliers from a training set is essentially the same as discarding those pixels for which the value of the reference population per floor is unknown.

One approach to this problem would be explicitly to incorporate information about numbers of levels, or equivalently building height, into the models. The ground reference pixel populations would be converted to populations per level, which would be used as the dependent variable to train the remote sensing estimation equation. This would then be applied to the image, and the resulting estimates of population per level would be back-transformed using the height information, to produce population estimates. Even very approximate region-by-region estimates of average building height, which could be incorporated as a multiplicative overlay, might bring about substantial improvement in population estimates such as those for Sydney. The "average building height" layer would be used to correct for widely distributed, lower profile, multilevel structures. In addition, an overlay can be constructed in which known major anomalous concentrations of population gleaned from ancillary sources, such as institutions and tower blocks, can be assigned to small regions and directly added to the population estimates from the linear model (this is the obverse of the binary mask for suppressing spurious population). One saving grace of large institutions and tower blocks is their visibility and regularity of shape, which makes such *ad hoc* adjustments quite feasible.

A less precise alternative to modelling population per level would be to employ more than one residential classification (see Langford *et al.* 1991; Lo 1995). This could be incorporated into the pixel-based methodology at the cost of considerable organizational complexity, with separate training sets being selected from within each residential stratum, and separate estimation equations being derived and applied. To the extent that high density comes about in the vertical dimension, this approach is essentially a discrete approximation to the explicit inclusion of height in the model. With the zone-based models used by Langford *et al.* (1991) and Lo (1995), only the discrete approach is feasible. But with pixel-based models there is no such restriction (see Table 8.1). Considering that some judgements about height would in any case probably be used to inform the multi-stratum classification, the more direct approach would be just as feasible to apply, and probably more effective.

One remote sensing approach to the measurement of building height is radar altimetry. Satellites exist with this capability, but in practice they have limited narrow swathe coverage and are designed principally for monitoring the sea surface rather than land. Alternatively, airborne scanning altimetry, such as LIDAR (Laser Induced Detection And Ranging), which has enormous potential for precision elevation mapping on land (see Chapter 3 by Baltsavias and Gruen for details on LIDAR, and Chapter 4 by Barnsley *et al.* on urban mapping, both in this book). Other potential approaches are the analysis of shadows (Shettigara and Sumerling 1998) and stereo image matching (Kim and Muller 1998).

8.5.3 *Other avenues of investigation*

Some other lines of enquiry of an *applied* nature might include:

- Exploration of the sensitivity of the results obtained to changes of scale or sensor resolution, by simulation and/or using data from other multi-spectral sensors, such as MSS (Multispectral Scanning Systems), SPOT (Système Pour l'Observation de la Terre) multi-spectral and new generation higher resolution sensors, with reference to such decision frameworks as those of Atkinson and Curran (1997) and Cowen and Jensen (1998).
- Combined use of more than one sensor, such as TM plus SPOT panchromatic, or TM plus night-time illumination (Imhoff *et al.* 1997; Sutton *et al.* 1997; Sutton, Chapter 14 in this book).
- Application of the procedure in other geographical, climatic and cultural settings.
- Applicability of the general approach to Synthetic Aperture Radar (SAR) imagery. Some research was carried out in the 1980s into the use of radar imagery for population estimation using selective imagery generated in Space Shuttle experiments (Harrison and Jupp 1989; Henderson and Xia 1997). However, this form of imagery has only recently become more generally available from orbital platforms.
- Application to the estimation of demographic measures other than total population. Some subpopulations, such as pre-school children or house-holds with high disposable income, may also have distinguishable spectral signatures.
- More *theoretical* investigations might include: modelling with simulated data in order to understand better the relationships between population and the physical properties of different surfaces and materials which underlie the pixel-level linear models identified (Forster 1980; Curran 1985). The use of generalized linear models, in particular models with spatially correlated Poisson error structures.

8.6 Conclusion

Models fitted at the level of individual TM pixels, utilizing the six untransformed TM bands for supervised classification and iterated regression modelling, have been shown to be capable of producing population estimates of comparable accuracy with the best and most highly tuned zone-based models. Pixel-based models use simpler and more robust mathematical formulations, provide greater spatially flexibility in application and are more amenable to spatially targeted refinement than zone-based models.

Three types of application have been discussed in this chapter. First, the method has been used to estimate and map small-area population density across large regions directly, using Landsat TM imagery supplemented by a broad land use/land cover classification and population counts of relatively small training areas. Second, the method has been applied to the task of disaggregating known populations of fixed zones to finer spatial resolutions, with greater accuracy than either choropleth or simple dasymetric redistribution methods. Third, the method can be used to detect and quantify population change between the dates of a remotely sensed image and a population census.

Many directions for further enquiry and potential refinement of this methodology have been identified. Of course, it is possible that, in developed nations at least, these ongoing efforts might be overtaken by the march of technological innovation, with some national census agencies now considering direct geo-coding of census counts. But even if these plans are realized, confidentiality restrictions will continue to constrain the spatial resolution at which data can be made publicly available, and hence spatial flexibility will be largely in the hands of the data providers rather than the end-users. The issue of timeliness and change detection between censuses will also remain. In these circumstances, it would seem that remote sensing will continue to have relevance to the problem of small-area population estimation.

8.7　Acknowledgements

Part of the research described in this chapter was supported through the provision of TM imagery by the Australian Centre for Remote Sensing and Geoimage Photographic Laboratory (P/L), ground reference population data by the Australian Bureau of Statistics and digitized census collection district boundaries by the Australian Survey and Land Information Group. Some tables and figures have been reproduced or adapted with permission from the American Society for Photogrammetry and Remote Sensing.

8.8　References

Arnold, R. A., Diamond, I. D. and Wakefield, J. C., 2000, The use of population data in spatial epidemiology. In P. Elliot, J. C. Wakefield, N. G. Best and D. J. Briggs (eds) *Spatial Epidemiology: Methods and Applications* (Oxford: Oxford University Press).

Atkinson, P. M. and Curran, P. J., 1997, Choosing an appropriate spatial resolution for remote sensing investigations. *Photogrammetric Engineering and Remote Sensing*, 63, 1345–51.

Booz Allen Hamilton, 2000, *Master Address File/Topographically Integrated Geographic Encoding and Referencing System (MAF/TIGER) Modernization Study* (Washington DC: United States Census Bureau, Geography Division).

Brugioni, D. A., 1983, The census: It can be done more accurately with space-age technology. *Photogrammetric Engineering and Remote Sensing*, 49, 1337–9.

Couloigner, I. and Ranchin, T., 2000, Mapping of urban areas: A multiresolution modelling approach for semi-automatic extraction of streets. *Photogrammetric Engineering and Remote Sensing*, 66, 867–74.

Cowen, D. J. and Jensen, J. R., 1998, Extraction and modeling of urban attributes using remote sensing technology. In D. Liverman, E. F. Moran, R. R. Rindfuss and P. C. Stern (eds) *People and Pixels: Linking Remote Sensing and Social Science* (Washington DC: National Academy Press), pp. 164–88.

Curran, P. J., 1985, *Principles of Remote Sensing* (New York: Longman).

Dempster, A. P., Laird, N. M. and Rubin, D. B., 1977, Maximum likelihood from incomplete data via the EM algorithm. *J. Royal Statistics Society, Series B*, 39, 1–38.

Elliot, P., Wakefield, J. C., Best, N. G. and Briggs, D. J. (eds), 2000, The use of population data in spatial epidemiology. *Spatial Epidemiology: Methods and Applications* (Oxford: Oxford University Press).

Fisher, P. F. and Langford, M., 1995, Modelling the errors in areal interpolation between zonal systems by Monte Carlo simulation. *Environment and Planning A*, 27, 211–24.

Fisher, P. F. and Langford, M., 1996, Modelling sensitivity to accuracy in classified imagery: A study of areal interpolation by dasymetric mapping. *Professional Geographer*, 48, 299–309.

Flowerdew, R. and Green, M., 1989, Statistical methods for inference between incompatible zonal systems. In M. F. Goodchild and S. Gopal (eds) *The Accuracy of Spatial Databases* (London: Taylor & Francis).

Forster, B. C., 1980, Urban residential ground cover using Landsat digital data. *Photogrammetric Engineering and Remote Sensing*, 46, 547–58.

Harrison, B. A. and Jupp, D. B. L., 1989, *Introduction to Remotely Sensed Data* (Canberra: Commonwealth Scientific Industrial Research Organization (CSIRO)).

Harvey, J. T., 1999a, *Estimation of population using satellite imagery*. Unpublished PhD thesis, University of Ballarat.

Harvey, J. T., 1999b, Modelling population associated with individual TM pixels. Paper given at Conference of the Annual Conference of the American Society for Photogrammetry and Remote Sensing, Portland (Bethesda, MD: ASPRS).

Harvey, J. T., 2000, Small area population estimation using satellite imagery. *Statistics in Transition*, 4, 611–33.

Harvey, J. T., 2002a, Population estimation models based on individual TM pixels. *Photogrammetric Engineering and Remote Sensing*, 68.

Harvey, J. T., 2002b, Estimating census district populations from satellite imagery: Some approaches and limitations. *International Journal of Remote Sensing*, 23, 2071–95.

Henderson, F. M. and Xia, Z. G., 1997, SAR applications in human settlement detection, population estimation and urban land use pattern analysis: A status report. *IEEE Transactions on Geoscience and Remote Sensing*, 35, 79–85.

Iisaka, J. and Hegedus, E., 1982, Population estimation from Landsat imagery. *Remote Sensing of the Environment*, 12, 259–72.

Imhoff, M. L., Lawrence, W. T., Stutzer, D. C. and Elvidge, C. D., 1997, A technique for using composite DMSP/OLS "city lights" satellite data to map urban area. *Remote Sensing of Environment*, 61, 361–70.

Karathanassi, V., Iossifidis, C. H. and Rokos, D., 1999, A thinning-based method for recognizing and extracting peri-urban road networks from SPOT panchromatic images. *International Journal of Remote Sensing*, 20, 153–68.

Kim, T. and Muller, J-P., 1998, A technique for 3D building reconstruction. *Photogrammetric Engineering and Remote Sensing*, 64, 923–30.

Langford, M. and Harvey, J. T., 2001, The use of remotely sensed data for spatial disaggregation of published census population counts. Paper given at Urban 2001, IEEE/ISPRS Joint Workshop on Remote Sensing and Data Fusion over Urban Areas, 7–8 November, Rome, Italy (Rome: IEEE/ISPRS).

Langford, M. and Unwin, D. J., 1994, Generating and mapping population density surfaces within a geographical information system. *The Cartographic Journal*, 31, 21–6.

Langford, M., Maguire, D. J. and Unwin, D. J., 1991, The areal interpolation problem: Estimating population using remote sensing within a GIS framework. In I. Masser and R. Blakemore (eds) *Handling Geographical Information: Methodology and Potential Applications* (London: Longman), pp. 55–77.

Lo, C. P., 1986, *Applied Remote Sensing* (Harlow: Longman).

Lo, C. P., 1995, Automated population and dwelling unit estimation from high-resolution satellite images: A GIS approach. *International Journal of Remote Sensing*, 16, 17–34.

Pollé, V. F. L., 1996, Planning urban services in developing countries in developing countries: Quantification of community service needs using remote sensing indicators. *ITC Journal*, 1996-1, 64–70.

Rindfuss, R. R. and Stern, P. C., 1998, Linking remote sensing and social science: The needs and challenges. In D. Liverman, E. F. Moran, R. R. Rindfuss and P. C. Stern (eds) *People and Pixels: Linking Remote Sensing and Social Science* (Washington DC: National Academy Press), pp. 1–27.

Shettigara, V. K. and Sumerling, G. M., 1998, Height determination of extended objects using shadows in SPOT images. *Photogrammetric Engineering and Remote Sensing*, 64, 35–44.

Staines, A. and Jarup, L., 2000, Health event data. In P. Elliot, J. C. Wakefield, N. G. Best and D. J. Briggs (eds) *Spatial Epidemiology: Methods and Applications* (Oxford: Oxford University Press).

Sutton, P., Roberts, D., Elvidge, C. and Meij, H., 1997, A comparison of night-time satellite imagery and population density for the continental United States. *Photogrammetric Engineering and Remote Sensing*, 63, 1303–13.

Webster, C. J., 1996, Population and dwelling unit estimates from space. *Third World Planning Review*, **18**, 155–76.

Yuan, Y., Smith, R. M. and Limp, W. F., 1997, Remodeling census population with spatial information from Landsat TM imagery. *Computers, Environment and Urban Systems*, **21**, 245–58.

9 Urban land use uncertainty

Bayesian approaches to urban image classification

Victor Mesev

9.1 Introduction

Whatever urban applications satellite images are used for, be it population mapping (Langford, Lo, Harvey, Chapters 6, 7, 8, respectively, in this book), land use layout, transportation, geodemographics (Harris, Chapter 10 in this book) or environmental pollution (Nichol, Chapter 11 in this book), the quality of their results is ultimately dependent on the quality of the initial image interpretation. Many urban methodologies and applications are derived from satellite images that are first classified into built-up land cover or residential land use. Although population and housing numbers have been estimated from raw radiometric values (Iisaka and Hegedus 1982; Lo 1986), the vast majority of research has been traditionally focused on classification methodologies: even pattern recognition systems rely initially on the identification of land covers.

Despite the plethora of classification techniques that have surfaced over the years, the accurate and consistent classification of urban surfaces from satellite images remains one of the most challenging and elusive. The difficulties of urban interpretation form the basis of this book and are frequently mentioned in most of the chapters. Suffice to say, the typically complex mixture and irregular arrangement of artificial and natural urban land cover types produce reflectance levels that are always the result of the interaction of more than one ground phenomenon. Allowing for variations in spatial resolutions, class descriptions and local conditions, the proportion of mixed pixels that represent typical urban areas frequently exceeds 90 per cent of the image. As a result, classifications of Earth observation satellite images representing urban areas have low accuracy levels compared with the physical environment, and as such have been traditionally consigned to somewhat large-scale distinctions of urban/non-urban land cover, built/non-built and, at best, categorical building types (Forster 1985; Haack *et al.* 1997). However, if urban images are ever to be applied to practical urban monitoring, management and planning activities, much more detail is needed on the structural morphology and functional use of land deemed to be urban.

Image classification techniques can be divided into two groups: "hard" and "soft". Hard classifiers are those that assign one, and only one, thematic class to each individual pixel, while soft classifiers relax this assumption and allow pixels multiple classes, sometimes at measurable proportionality. Leaving soft classifiers aside for the moment, this chapter will examine one popular hard per-pixel classification technique – the maximum likelihood (ML) discriminant function – and in particular the Bayes' modification of class membership probabilities. Under standard ML classifiers each class has an equal probability of occurrence, but the Bayes' modification allows bias to influence the classification in favour of classes with known areal estimation. To test the merits of the approach the intention is to utilize ancillary data, first, to spatially segment a satellite image representing an urban area and, then, within each segment, to identify and update the prior probabilities for separate ML classifications. Note that ancillary data are any data that are beyond the spectral domain, and include sources such as population censuses and delivery addresses. The vehicle for storing and handling these exogenous data sets is invariably a geographical information system (GIS).

The Bayes' modification of the ML discriminant function in multi-spectral image classification is not new. Early remote sensing texts, such as Swain and Davis (1978) and Haralick and Fu (1983), reported the statistical derivation of the technique, and so too did Haralick *et al.* (1973) on the use of spatial information in a single band. Kettig and Landgrebe (1976) developed a multivariate method, and Strahler (1980) demonstrated its functionality using environmental indicators. Contributions from Skidmore and Turner (1988) extracted class probabilities from grey-level frequency histograms, Maselli *et al.* (1992) formalized a parametric/non-parametric link between the ML discriminant function and prior probabilities, Conese and Maselli (1992) evaluated error matrices and Foody *et al.* (1992) investigated the potential of posterior probabilities in continuous class memberships. However, research on urban applications seems to have been largely neglected, undoubtedly because of the frequent heterogeneity of urban land use composition. Yet, it is exactly when there is uncertainty in class membership that prior/posterior probability estimation should have theoretically the most influence (Mather 1985; Mesev 1998; Mesev *et al.* 2001). This chapter builds on previous research in two directions – first, in further exploring the performance of prior probabilities for improving urban classifications and, second, assessing whether prior probability classifications within a spatially segmented image will result in higher thematic accuracy. The reasoning behind segmentation is to reduce spectral variance and increase homogeneity. If prior probabilities are to succeed and increase classification accuracy they need to operate within more localized parameters than have hitherto been witnessed. Local, or intra-urban (within urban), segmentation represents a more precise class membership from ancillary information, and therefore more accurate class prior probabilities.

In addressing these two objectives, the chapter contributes to research and development discussions on the closer interaction between image processing and GIS (Mesev 1997).

As mentioned, the ML-modified classifier is a per-pixel algorithm, which is conceptually straightforward and relatively simple to implement (Tom and Miller 1984) (see also Aplin, Chapter 1 in this book for comparisons with per-parcel classifications). This chapter does not attempt to challenge the well-established literature on Bayes' modifications to the ML classifier. Instead, it aims to examine the possibility of enhancing the performance of prior probabilities by reducing uncertainty between spectrally similar class memberships and creating more favourable conditions for their calculation. What is meant by "favourable", in terms of spatially segmenting a remotely sensed image, is the use of reliable urban data and thereby reducing spectral variation and generating class memberships that are more precise (Hutchinson 1982). Alternative research into modifying classifiers takes into consideration contextual, textural (Myint, Chapter 5 in this book) and spatial properties (Barnsley and Barr 1997) inherent in the image, while others implement statistical operators such as neural networks and fuzzy set principles.

In contrast, soft classifiers, including those employing mixture models and fuzzy set theory, allocate proportions of one or more classes per pixel. As pixels representing urban areas are virtually all composed of mixed surfaces, the difference between the two types of classifier is a matter of scale, generalization and classification scheme. The focus in this chapter is on improving the performance of the popular Bayes' modified ML classifier, not to debate the continuum of pixel representation (Fisher 1997). Besides, given the severe spatial heterogeneity in the arrangement of urban structures, there is reason enough to move away from detail (soft classifiers) in favour of more aggregated or "averaged" pixels (hard classifiers) (see Baltsavias and Gruen, Aplin, Chapters 3, 2, respectively, in this book). In other words, the breakdown of mixed pixels may produce too much information that cannot be easily categorized within a workable scheme: a situation of "not seeing the wood for the trees". The three urban land uses to be classified are variations of residential density – high, medium and low – and are representative of variations in proportions of buildings-to-vegetation ratios. For instance, pixels representing mostly, if not all, buildings are classified as high residential density and pixels representing sizeable proportions of vegetation in conjunction with buildings as low residential density. Moreover, the three residential density classes are comparable with information held by the UK Population of Census on typical British housing, namely detached, semi-detached and terraced. This association creates a situation where demonstration of the Bayes' modification of the ML classifier can proceed within the framework of census-derived prior probabilities.

9.2 Modified ML classification

As a background, the ML algorithm in remote sensing classification is parametric and depends on each class being represented by a Gaussian probability density function completely described by the mean vector and variance–covariance matrix across all available spectral bands, and, if possible, ancillary information. Given these parameters, it is possible to compute the statistical likelihood of a pixel vector being a member of each spectral class (Thomas *et al.* 1987; Besag 1986). The objective is to assign the most likely class w_j from a set of N classes, $w_1, \ldots, w_N$, to any feature vector $\mathbf{x}$ in the image. A feature vector $\mathbf{x}$ is the vector $(\mathbf{x}_1, \mathbf{x}_2, \ldots, \mathbf{x}_M)$, composed of pixel values in M features (in most cases, spectral bands). The most likely class w_j for a given feature vector $\mathbf{x}$ is the one with the highest posterior probability $\Pr(w_j|\mathbf{x})$. Therefore, all $\Pr(w_j|\mathbf{x})$, $j \in [1, \ldots, N]$ are calculated, and the w_j with the highest value is selected. The calculation of $\Pr(w_j|\mathbf{x})$ is based on Bayes' theorem:

$$\Pr(w_j|\mathbf{x}) = \frac{\Pr(\mathbf{x}|w_j) \times \Pr(w_j)}{\Pr(\mathbf{x})} \tag{9.1}$$

On the left-hand side is the *posterior* probability that a pixel with feature vector $\mathbf{x}$ should be classified as belonging to class w_j. The right-hand side is based on Bayes' theorem, where $\Pr(\mathbf{x}|w_j)$ is the conditional probability that some feature vector $\mathbf{x}$ occurs in a given class: in other words, the probability density of w_j as a function of $\mathbf{x}$ (Besag 1986). Supervised classifications, such as the ML, derive this information from training samples. Often, this is done parametrically by assuming normal class probability densities and estimating the mean vector and covariance matrix. Alternatively, it is possible to use Markov random fields (Berthod *et al.* 1996), or non-parametric methods, such as k-Nearest Neighbour (kNN). "Standard" kNN methods directly implement a decision function based on the number of training pixels per class proportional to the prior probability (Fukunaga and Hummels 1987; Therrien 1989). This is the prior probability of the occurrence of w_j irrespective of its feature vector, and as such is open to estimation by prior knowledge external to the remotely sensed image.

External prior knowledge will typically include information on the distribution and relative areas covered by each class in feature space. This ancillary information is most readily generated from urban data derived from a GIS (Harris and Ventura 1995; Mesev 1998). It therefore follows that the accuracy of class priors is at best equal to the quality of GIS prior knowledge. In image classification terms, prior probabilities can be visualized as a means of shifting decision boundaries to produce larger volumes in M-dimensional feature space for classes that are expected to be large and smaller volumes for classes that are expected to be small. The denominator in Equation (9.1), Pr(**x**), is the unconditional probability density that is used to normalize the numerator such that:

$$\Pr(\mathbf{x}) = \sum_{j=1}^{N} \Pr(\mathbf{x}|w_j) \times \Pr(w_j) \qquad (9.2)$$

Typically, ML classifiers assume prior probabilities to be equal and assign each Pr(w_j) a value of 1.0. However, variations in prior probabilities can be an important remedy for the problem of spectrally overlapping classes. If a feature vector **x** has probability density values that are significantly different from zero for several classes, it is not inconceivable for that pixel to belong to any of these classes. When selecting a class solely on the basis of its radiometric values at various spectral channels, a large probability of error frequently results (Conese and Maselli 1992; Steele *et al.* 1998). The use of appropriate prior probabilities, based on reliable ancillary information, is one way to reduce this uncertainty in class assignments. Moreover, it would seem intuitively more sensible to suggest that some classes are more likely to occur than others.

9.2.1 Local/individual/global prior probabilities

At this stage of the discussion, it is important to differentiate between "global", "individual" and "local" priors. Many proprietary software packages allow the use of global prior probabilities, where the user is expected to estimate them simply by using information on the anticipated (relative) class areas. The improvement on classification is often limited (Mather 1985). At the other extreme, a vector of prior probabilities can be calculated for each and every individual pixel. This, however, is pointless, because if the correct prior probabilities for each individual pixel were known beforehand, the classification would not be necessary (Mesev *et al.* 2001).

Given these problems, a compromise scale somewhere between global and individual can be derived by, first, subdividing the image into strata, or segments, according to ancillary context data and, then, finding the local prior probability vector for each stratum. In the case study, extraneous data are used to stratify and segment a satellite image according to some contextual rules to produce housing classes. Housing data are used from the 1991 UK Census of Population at the smallest collection unit, the ED (Enumeration Districts, which represent an average of around 250 house-holds within urban areas), to infer local prior probabilities of housing type.

9.3   Modification of prior probabilities

9.3.1   *Study areas*

Spatially stratified and prior probability modified Bayes' ML classifiers are applied to images of four cities and towns in the south-west of England. These are Bristol, Swindon, Bath and Taunton, with populations of 584,289, 164,245, 88,495 and 57,028, respectively. The objective is to classify a Landsat ETM+ (Enhanced Thematic Mapper Plus) multispectral image (Path 203/Row 024), taken on 30 April 2000, into three residential density categories – low, medium and high – which are associated with detached, semi-detached and terraced, respectively, for each of the four settlements. Class area estimates for each urban land use generated with Bayes' modifications are then compared with classifications generated using the standard, equal, prior probability ML technique. Any improvements in accuracy will be measured by a convergence toward known housing proportions from the 1991 Population Census. The temporal differences between the Landsat image and the population census are unavoidable but do not represent a weakness in the methodology. Remember, the aim is to compare unequal with equal priors, not the exact extent of the housing morphology. The software used comprised Interactive Data Language (IDL)-modified Research Systems Inc. (RSI) ENVI™ programs.

9.3.2   *Methodology*

In the UK, housing is readily differentiated into three categories, detached, semi-detached and terraced. These are essentially surrogates for the target classes of low-, medium- and high-density residential land use, respectively. Although this relationship is in part dependent on the size of buildings and size of gardens, there is still a strong correlation between types of properties and densities of properties. In Table 9.1, typical housing densities for each category are calculated from the 1991 UK Census of Population, and

Table 9.1 Housing density categories for each of the four settlements in south-west England

Settlement	Housing density categories (dwellings/ha)		
	Low density	*Medium density*	*High density*
Bristol	<10	11–24	25–100
Swindon	<12	13–70	71–173
Bath	<12	13–28	29–128
Taunton	<13	14–48	49–175
Average	<13	14–43	44–136

expressed as the number of dwellings per hectare (see Section 9.3.3). These densities are used to calibrate the prior probabilities for each land use class at the local level. For prior probabilities to function optimally they need to function within inclusive feature space and derive mutually exclusive classes. This essentially means that, for the classification of mutually exclusive housing types, an image must only be composed of housing segments. The assumption is that new, commercial, high spatial resolution satellite sensor images can be routinely segmented into urban and non-urban, as well as housing and non-housing, with reliable accuracy. Within the housing feature space, prior probabilities of the surrogate housing types – detached, semi-detached and terraced – may be generated by census data and inserted into the ML classifier to produce the spectral classes of low-, medium- and high-density housing. This is exemplified by the city of Bristol in Figure 9.1, which shows the Landsat ETM image with housing proportions calculated at selected individual census ED (tract) level.

Prior probabilities are determined as follows. Consider z_k as the census variable "housing type" (where k: 1 = detached, 2 = semi-detached and 3 = terrace). When stratified into exclusively housing feature space, the three classes will have pixels with feature values x_i, where $x_i, \ldots, x_A$ are not necessarily mutually exclusive. The objective is to find the probability that a random pixel (within the housing type stratum of the image) will be a member of a spectral class w_j (where j: 1 = low, 2 = medium, 3 = high), given its density vector of observed measurements x, in m-dimensional feature space *and* that it belongs to ancillary class z_k, described as:

$$\Pr(w_j|x, z_k) \tag{9.3}$$

It is also assumed that the effects of z_k are external to the original generation of the mean vector and covariance matrix of w_j. As a result the likelihood function $\Pr(w_j|x)$ is unaltered by the introduction of z_k, but is simply modified by the conditional probability:

$$\Pr(w_j|z_k) \tag{9.4}$$

This is a process of identifying the association between spectral class w_j with census variable z_k. For example, the spectral class labelled "low-density housing" would be directly associated with a conditional probability of the census variable "detached dwellings". In effect, w_1 is weighted by the probability of z_1 producing the prior probability of $\Pr(w_1)$. In the empirical examples, prior probabilities of each of the three dwelling types are assumed to exist in inclusive m-dimensional feature space, so that $\Pr(w_1) + \Pr(w_2) + \Pr(w_3) = 1.0$. The probability densities $d_{i1} = \Pr(x_i|w_1)$, $d_{i2} = \Pr(x_i|w_2)$, $d_{i3} = \Pr(x_i|w_3)$ are known for each pixel. Let l_{i1} be shorthand for the posterior probability $\Pr(w_1|x_i, z_1)$ that pixel i belongs to class w_1, and let p_j

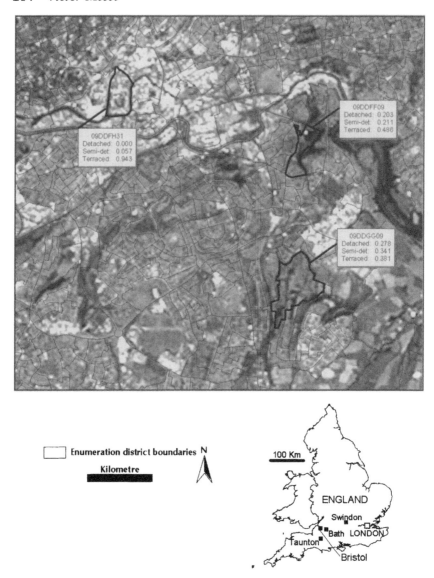

Figure 9.1 Landsat ETM+ image of the central and south-east areas of Bristol,
England. Three examples of local prior probabilities of housing are
generated from the 1991 UK Census of Population, and represented at
the ED level.

be shorthand for the prior probabilities. The Bayes' modified ML is now
represented as:

$$l_{i1} = \frac{d_{i1}p_1}{d_{i1}p_1 + d_{i2}p_2 + d_{i3}p_3} \qquad (9.5)$$

Likewise, $l_{i2} = \mathrm{Pr}(w_2|\mathbf{x}_i, z_2)$ and $l_{i3} = \mathrm{Pr}(w_3|\mathbf{x}_i, z_3)$ may also be found, and, of course, the sum of the three posterior probabilities equals 1.0:

$$l_{ij} = \frac{d_{ij}p_j}{\sum_{j=1}^{3} d_{ij}p_j} \qquad (9.6)$$

9.3.3 Implementation

Before the modified ML classifier is implemented, a series of hierarchical segmentations are carried out in order to partition the image and generate the housing stratum from which the three housing types are ultimately derived. Segmentations are applied within standard unsupervised classifications using the ISODATA (Iterative Self-Organizing DATA analysis) algorithm. These produce generalized urban and non-urban strata, from which the urban stratum is subdivided into built-up and non-built-up. The built-up stratum is then segmented into housing and non-housing. Using the housing stratum, prior probabilities for each housing type are calculated at the local level. This means that area proportions for each of three housing types (detached, semi-detached and terraced) are calculated for each of the census EDs that represent a settlement. In Figure 9.1, EDs (e.g., 09DDFH31) from the city of Bristol are illustrated along with their respective areal proportions of housing types. These proportions are normalized to create a probability distribution, and modified to take into account their relative size ratios (average prior probabilities are given in Table 9.2). The size ratio transformation helps to preserve the relative areal proportions of each housing type, where for instance detached housing occupies a larger physical space than terrace dwellings. Using stereoscopic aerial photographs, twenty samples of dwelling type sizes were generated and average relative size ratios between dwelling types were established. The ratios stabilized at 1 detached dwelling to 1.5 semi-detached and 1 detached to 2.25 terraced. Although

Table 9.2 Average prior probabilities for the three housing types derived from census estimates (with scaling factors)

Settlement	Average census prior probabilities			
	Low density	*Medium density*	*High density*	*Total*
Bristol	0.0946	0.2786	0.6268	1.0
Swindon	0.1545	0.2659	0.5795	1.0
Bath	0.2472	0.2374	0.5154	1.0
Taunton	0.1044	0.3022	0.5935	1.0
Scaling	1.00	1.50	2.25	–

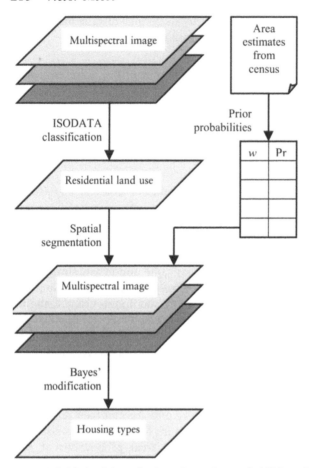

Figure 9.2 Methodology for inserting prior probabilities within a spatially segmented and Bayes-modified ML classification.

these are approximations, they are still more realistic than assuming absolute 1 : 1 linear relationships. The entire classification process is illustrated in Figure 9.2.

9.3.4 Results

First of all, as a caution, the modification of prior probabilities, especially in complex urban landscapes does not produce drastic differences in classification. That is because prior probabilities are only tweaking the decision rule boundaries of the ML classifier, and not completely altering them. However, any improvements in urban classification, no matter how small, are still valuable improvements.

In the cases of the four settlements, prior probabilities for all three housing types are entered into the ML classifications at the stratified census ED (local) level. Results in Figures 9.3 and 9.4 (see colour plates) are based on comparisons of class area estimates of classifications of the three housing types generated by equal and stratified unequal prior probabilities. Figure 9.3 shows area estimates for most urban land use classes produced from the Bayes-modified ML classifier to be closer to those derived from the size ratio-transformed census figures. The difference is slight, as expected, and for the majority of cases in the same direction (overestimation or underestimation). Total absolute error for all settlements is consistently lower under conditions of unequal as opposed to equal prior probabilities. However, in terms of housing, there are considerable variations between types within individual settlements and between the four settlements. No one housing type has consistently lower area estimation error, but there is some evidence to suggest that high-density housing is underpredicted (i.e., less pixels classified) and conversely low-density housing is overpredicted (i.e., more pixels classified). The reason for this may be case-specific and lie in the highly concentrated nature of British housing in central areas of towns (but see Chapter 8 in this book by Harvey for similar results for cities in Australia). The spatial extent of individual houses around the central core may sometimes be much smaller than the spatial resolution of the satellite sensor. Anyway, what becomes apparent from these results is that classifications are highly site-specific, and they underline the immense problems that arise when subresidential classifications are attempted. For example, consider the case of Bristol in Figure 9.4 (colour plates) where only the central and south-eastern part of the city is shown for clarity. A visual comparison of urban land use coverage reveals similar distributions of the three housing types between equal priors (Figure 9.4a, colour plates) and unequal priors (Figure 9.4b, colour plates). However, on closer scrutiny, slightly more pixels have been classified as detached housing (low-density land use) under conditions of equal prior probabilities, particularly across the more affluent parts of the south-east of the study area. At the same time, slightly more pixels are classified as terraced under unequal prior conditions. In both cases, class estimates from unequal (stratified and modified) priors are closer to census estimates, and this is indicative of the flexibility of modified prior probabilities and their ability to incorporate spatial information from beyond the spectral domain. Individual census tracts can further highlight differences in classification between equal and unequal priors. The census tracts 09DDFF09 and 09DDGG09 demonstrate how lack of prior knowledge has resulted in the misclassification of detached dwellings. In the case of the census tract (09DDFH31) within the city centre (top left of the study area) detached housing is again classified using equal priors despite the fact that none are recorded in the census. When this zero prior probability is included within the stratified and modified method fewer pixels are classified as detached.

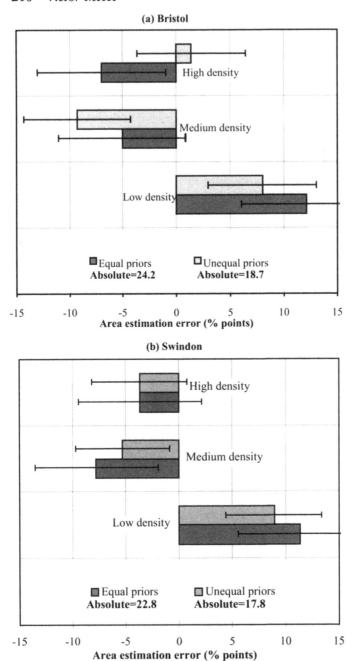

Figure 9.3 Area estimation results from ML classification using equal and unequal prior probabilities on Landsat ETM image representing four settlements in south-west England. Area estimation (shaded bars) are calculated from census numbers and standard errors are shown by line bars.

(c) Bath

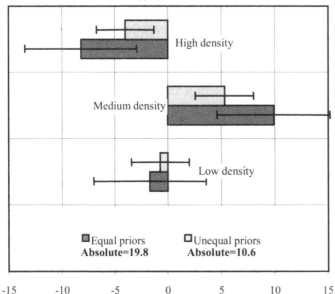

(d) Taunton

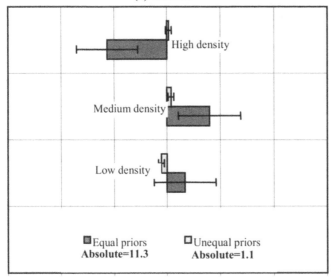

Admittedly, the testing procedure is essentially circular: classifications using census-derived prior probabilities are tested against census distributions. However, the objective of this chapter is not exclusively to test equal with unequal priors, but also to observe the relative performance of the Bayes' classifier across the three land use types. In this sense, what the results reveal is that total absolute error is lower under modified prior probabilities, but that no one housing type has consistently lower area estimation error. However, there is some evidence to suggest that high-density housing is underestimated (less pixels classified) and conversely that low-density housing is overestimated (more pixels classified). Further accuracy assessments of other settlements, including one with a detailed accuracy evaluation, can be found in Mesev (1998), with all indicating slight to moderate improvements.

9.4   Conclusions

The Bayes' classifier has a long history in environmental remote sensing, but a much shorter one in urban mapping. This is the unfortunate consequence of dealing with a range of surfaces that can exhibit variable levels of complex spatial heterogeneity. Indeed, it is precisely when class memberships are uncertain and spectrally overlapping that probabilistic and flexible ML modifications are indispensable. The importance of the Bayes' classifier in urban areas is, however, not paralleled by the level of research. There seems to be a lack of methodological development with extraneous urban data and, more specifically, with census data, the most obvious source for prior probability modification.

This chapter has contributed to the development of the Bayes' classifier for urban applications in two ways. The first is on reinforcing links between socio-economic data from the census with spectral patterns inherent in a satellite image (in the same vein as in the previous three chapters). Area estimates of separate, yet spectrally similar classes are readily available from sources such as the Census of Population, and, when accommodated within the Bayes' classifier, can considerably improve classification accuracy. The second contribution is on spatial segmentation, which is essential for the accurate calculation of local, not global, priors. Improvements in classification accuracy have been reported in all four settlements. However, these are relatively small and further testing of the methodology is essential on sensor data at much finer spatial resolutions to see whether standard per-pixel algorithms can contribute to the interpretation of high-volume, more detailed representations of urban morphology. One way to operationalize the methodology in a practical sense is to link probabilistic information from modified ML classifications directly within a GIS. For instance, posterior probabilities of per-pixel class membership may be stored and updated in a GIS database, and used to resolve land use planning and land use change queries. However, this development is still at the theoretical stage. The ML classifier is simple, yet robust enough to accommodate modifications.

9.5 References

Barnsley, M. J. and Barr, S. L., 1997, Distinguishing urban land use categories in fine spatial resolution land cover data using a graph-based, structural pattern recognition system. *Computers, Environment and Urban Systems*, 21, 209–26.

Berthod, M., Kato, Z. Z., Yu, S. and Zerubia, L., 1996, Bayesian image classification using Markov random fields. *Image and Vision Computing*, 14, 285–95.

Besag, J., 1986, Towards Bayesian image analysis. *Journal of the Royal Statistical Society*, 48, 259–302.

Conese, C. and Maselli, F., 1992, Use of error matrices to improve area estimates with maximum likelihood classification procedure. *Remote Sensing of Environment*, 40, 113–24.

Fisher, P., 1997, The pixel: A snare and a delusion. *International Journal of Remote Sensing*, 18, 679–85.

Foody, G. M., Campbell, N. A., Trodd, N. M. and Wood, T. F., 1992, Derivation and applications of probabilistic measures of class membership from the maximum-likelihood classification. *Photogrammetric Engineering and Remote Sensing*, 58, 1335–41.

Forster, B. C., 1985, An examination of some problems and solutions in monitoring urban areas from satellite platforms. *International Journal of Remote Sensing*, 6, 139–51.

Fukunaga, K. and Hummels, D., 1987, Bayes error estimation using Parzen and k-NN procedures. *IEEE Transactions on Pattern Analysis and Machine Intelligence*, 9, 634–43.

Haack, B., Guptill, S., Holz, R., Jampoler, S., Jensen, J. R. and Welch, R., 1997, Urban analysis and planning. In W. Philipson (ed.) *Manual of Photographic Interpretation*. (Bethesda, MD: American Society for Photogrammetry and Remote Sensing), pp. 517–53.

Haralick, R. M. and Fu, K., 1983, Pattern recognition and classification. In R. N. Colwell (ed.) *Manual of Remote Sensing* (Falls Church, VA: American Society for Photogrammetry and Remote Sensing).

Haralick, R. M., Shanmugam, K. and Dinstein, I., 1973, Textural features for image classification. *IEEE Transactions on Systems, Man, and Cybernetics*, SMC-3, 610–21.

Harris, P. M. and Ventura, S. J., 1995, The integration of geographic data with remotely sensed imagery to improve classification in an urban area. *Photogrammetric Engineering and Remote Sensing*, 61, 993–8.

Hutchinson, C. F., 1982, Techniques for combining Landsat and ancillary data for digital classification improvement. *Photogrammetric Engineering and Remote Sensing*, 48, 123–30.

Iisaka and Hegedus, 1982, Population estimation from Landsat imagery. *Remote Sensing of Environment*, 12, 259–272.

Jensen, J. R. and Cowen, D. C., 1999, Remote sensing of urban/suburban infrastructure and socio-economic attributes. *Photogrammetric Engineering and Remote Sensing*, **65**, 611–22.

Kettig, R. L. and Landgrebe, D. A., 1976, Classification of multispectral data by extraction and classification of homogeneous objects. *IEEE Transactions on Geoscience and Electronics*, **GE-14**, 19–26.

Maselli, F., Conese, C., Petkov, L. and Resti, R., 1992, Inclusion of prior probabilities derived from a nonparametric process into the maximum-likelihood classifier. *Photogrammetric Engineering and Remote Sensing*, **58**, 201–7.

Mather, P. M., 1985, A computationally-efficient maximum likelihood classifier employing prior probabilities for remotely sensed data. *International Journal of Remote Sensing*, **6**, 369–76.

Mesev, V., 1997, Remote sensing of urban systems: hierarchical integration with GIS. *Computers, Environment and Urban Systems*, **21**, 175–187.

Mesev, V., 1998, The use of census data in urban image classification. *Photogrammetric Engineering and Remote Sensing*, **64**, 431–8.

Mesev, V., Gorte, B. and Longley, P. A., 2001, Modified maximum-likelihood classification algorithms and their application to urban remote sensing. In J.-P. Donnay, M. J. Barnsley and P. A. Longley (eds) *Remote Sensing and Urban Analysis* (London: Taylor & Francis), pp. 72–94.

Skidmore, A. K. and Turner, B. J., 1988, Forest mapping accuracies are improved using a supervised nonparametric classifier with SPOT data. *Photogrammetric Engineering and Remote Sensing*, **54**, 1415–21.

Steele, B. M., Winne, J. C. and Redmond, R. L., 1998, Estimation and mapping of misclassification probabilities for thematic land cover maps. *Remote Sensing of Environment*, **10**, 192–202.

Strahler, A. H., 1980, The use of prior probabilities in maximum likelihood classification of remotely sensed data. *Remote Sensing of Environment*, **10**, 135–63.

Swain, P. H. and Davis, S. M., 1978, *Remote Sensing: The Quantitative Approach* (New York: McGraw-Hill).

Therrien, C. W., 1989, *Decisions, Estimation and Classification* (Chichester, UK: John Wiley & Sons).

Thomas, I. L., Benning, V. M. and Ching, N. P., 1987, *Classification of Remotely Sensed Images* (Bristol, PA: IOP).

Tom, C. H. and Miller, L. D., 1984, An automated land use mapping comparison of the Bayesian maximum likelihood and linear discriminant analysis algorithms. *Photogrammetric Engineering and Remote Sensing*, **50**, 193–207.

10 Population mapping by geodemographics and digital imagery

Richard Harris

10.1 Introduction

Urban image classification is an inferential and reductionist science. Satellite and airborne sensors gather enormous quantities of (digital) data, measuring radiation at one or more divisions of the electromagnetic spectrum, subject to atmospheric conditions or other contingencies. Those data are then analysed and sorted using a toolbox of (frequently aspatial) analytical techniques into a much smaller number of land cover classes. Often, however, it is not really land cover or urban form that is of interest to the end-user but land use or urban function. Problematically, land cover classes are not equivalent to land use because use serves social purpose and this rarely directly maps to taxonomies of physical measurement. Land cover has been described as first-order information; land use as second-order, inferred from the first (Barr and Barnsley 1999). Even supposing there was a direct and one-to-one correspondence between land cover and land use regarding urban form and urban function, knowledge about land usage stills casts nothing more than a penumbra on the most important component of any urban area: the people and their socio-economic and behavioural character-istics. Yet, despite these apparent deficiencies, urban remote sensing (RS) has been shown to be a valuable aid for urban analysis and planning, and its usage is expected to increase (Donnay *et al.* 2001).

Although the land cover/land use differential is widely recognized in RS, it is rarely extended to emphasize the differences between land uses and land users. Arguably the latter fall outside the immediate remit of urban RS and image classification. However, to move beyond partial or myopic under-standings of urban systems requires greater integration of information about both physical and social environments: about land cover, land use, land users and their interactions. In this vein there has been a drive toward RS-GIS (Geographical Information System) integration and a sense of "best of both worlds" (see Donnay *et al.* 2001). Progress has also been made in developing explicitly spatial techniques that more directly infer land use from high spatial resolution imagery (by pattern recognition, for example:

Barnsley *et al.*, Chapter 4 in this book). Notwithstanding these developments, Longley and Mesev (2001: 164) raise a caveat:

> [although] remote sensing has always been a data-rich subject [...]
> irrespective of increases in satellite precision, improvements in the
> detection of built form will provide only limited perspective upon the
> functioning of urban settlements. By contrast, socio-economic informa-
> tion provides an imprecise representation of form [...] [but] tells us
> rather more about the activity patterns which characterise the function-
> ing of settlements.

The quintessential reason that socio-economic information gives only
imprecise representation of form is because the data are aggregated into
coarse and often arbitrary zones (Openshaw 1985, but see plans for the
2001 UK Census data dissemination: Martin 1998). While developments in
RS technology allow panchromatic images to be produced at one metre
precision with the possibility of update at near-weekly intervals, no
commensurate improvements have been made to the spatial and temporal
resolution of the various sources of official, socio-economic data in the UK,
though analogous situations also exist in other countries. This is partly
because direct sources of personal, socio-economic and behavioural information are subject to more stringent data protection legislation than
remote observation (Curry 1998).

The urban analyst faces a difficult choice. On the one side there are
extensive, up-to-date and "scientific" sources of data that are qualitatively
shallow and offer only indirect information on human populations. On the
other there are more direct sources of socio-economic (areal) information
that are coarsely aggregated, infrequently updated and only describe the
characteristics of areas on average. It is extremely unusual for socio-
economic data sets to include, in addition to the average, further information about population diversity (or variance). This situation tends to force
the erroneous assumption of population homogeneity within areas.
Consequently, localities can be labelled as wholly deprived or entirely
affluent, despite this totality being in apparent contradiction to the
fragmentary nature of (post-) modern city populations at subneighbourhood
scales (Soja 2000).

The trend of increasing precision is not limited solely to raster-based data
products. Numerous sources of predominantly commercial, vector "framework data" (Rhind 1997) have emerged in recent years, such as the Address-
Point and Code-Point products from the Ordnance Survey (OS) of Great
Britain. These offer precision of 0.1 m and 1 m, respectively. Harris and
Longley (2000) combine these data sets with a SPOT (Système Pour
l'Observation de la Terre) image, modelling residential densities and
identifying high-rise buildings in the city and county of Bristol, England
(see also Mesev and Longley 1999; Longley and Mesev 2001). The analysis

is limited, however, by the amount of socio-economic information that can be inferred from the two OS products. Both are really intended to provide only a geographical register of postal mail delivery points (at a property address or unit postcode level, respectively: see Harris and Longley 2000). Yet, as such, they provide the framework that can be used to geo-reference and model alternative sources of data that share a postal address field.

The objective of this chapter is to chart a symbiotic relationship between urban geodemographic analysis and image classification. Urban geodemographics is defined in a broad sense as "the analysis of socio-economic and behavioural data to investigate the geographical (or areal) patterns that both structure and are structured by urban forms and functions". The underlying logic is that articulated by Harris and Longley (2000): that RS brings a potentially strong, scientific framework for socio-economic analysis of urban form and monitoring urban change, but that potential will only be realised fully when urban RS is integrated with new sources of socio-economic data. The eventual aim is to increase understanding of the complex interplay between land cover, land use and land users. This chapter makes some tentative steps in that direction, emphasizing some simple techniques and benefits that arise from movement towards RS-GIS integration. Code-Point, a commercial source of "lifestyles" data and a simple image classification are treated as the primary case studies for a study region of Bristol, England. The region was chosen to facilitate comparison with earlier research described by Mesev (1998), Longley and Harris (1999), Harris and Longley (2000), Longley and Mesev (2000) and Harris and Longley (2001). This chapter is largely an extension of the work described in the first and third of those references, with an increased emphasis on "unconventional" but disaggregate sources of socio-economic data that come closest to matching the spatial and temporal precisions achieved by modern RS platforms. Such data, while often highly detailed and spatially extensive (compared with official sources of socio-economic data, at least), are primarily intended for direct marketing, not for spatial analysis *per se*. Nevertheless, this chapter follows and elaborates on Longley and Harris's (1999) suggestion that such data offer a viable extension to more traditional, socio-economic population mappings (for other recent representations of geographical distributions in urban areas (see Coombes and Raybould 2000; Martin *et al.* 2000; Thurstain-Goodwin and Unwin 2000; Baudot 2001; Donnay and Unwin 2001; Unwin and Fisher 2001; Weber 2001).

10.2 Urban geodemographics

A typical example of (census-based) geodemographic analysis is given by a retail company that takes their client list and stratifies it into different types of consumer. The grouping begins by linking the address of each client to a predetermined classification of the type of area that address is found in. As a consequence, the clients are divided into groups not on the basis of their own, individual characteristics *per se* but according to some sort of social average for the area in which they live. The area classification would be purchased from a third-party data vendor that has produced a statistical amalgam of small area, census data for the n output areas in the country concerned (e.g., $n \approx 150,000$ for 1991 census enumeration districts in England and Wales). The geodemographic classification is produced by the data vendor grouping, on a like-with-like basis, the n areas into a much smaller number of k classes. Commonly k is in the range from about 10^1 to 10^2, depending on the application. The retail company completes its analysis by comparing the proportion of their clients in each of the k classes with the corresponding proportions for all consumers within the firm's catchment area (or some other suitable measure). The comparison allows the retail company to infer information about their core customer and market accordingly (see Birkin 1995).

There is nothing exclusively urban about the procedure described above. However, the origins of geodemographic analysis do lie with both early twentieth century mappings of poverty in Victorian London (Booth 1902–3) and also the models of urban morphology developed by what is now known as the Chicago School of human ecologists (see, for example, Park and Burgess 1925). Geodemographics has been defined as "the analysis of people by where they live" (Sleight 1997: 16). More correctly, it is the analysis of people based on a classification that defines the type of area in which they live. There are similarities between conventional, geodemographic classifications and measures of deprivation for policy research, for example. Although the former favours cluster analysis and area classification on a nominal/ordinal scale and the latter favours regression analysis and classification on an interval/ratio scale, the result is still comparable: a single value is ascribed to an area, and by inference to its population. Image classification is frequently undertaken in a similar manner: a finite but large number of small areas (pixels) are reduced into a much smaller number of (usually) mutually exclusive land cover or land use classes. Again, there are differences in technique (though geodemographic classifications have been developed using techniques more familiar to RS practitioners: specifically probabilistic methods and neural classifiers – Flowerdew and Leventhal 1998; Openshaw and Wymer 1995), but both result in area-based taxonomies.

Note how conventional geodemographics gives primacy to areas as the starting point for analysis. In the example of the retail company the area classification was built first (by a third party) and then individual client data were appended for analysis. This is a top-down approach that uses the type of area, as defined by the classification, as a surrogate variable for explaining individual preferences, characteristics and behaviours. By coupling the sorts of data looked at in Sections 10.3.2 and 10.4.2 of this chapter with new techniques of geocomputation (Longley *et al.* 1998;

Openshaw and Abrahart 2000) it is possible to reverse the methodology, building up from individual data in a way that can lead to more meaningful zone designs and area delimitation (Harris 2001). In other words, it is the aggregate or areal patterns that are explained by reference to the individual data, not vice versa. The advantages of adopting a bottom-up and more geographically sensitive approach are in some ways analogous to those achieved by moving beyond aspatial, pixel-by-pixel classifications in RS and adopting more geographically considered techniques.

10.3 Using image classification and other data to improve geodemographic representation

Notwithstanding the different sources of data (satellite and aerial photography vs. social survey and census), a main contrast between image and geodemographic classification is the emphasis: the former is on landscape, the latter on land users (people). All settlements are made up of both so, in principle, the two approaches might be regarded as complementary. Figure 10.1a shows a conventional, census-based and geodemographic mapping of the population within the Bristol study region (see inset to

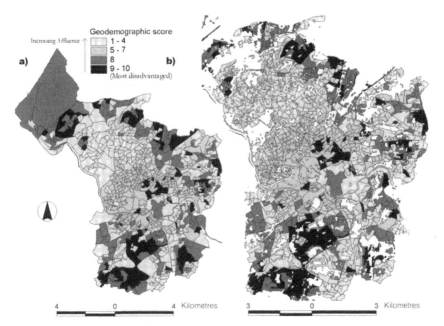

Figure 10.1 Geodemographic-based population maps of Bristol, England: (a) with 1991 Census zone enumeration district boundary outlines; (b) with improved boundary representation: non-residential areas demarcated (see Section 3.1).

Figure 10.4 to locate Bristol within England). The classification shown is the post-1991 UK Census Super Profiles typology for enumeration districts at the level of aggregation where $k = 10$ (plus an eleventh group for "residual" areas) (Batey and Brown 1995). The classes are arranged broadly along an ordinal scale from "Affluent Achievers" (Group 1) to "Have-Nots" (Group 10). In Figure 10.1a the census zones have been sorted into quartile groups based on their geodemographic classification. What is shown is essentially a choropleth map of average wealth for the census areas of Bristol (Voas and Williamson 2001). There are three well-known limitations to maps such as Figure 10.1a:

- First, it is implied that populations are uniformly distributed geographically within each zone, or, rather, that all census areas are fully occupied by residents (there are apparently no non-residential spaces within zones).
- Second, it is also implied that populations are socio-economically uniform within each areal division. "One size fits all" where all residents within an area are assigned the same label.
- Third, the maps suggest that all change (in geodemographic condition) occurs only at the borders between zones.

10.3.1   *Improving the boundary representation*

In response to the first limitation, a simple modification to Figure 10.1a is made by using a standard GIS function to overlay a classified RS image upon the census-based mapping. The land cover theme, intended for demonstrative purposes only, is derived from the Cities Revealed set of high spatial resolution, aerial photography. The theme has two broad classes: built and not built (Figure 10.2). In effect, the RS classification is used as a template to cut out from the original mapping those areas indicated by the RS data to be non-residential. The result is shown in Figure 10.1b and gives a more realistic, planar view of the residential extent of Bristol. It remains far from perfect, however, as witnessed by the M32 motorway appearing from the centre to the north-east of the image (while there is considerable traffic congestion in and out of Bristol during peak hours it is not, as yet, so great that drivers spend their entire day residing on a motorway!). The cartographic product can be refined further by arguing that as the distance from the centre of a residential unit postcode increases so does the likelihood that the apparently built-up/residential areas are, in fact, misclassified (Figure 10.3). The M32 motorway is now correctly identified as unlikely to be residential.

In the UK, unit postcodes define small groups of neighbouring mail delivery points, either residential or commercial properties. Residential postcodes typically comprise between twelve and sixteen dwellings on average (Shepherd and Ming 1998). Unit postcodes are really lists of

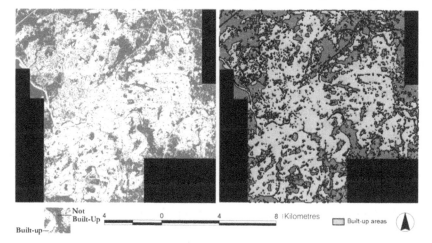

Figure 10.2 Cities Revealed aerial photograph classified into "built" and "not built" land classes.

Figure 10.3 A Code-Point-based model used to determine which areas are residential within the built/not built image classification.

properties and have no certain or absolute geographical boundaries. However, there are two sources of data available that define a geographical centroid for unit postcodes. The first is the Postzon file which has a 100-m resolution in England and Wales (10 m in Scotland) and is accessible to registered users at Manchester Information and Associated Services (www.mimas.ac.uk). The second, more accurate and more precise, is the Code-Point product which is 1-m precise in the majority of cases (and is described in detail by Harris and Longley 2000). Applying Code-Point data to calculate distances from the centre of residential postcodes produces a simple measure of (un)certainty in using the built-up/not built-up theme to identify residential areas. Accurate measures of residential land areas are important. They are the basis (specifically, the denominator) for calculating residential population densities, an increasingly used measure of urban sustainability (Jenks *et al.* 1996; Harris and Longley 2000; Rogers and Power 2000).

10.3.2   *Improving the representation of population heterogeneity*

The RS-derived changes to the geodemographic mapping may produce more realistic delimitation of the residential extent of Bristol. They do not, however, resolve the second or third limitations of the original map: Figure 10.1b still implies socio-economically uniform populations and only changes at the boundaries of the administrative zones. It is possible, of course, to remodel census or other zonal geographies by using various surface modelling techniques to create near-continuous representation of (changing) population density across a study region (Bracken and Martin 1989; Donnay and Unwin 2001; Tate 2000; Atkinson 2002). Often this is the only way to "disaggregate" coarse zone-based data given the confidentiality strictures of governmental data dissemination that preclude direct access to the original source data (i.e., the "raw" data prior to aggregation). In contrast, private commerce and retailing increasingly have moved toward "one-to-one" and direct marketing, requiring more detailed knowledge about the consumer behaviours and preferences of individuals and/or their households. The types of data collected to provide such knowledge are often grouped under the heading of "lifestyles" information. These sorts of non-aggregate data can be used as the basis for measuring diversity and socio-economic variations within administrative zones, although having access to individual or household-level data sets raises issues of confidentiality and protection of a survey respondent's right to privacy.

There are two main limitations to using lifestyle data for socio-economic research. The first is one of access and the possibly prohibitive cost of acquiring the data other than through the goodwill of data vendors. The second is ascertaining the quality of the data. It is known that various sample and response biases can affect composite lifestyle data sets. An example is the collection of lifestyle information through postal shopping surveys that tend to preclude young (male) adults but otherwise leave it as the recipient's choice whether to return the questionnaire or not: the respondents are broadly self-selecting (Longley and Harris 1999). What is more difficult to quantify is whether these biases have any real effect on the

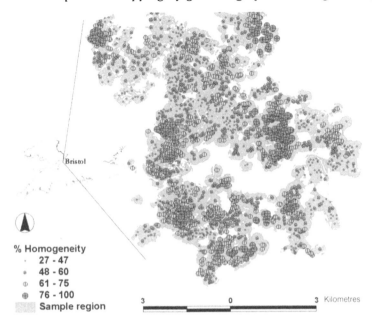

% Homogeneity
- 27 - 47
- 48 - 60
- 61 - 75
- 76 - 100
- Sample region

Figure 10.4 Indicators of housing stock used to measure small-area (percentage) homogeneity. (Inset reproduced from Ordnance Survey map data by permission of Ordnance Survey, Crown Copyright 2001.)

data quality, whether the quality varies from place to place and so forth. This is because there is no real benchmark to gauge the data against, other than at a coarsely aggregate level (coarse both spatially and temporally). Despite these difficulties, research has suggested that geographically sensitive aggregation of lifestyle data can produce results that at least appear consistent with information obtained from other data sources (Frost and Harris 2001).

One way of handling lifestyle data to improve the robustness of calculations based on them is to consider each record, each individual or household within its immediate, geographical context, rather than in isolation or as an atomistic entity. For example, it is a relatively simple procedure to find the n_i nearest neighbours around a point i and group them together to form a neighbourhood cluster with class size of m where:

$$m = p_{0i} + p_{1i} + p_{2i} + \cdots + p_{ni} \tag{10.1}$$

and p_{0i} is the population at point i, p_{1i} is the population at the (first) nearest neighbour to i, p_{2i} is the population at the second nearest neighbour and so forth to the nth neighbour. Figure 10.4 was derived using such a procedure, grouping each of 2,703 lifestyle records for a sample area within the Bristol region into neighbourhood clusters where $m = 50$. Note that those clusters overlap, with most households belonging to more than one group. Having grouped the households it is possible to calculate, in this case, the most frequently occurring property type within the cluster and the proportion of households residing in that type of property. In Figure 10.4 that proportion is assigned back to each point i (around which the cluster was formed).

For reasons of data privacy the precise geo-reference of each household is not actually known for this analysis of the lifestyle data set; instead, the household can be located at a unit postcode level. Consequently, prior to clustering, the 2,703 household records were grouped to 1,936 Code-Points. Every point i in Figure 10.4 therefore represents the centroid location of one of 1,936 residential postcodes within Bristol. The analytical procedure creates as many clusters as there are Code-Points since the search procedure is undertaken for each point in turn. The Code-Point data suggest that within the 1,936 postcodes there is a total of 43,884 residential properties. This implies that the 2,703 lifestyle records constitute a 6 per cent sample of residential properties within the trial region. The number of postcode centroids in each group n will vary from cluster to cluster as it is m (the class size) that is held constant. The value of n is governed by both the variable size of postcode populations (number of residential delivery points per unit postcode) and the local sample and response rates for the national shopping survey from which the lifestyle information was sourced. For the 1,936 neighbourhood clusters the mean value of the furthest (nth) point from i is 422 m, with a minimum value from i to n of 196 m and a maximum of 1,153 m. The mean distance from i to the nth/2 member of each cluster (i.e., the member at the median distance) is 278 m but ranges from 107 m to 918 m across the set of 1,936 clusters. What Figure 10.4 shows is a simple measure of small-area homogeneity, as determined from the housing stock, within the immediate vicinity of each Code-Point centroid. The geographical patterns revealed from the bottom-up analysis are not constrained to fit a predetermined zoning scheme primarily designed for other administrative purposes. It is likely that other lifestyle variables, in particular household income, offer better indicators of geodemographic diversity at fine scales. The reason for choosing property type here will become evident in Section 10.4.2.

10.4   Using geodemographic mappings for image classification

In Section 10.3.1 it was shown how a simple RS-based classification of built and not built-up land areas could be used to improve the geodemographic representation of urban areas in Bristol. In part this was a rather routine observation that rested on using only the most elementary functions of a typical GIS. Yet, if the simplicity of the procedure is accepted then there can be little excuse for not modifying existing (vector) boundary files to produce more convincing representations of urban forms. With increased attention being given in the UK to rapid and easy access to the wide range of "neighbourhood statistics" produced for standard administrative zones such as political wards (www.statistics.gov.uk), it is important to consider also how such data can be mapped in ways that are relatively easy to undertake but also produce a convincing and meaningful cartographic end product (see also Coombes and Raybould 2000).

Can the argument be inverted? Can geodemographic approaches in the broadest sense be applied in relatively simple ways to inform and improve methods for image classification? A precedent was set by Mesev (1998) who outlined a threefold strategy for incorporating property data from the 1991 UK Census of Population within supervised classifications of RS imagery. The idea, first, is to determine the extent of the built environment using satellite observation but guided by census mappings to select training samples. The training samples are chosen on the basis that the population-weighted centroids attached to UK small area census data are most likely to be at the centre of the greatest concentration of residential land use with each census zone.

Second, the census data are also used to modify the prior probabilities attached to each residential density category within the Maximum Likelihood (ML) image classifier. Put simply, if the census data suggest that a greater proportion of the land area ought to be classified as "terraced" as opposed to, say, "semi-detached", then the former land class is given an increased likelihood of being assigned to a pixel during the classification. A problem with the procedure is that it tends to conflate the differences between land cover and land use, treating the mutually exclusive property types as though they are also mutually exclusive in terms of their spectral signature, which they are not (Harris and Longley 2001). Better results are achieved fitting the prior probabilities "backward" by iteration until the proportion of the land area (number of pixels) assigned to each density category reaches a best fit with that suggested by the census to be correct (Mesev *et al.* 2001).

The third stage in Mesev's (1998) strategy is to remodel the polygonal census boundaries and data into near-continuous raster surfaces using the algorithms developed by Bracken and Martin (1989) (see also Martin and Bracken 1991). These surfaces are then used to help validate the results of the image classification. They offer a marked improvement over traditional choropleth maps by giving a more attuned differentiation of residential from non-residential land use. They also avoid implying that all change occurs only at the boundaries of administrative zones. To summarize, Mesev's (1998) strategy for mapping population densities is to use:

- census centroids to direct the selection of training samples;
- census data to modify Bayesian probabilities within a ML classifier;
- census surfaces to validate the results of the classification.

10.4.1   *From census to code-point*

Unfortunately, a major limitation to Mesev's strategy is the data source. Even when remodelled to provide better discrimination of residential and non-residential areas, the census data permit at best a raster grid size of 200 m × 200 m. This is hardly commensurable with the new sources of submetre precision RS imagery! The assumption that each zone centroid is a "high information point" with the greatest mass of population and thus the best place to select training samples is also problematic: the points were determined by eye at the census agency with a resolution of about 10 m in urban areas. To improve the precision, Harris and Longley (2000) modified Mesev's strategy to incorporate OS Code-Point data and reduced the raster surface of residential land area to grid dimensions of 50 × 50 m. Using Code-Point data also has the advantage that the centroid location (high information point) for unit postcodes ought always to be at the location of a residential property. This is because the centroid listed in the Code-Point file is defined as the "Address-Point co-ordinate of the nearest delivery point to the calculated mean position of the delivery points in the unit" (Ordnance Survey 1997). The Address-Point file gives a national grid reference for each mail delivery point in the UK, to a precision of 0.1 m in the majority of cases (Ordnance Survey 1996). The Code-Point centroid does not necessarily define the geographical centre of a postcode. More usefully, it actually defines the location of the residential property that is nearest to the centre. As a consequence Code-Point coordinates can be used as the basis for collecting training samples of properties. The same is not true of census centroids, which can be located within the middle of streets, gardens, parkland or so forth.

It was stated, however, that the Code-Point dataset is a geographical register of mail delivery points and, as such, provides no direct information on the type of property found at each postcode centroid. To some extent this information can be inferred by looking at the number of residential delivery points (the density) per unit postcode: a high number suggests a postcode where apartments are dominant; a low number suggests detached properties (see also Bibby and Shepherd 1999). Alternatively, where the Address-Point data are available it is possible to search on the address field for clues about the type of properties present: Flat 15 or Royal York Terrace, for example, provides strong circumstantial evidence! Unfortunately, in most cases neither the Address-Point nor Code-Point data will be so revealing. A third option is to incorporate the lifestyle information discussed in Section 10.3.2 as part of the selection strategy.

10.4.2   *Using lifestyle data for image classification*

From an image classification perspective, an interesting feature of some lifestyle data sets is that they contain detailed information on, for example, property types, including when they were built and what modifications have been made to them. Moreover, the data are often sourced from annual and national shopping surveys, so what is (in principle) available is an up-to-date and national sample of properties nationwide. From a social analytical perspective there may well be problems with the data, with, perhaps, certain socio-economic groups over- or underrepresented in the sample (see Longley and Harris 1999). However, such problems do not preclude using the lifestyle information to identify (Code-) points where, say, terraced properties built before 1900 are found, and then using this information as the basis for selecting training samples for image classification (see also Harris and Longley 2001). The argument for doing so is basically the same as that made for using Code-Point instead of census centroids as the basis for directing the selection of training samples, except by now appending the lifestyles data to the Code-Point file it is known not only that there is a residential property at the point, but also what sort of property it is likely to be. The lifestyle data can be appended to the Code-Point file by virtue of the unit postcode which can be used as the common attribute or join field. (In fact, the full postal address of each survey respondent is known to the lifestyle vendor but was degraded to the unit postcode level for reasons of data protection. In principle, however, it is possible to achieve greater precision and accuracy by exactly matching each lifestyle record to a unique Address-Point. Unless the respondent has lied it is then known, with certainty, what sort of property is found at each point, how old it is and so forth.)

The lifestyle data can also be used at the post-classification stage to validate the results of an image analysis. To recall, Figure 10.4 showed a simple measure of the uniformity of property types around and inclusive of a selection of Code-Points within the Bristol study region. In order to derive this measure it was first necessary to identify which property type was dominant within each neighbourhood cluster and thence calculate its share as a proportion of all property types within the group. Rather than showing the calculations for all property types as in Figure 10.4, it is possible to isolate any one type, showing the areas where it likely dominates and to what extent. In Figure 10.5, semi-detached properties have been selected.

Like the mapping shown in Figure 10.3, Figure 10.5 permits an external validation of an RS image; in this case one that is classified into dwelling-type categories. Despite the presumed vagaries of the lifestyle data, we can still reasonable assume that where a high proportion of semi-detached properties is indicated in each neighbourhood cluster then this does indeed reflect the actual housing stock on the ground. It would be surprising if an image stratified into classes of dwelling type did not reflect the geographical patterning suggested by the socio-economic data and would, in such a circumstance, suggest the classification to be errant. The logic suggests a way of extending Mesev's (1998) schema not only to validate an image classified into a built/not-built dichotomy using external, socio-economic data, but also a more extensive typology divided into dwelling

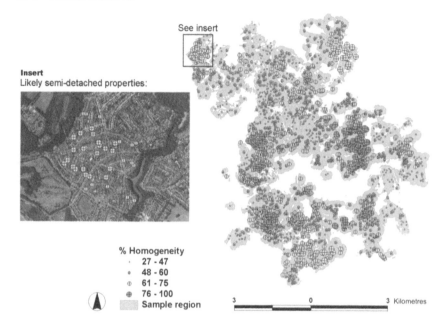

Figure 10.5 Combining the lifestyles and Code-Point population maps to select, from the Cities Revealed image, training samples with high likelihood of being semi-detached properties.

types or density classes. Such a procedure might begin to avoid the somewhat self-referential and tautological, internal validation exercises that characterize some urban RS research in the interest of expediency and the absence of other ways of confirming the classification accuracy.

To reiterate, Mesev's (1998) paper advocated the use of ancillary census data in urban image classification and outlined a strategy to do so. Since that time increased spatial and temporal precisions in the supply of RS imagery have not been met by accordant improvements in official or semi-official sources of socio-economic data within the UK. There is now a marked discrepancy in the resolution of remotely sensed and social survey data. As a consequence, though the logic of Mesev's strategy remains valid, its dependence on census data renders its original formulation somewhat anachronistic. Fortunately, new sources of "unconventional" geodemographic data are presenting new opportunities to update the basic premise and further the integration of RS-GIS. Following Mesev (1998) and Harris and Longley (2000) the following strategy for urban image classification (in the UK) is suggested, incorporating both geodemographic information and RS imagery as the basis for mapping population distributions and densities (see also Mesev, Chapter 9 in this book):

- Using Code-Point data and a dichotomous (built/not built) image classification as a first filer to differentiate residential from non-residential land uses (Figures 10.2 and 10.3).
- Using lifestyle data to direct the selection of training samples (Figure 10.5).
- Using census or other reliable survey data to modify, by iteration, the Bayesian probabilities attached to each land use class within an ML classifier (Mesev *et al.* 2001). Lifestyle data are less suited to this task since the samples are not necessarily representative of the population at large.
- Using bottom-up, lifestyle-based population mappings as an aid to validating the image classification.

10.5 Conclusions

As this chapter has progressed so the population mapping procedures discussed have become increasingly "data-rich". We commenced with a standard census boundary file and a simple, RS-derived classification. It was suggested that an overlay of the latter on the former could be used to improve the representative ability of the census boundary file to describe the urban extent of Bristol. Subsequently, Code-Point data were appended to firmly establish which built-up areas were residential (and, by extension, which were predominantly commercial). It was shown that lifestyle data can be used to guide the selection of training samples for image classification and that this might yield more accurate classification of residential areas into high- or low-density population classes. Finally, it was also suggested that bottom-up modelling of the lifestyle data could be used as the basis for validating the image classification. To date, the effectiveness of this strategy has yet to be fully ascertained, though some initial research has been presented by Harris and Longley (2000), and Harris and Longley (2001).

It should be added that RS imagery and social classifications are complementary in ways that do not require any necessary move toward physically integrating the data. Each provides a different perspective on the "layout" and workings of urban systems, and in a qualitative sense that alone may be sufficient. A simple example is given by the commercial website multimap.com. There it is possible to enter a unit postcode and immediately retrieve an OS-based map, a geodemographic classification of the postcode and an aerial view of the immediate area. Having the aerial image alongside the geodemographic classification is an extremely useful aid to contextualizing or understanding (literally, visualizing) the social data. In fact, vendors of geodemographic classifications have often included a typical street scene or pictorial property type to help describe the uniqueness of each geodemographic class (see, for example, Birkin 1995). With the advances in technology and geographical data processing it is now possible to produce an actual (overhead) image of nearly any postcode queried by the user. Such information could provide a useful adjunct to the sorts of neighbourhood data being made publicly available through the websites of national statistics agencies (e.g., the UK's Neighbourhood Statistics Service at www.statistics.-gov.uk/neighbourhood/home.asp).

The subtext to this chapter is that there are benefits to be gained from RS-GIS integration, even if the "integration" is very much at an *ad hoc* or basic level. Unfortunately, greater integration and thence greater understandings of urban environments continue to be held back by the disparate resolutions of RS-based and GIS (i.e., social survey) data. New data frameworks such as the UK's National Digital Framework (NDF) or the US National Spatial Data Infrastructure (NSDI) offer an important step to resolving the "precision gap". Yet, while the NDF may well be a "definitive, consistent and maintained framework for the referencing of geographical information in Great Britain" (Ordnance Survey 2001), it does not, in itself, provide the spatially and temporally precise socio-economic data that are required to begin unlocking the complex workings of urban systems within the context of urban RS.

10.6   Acknowledgements

The census boundary files are Crown Copyright (Economic & Social Research Council/Joint Information Systems Committee (ESRC/JISC) purchase). The lifestyle data are the copyright of Claritas UK (http://www.claritas.co.uk/) and are used with permission. The Code-Point® and Cities Revealed® data were made available as part of the (UK) Natural Environment Research Council Urban Regeneration and the Environment programme, grant GST/02/2241 (urgent.nerc.ac.uk/ and www.casa.ucl.ac.uk/urgent/). Code-Point is Crown Copyright (www.ordsvy.gov.uk). Cities Revealed is the copyright of the GeoInformation Group (www.crworld.co.uk). The initial classification of the Cities Revealed image (Figure 10.2) was undertaken by Alan Steel at the Department of Geography, University of Wales Swansea. The inset to Figure 10.4 was reproduced from Ordnance Survey map data by permission of the Ordnance Survey and is Crown Copyright. The views expressed are those of the author alone. Responsibility for any errors that have arisen from the analysis or discussion of the data sets also lies solely with the author.

10.7   References

Atkinson, P. M., 2002, Surface modelling: What's the point? *Transactions in GIS*, 6, 1–4.

Barr, S. L. and Barnsley, M. J., 1999, A syntactic pattern recognition paradigm for the derivation of second-order thematic information from remotely sensed images. In P. M. Atkinson and N. J. Tate (eds) *Advances in Remote Sensing and GIS Analysis* (Chichester, UK: John Wiley & Sons), pp. 167–84.

Batey, P. and Brown, P., 1995, From human ecology to customer targeting: The evolution of geodemographics. In P. A. Longley and G. Clarke (eds) *GIS for Business and Service Planning* (Cambridge: GeoInformation International), pp. 77–103.

Baudot, Y., 2001, Geographical analysis of the population of fast-growing cities in the third world. In J-P. Donnay, M. J. Barnsley and P. A. Longley (eds) *Remote Sensing and Urban Analysis* (London: Taylor & Francis), pp. 225–41.

Bibby, P. and Shepherd, J., 1999, Monitoring land cover and land use for urban and regional planning. In P. A. Longley, M. F. Goodchild, D. J. Maguire and D. Rhind (eds) *Geographical Information Systems*, 2nd edn (New York: John Wiley & Sons), pp. 953–65.

Birkin, M., 1995, Customer targeting, geodemographics and lifestyle approaches. In P. A. Longley and G. Clarke (eds) *GIS for Business and Service Planning* (Cambridge: GeoInformation International), pp. 104–49.

Booth, C., 1902-3, *Life and Labour of the People of London*, 3rd edn (London: Macmillan).

Bracken, I. and Martin, D., 1989, The generation of spatial population distributions from census centroid data. *Environment and Planning A*, **21**, 537–43.

Coombes, M. and Raybould, S., 2000, Policy-relevant surfaced data on population distribution and characteristics. *Transactions in GIS*, **4**, 319–42.

Curry, M. R., 1998, *Digital Places: Living with Geographic Information Technologies* (London: Routledge).

Donnay, J-P. and Unwin, D., 2001, Modelling geographical distributions in urban areas. In J-P. Donnay, M. J. Barnsley and P. A. Longley (eds) *Remote Sensing and Urban Analysis* (London: Taylor & Francis), pp. 205–24.

Donnay, J-P., Barnsley, M. J. and Longley, P. A. (eds), 2001, *Remote Sensing and Urban Analysis* (London: Taylor & Francis).

Flowerdew, R. and Leventhal, B., 1998, Under the microscope. *New Perspectives*, **18**, 36–8.

Frost, M. and Harris, R., 2001, Indicators of deprivation for policy analysis in GIS: Getting within wards. In D. B. Kidner and G. Higgs (eds) *Proceedings of the GIS Research UK 9th Annual Conference* (Cardiff: University of Glamorgan), pp. 254–5.

Harris, R., 2001, The diversity of diversity: Is there still a place for small area classifications? *Area*, **33**, 329–35.

Harris, R. J. and Longley, P. A., 2000, New data and approaches for urban analysis: Modelling residential densities. *Transactions in GIS*, **4**, 217–34.

Harris, R. J. and Longley, P. A., 2001, Data-rich models of the urban environment: RS, GIS and 'lifestyles'. In P. Halls (ed.) *Innovations in GIS 8: Spatial Information and the Environment* (London: Taylor & Francis), pp. 53–76.

Jenks, M. Burton, E. and Williams, K. (eds), 1996, *The Compact City: A Sustainable Urban Form?* (London: E & FN Spon).

Longley, P. A. and Harris, R. J., 1999, Towards a new digital data infrastructure for urban analysis and modelling. *Environment and Planning B*, **25**, 855–78.

Longley, P. A. and Mesev, V., 2000, On the measurement and generalisation of urban form. *Environment and Planning A*, **32**, 473–88.

Longley, P. A. and Mesev, V., 2001, Measuring urban morphology using remotely sensed imagery. In J-P. Donnay, M. J. Barnsley and P. A. Longley (eds) *Remote Sensing and Urban Analysis* (London: Taylor & Francis), pp. 163–83.

Longley, P. A., Brooks, S. M. and McDonnell, R., 1998, *Geocomputation: A Primer* (Chichester, UK: John Wiley & Sons).

Martin, D., 1998, 2001 Census output areas: From concept to prototype. *Population Trends*, **94**, 19–24.

Martin, D., Tate, N. J. and Langford, M., 2000, Refining population surface models: Experiments with Northern Ireland census data. *Transactions in GIS*, **4**, 343–60.

Mesev, V., 1998, The use of census data in urban image classification. *Photogrammetric Engineering and Remote Sensing*, **64**, 431–8.

Mesev, V. and Longley, P. A., 1999, The rôle of classified imagery in urban spatial analysis. In P. M. Atkinson and N. J. Tate (eds) *Advances in Remote Sensing and GIS Analysis* (Chichester, UK: John Wiley & Sons), pp. 185–206.

Mesev, V., Gorte, B. and Longley, P. A., 2001, Modified maximum-likelihood classification algorithms and their application to urban remote sensing. In J.-P. Donnay, M. J. Barnsley and P. A. Longley (eds) *Remote Sensing and Urban Analysis* (London: Taylor & Francis), pp. 71–94.

Openshaw, S., 1984, *The Modifiable Areal Unit Problem*, Concepts and Techniques in Modern Geography 38 (Norwich, UK: Geo Books).

Openshaw, S. and Abrahart, R., 2000, *GeoComputation* (London: Taylor & Francis).

Openshaw, S. and Wymer, C., 1995, Classifying and regionalizing census data. In S. Openshaw (ed.) *Census Users' Handbook* (Cambridge: GeoInformation International), pp. 239–69.

Ordnance Survey, 1996, *AddressPoint Sample Data User Guide* (Southampton, UK: Ordnance Survey).

Ordnance Survey, 1997, *DataPoint Sample Data User Guide* (Southampton, UK: Ordnance Survey).

Ordnance Survey, 2001, http://www.ordsvy.gov.uk/business/dnf/whatisit.htm (downloaded 22/8/01).

Park, R. E. and Burgess, E. W., 1925, *The City: Suggestions for Investigations of Human Behavior in the Urban Environment* (Chicago: University of Chicago Press).

Rhind, D., 1997, *Framework for the World* (Cambridge: GeoInformation International).

Rogers, R. and Power, A., 2000, *Cities for a Small Country* (London: Faber & Faber).

Shepherd, J. and Ming, D., 1998, *Postcodes into Geography* (London: Geographic Information Services).

Sleight, P., 1997, *Targeting Customers: How to use Geodemographic and Lifestyle Data in Your Business*, 2nd edn (Henley, UK: NTC Publications).

Soja, E. W., 2000, *Postmetropolis: Critical Studies of Cities and Regions* (Oxford: Blackwell).

Tate, N. J., 2000, Surfaces for GIScience. *Transactions in GIS*, **4**, 301–3.

Thurstain-Goodwin, M. and Unwin, D., 2000, Defining and delineating the central areas of towns for statistical monitoring using continuous surface representations. *Transactions in GIS*, **4**, 305–17.

Unwin, D. and Fisher, P., 2001, *Virtual Reality in Geography* (London: Taylor & Francis).

Voas, D. and Williamson, P., 2001, The diversity of diversity: A critique of geodemographic classification. *Area*, **33**, 63–76.

Weber, C., 2001, Urban agglomeration delimitation using remotely sensed data. In J.-P. Donnay, M. J. Barnsley and P. A. Longley (eds) *Remote Sensing and Urban Analysis* (London: Taylor & Francis), pp. 145–59.

11 GIS and remote sensing in urban heat islands in the Third World

Janet Nichol

11.1 Introduction

Knowledge of the magnitude and morphology of urban heat islands was greatly advanced by Chandler (1965) working in London, who measured minimum air temperatures across the city, and noted an increase toward the city centre, with a steep temperature gradient at the rural–urban boundary. Due to the relationship between a city's ambient air temperature and other factors of the living environment such as air pollution, wind and levels of human comfort, many subsequent studies have taken place. The usual method of recording urban heat islands is to measure screen-level air temperature by vehicle traverse across the city by night; the main disadvantages being incomplete spatial coverage of such sample data and the lack of time synchronization across the traverse.

The first remote sensing studies of urban heat islands were undertaken by satellites launched during the 1970s including the Television Infrared Observation Satellite (TIROS) (Rao 1972), National Oceanic and Atmospheric Administration's (NOAA) Advanced Very High Resolution Radiometry (AVHRR) (Carlson and Boland 1978) and the Heat Capacity Mapping Mission (HCMM)[1] (Price 1979). Carlson and Boland validated the satellite-derived heat island by demonstrating a close relationship between the conventional heat island measured by screen-level air temperature and the surface temperature measured by remote radiometers. Since heat islands are caused by replacement of natural evaporative and porous land surfaces by non-evaporative man-made surfaces, cities are identifiable on low resolution satellite sensor images for their temperature contrasts, as much as for their optical differences with surrounding rural areas. However, the nature and intensity of the rural–urban temperature difference is dependent on the respective land cover types, which may vary between climatic regions at global scale. Figure 11.1 corresponds to two subscenes of

1 HCMM was an experimental satellite operating in the 10.5–12.5-μm region, having a 500-m spatial resolution, which was launched in 1978 and shut down in 1981.

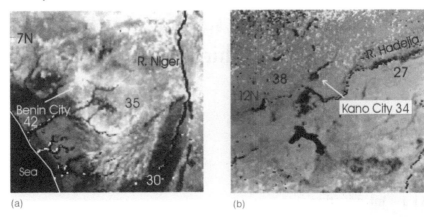

Figure 11.1 Thermal imagery of West Africa from the NOAA AVHRR 1-km land
data set: (a) humid forest zone at 6–7°N, (b) semi-arid savanna zone at
12°N.

the same day-time NOAA AVHRR Channel 4 (10.3–11.3-μm) image of
West Africa during the January 1992 dry season. It suggests that Benin City
in the humid forested zone (Figure 11.2) constitutes a heat island with urban
temperature greater than rural $(T_u > T_r)$ (lighter tone) whereas Kano, in the
savanna zone farther north, constitutes a "heat sink", or cool island
$(T_u < T_r)$ (darker tone).[2]

Present knowledge of urban climate is largely based on studies of Western
cities. Moreover, there are few remote sensing investigations of the thermal
characteristics of tropical and Third World cities. Transferability of knowledge to tropical cities may be compromised due to differences in solar
irradiance resulting from a higher solar zenith angle in the tropics, as well as
differences in building style and materials, and planning layouts and designs.
Jauregui (1984), for example, noted that in Third World tropical cities a
larger proportion of streets are unpaved, there is a relatively small
proportion of green space and the vertical extent of urban canyons may
be shallow. A comparison between North American, European and tropical
cities gave approximate values for heat island intensity $\Delta T(u - r)$ of 8, 6
and 4°C, respectively. The objectives of heat island studies also differ, since
in tropical cities day-time temperatures more often exceed the threshold for
human comfort[3] and a major concern of urban climatology here is to
promote, rather than prevent heat loss. Therefore the phenomena of interest
are those factors that reduce or delay the receipt and absorption of solar

2 Carnaham and Lawson (1990) have noted a heat sink, occurring under unusual climatic
   conditions, for a temperate zone city in the USA.
3 Human comfort is a combination of temperature and humidity that determines human
   response to climate stress (see http://ecep1.usl.edu/ecep/comfort/a/a.htm). IPCC (2001) refers
   to this concept as a heat index.

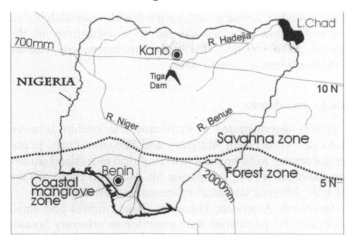

Figure 11.2 The location of Kano in northern Nigeria.

radiation by urban structures, creating a heat sink, frequently observed in tropical cities. The distribution and intensity of such heat sinks in Singapore have been studied in the field, and from thermal satellite sensor images by Nichol (1996a, b). She demonstrated that in Singapore, a humid tropical city near the equator, average day-time air temperatures in high-rise built areas were 2°C cooler than in open suburban areas. Significant control of unwanted heating effects during the day-time has been successfully accomplished here by strategic planning of building height and orientation combined with informed tree planting programmes.

Apart from applications to urban environmental quality, the Inter-Government Panel on Climate Change (IPCC) has now recognized the need to include heat island measurements in climate change models (IPCC 2001). In the rural tropics 85 per cent of incoming solar radiation is used for evapotranspiration (Chang 1965). The dispersion of a much greater proportion of this energy into the urban canopy layer as sensible heat coupled with the generally lower albedo of urban surfaces results in convectional instability within the urban boundary layer with higher precipitation over the city and for some distance downwind. Population growth figures for tropical countries make such changes particularly significant. Notably, the urban populations of sub-Saharan Africa and East Asia and the Pacific more than doubled between 1980 and 1999 (UNEP 2001).

11.2   Methods for studying the urban heat island

Conventional measurements of heat island intensity $\Delta T(u - r)$ consist of night-time observations of screen-level air temperature by vehicle traverse, ideally corresponding to the time of minimum temperature. The difficulty of synchronizing data collected across a city, coupled with unavailability of such data for a wide range of cities has led to investigations of the relationship between $\Delta T(u - r)$ and two other parameters, city size by population and thermal remotely sensed data.

11.2.1 City size by population

Although Oke (1982) demonstrated a significant relationship between population and $\Delta T(u - r)$ in North American and European cities, population alone does not appear to be applicable to empirical heat island studies worldwide. Thus, Jauregui (1984), comparing his data with those of Oke (1982) obtained four different statistical relationships between $\Delta T(u - r)$ and population for North American, European, large tropical and small tropical cities. Additionally, population data given for an arbitrary boundary defining an urban area may be difficult to relate to air temperature in the vicinity of a weather station.

11.2.2 Remotely sensed measurements of surface temperature

The main advantage of remotely sensed thermal data for studying the urban heat island lies in the ability to provide a time-synchronized dense grid of temperature data over a whole city, but there are numerous constraints:

- Since the satellite-derived heat island is based on surface temperature (T_s), the optimum usefulness of these data depends on defining their relationship to a more conventional view of the urban heat island such as is defined by screen-level air temperature (T_a) during either day or night, according to the time of imaging. Although, in general, near-surface climates are known to be intimately related to those of the active surface (Chandler 1965; Price 1979; Carlson and Boland 1981; Goldreich 1985), Roth *et al.* (1989) warn against regarding remotely sensed heat islands as synonymous with those based on air temperature measurements at near ground level. However, Nichol has demonstrated that for the particular conditions of high-rise housing estates in Singapore, T_s is significantly related to T_a. Very low wind regimes near the equator as in Singapore promote a strong relationship between T_s and T_a, making remotely sensed T_s an acceptable proxy for ambient air temperatures (Nichol 1996a, b).
- Many satellite-based sensors only record during the day-time when densely built areas of high building mass, which heat slowly and provide shade, may constitute a heat sink (Nichol 1996a, b; Tso *et al.* 1994). This is also true for tropical arid zone cities (see Figure 11.1b).
- Differences in T_s are largest during the day-time, thus satellite-derived heat islands are more pronounced than those based on air temperature, for which the greatest differences are at night.

- Satellites only record the temperature of horizontal surfaces, which may only approach the complete radiating surface in rural areas. The effective (active) surface area of a city, especially of high-rise areas, is much larger than the equivalent countryside of the same size (Oke 1987; Voogt and Oke 1996). However, in tropical regions this problem is mitigated to some extent by the high Sun angle, which makes building roofs relatively more important than walls in terms of surface energy exchange.

- Timing of the satellite overpass may not be optimal for detecting temperature differences. Thus the Landsat overpass, at 9.15 a.m. to 10.15 a.m. local time is near the thermal crossover time whereas AVHRR at 2.30 p.m. is near the day-time maximum of $\Delta T(u - r)$.

- Satellites may provide an instantaneous view of a whole city, but cannot provide a diurnal profile of temperature changes, which may be required for modelling and visualization of the urban infrastructure in two and three dimensions for input to urban planning strategies. Supplemental data from ground-based studies at intervals during the day is required (e.g., Nichol 1996a, b; Voogt and Oke 1996).

- Although satellite-derived radiance values can readily be converted to equivalent black body temperatures (T_b) using Planck's law (Malaret *et al.* 1985), this underestimates T_s if corrections for emissivity differences according to the type of land cover are not carried out. A clay loam soil of emissivity 0.92 and vegetation of emissivity 0.98, both with a true kinetic temperature of 27°C, will have radiant temperature values of 20.8°C and 25.5°C, respectively.

- These data, which are equivalent to the satellite measured T_s, can only be considered accurate in clear, dry atmospheres, and a further correction using atmospheric data should be made if absolute temperatures are desirable.

- Low spatial resolution constitutes a further source of inaccuracy, particularly in high-spatial-frequency urban terrain, since a low proportion of "pure" pixels are present in the image and the climate of the urban canopy layer may be discontinuous between structures especially during times of solar insolation (i.e., during the day).

If the above constraints can be minimized, remotely sensed data may be used as a surrogate for air temperature, and thus used by urban planners to indicate the need for new or revised urban design or landscaping policies for mitigating the adverse thermal effects of building materials, geometry or high building mass.

11.3   Analysing the urban thermal environment within cities

The IPCC summary document (IPCC 2001) predicts a very likely increase in the heat index over most land areas during the twenty-first century. For cities in relatively developed tropical countries such an increase may be less serious than in Third World cities where urban growth is often unplanned and air conditioning unavailable. For the latter, the adoption of a few simple and inexpensive planning principles may help provide a better living and working environment for a large segment of the world's inhabitants.

11.3.1   Case study of Singapore: a humid tropical city

Singapore was the first developed country on the equator (Figure 11.3). It is a highly urbanized city-state with a number of densely built population and commercial centres in a matrix of semi-urban land cover types. It is therefore difficult to establish a precise background temperature for comparison with values in the city centre, though night-time traverses conducted by the Singapore Meteorological Service (1982) established a heat island magnitude of 5°C for the warm July situation and 2–3°C for the cooler January monsoon period. The particular lifestyle that has evolved in this small and densely populated, high-rise city is dependent on the creation of a comfortable living environment. Temperatures in Singapore exceed the levels of human comfort for 40 per cent of the hours of any one year. Greening campaigns have been adopted since 1967 in order to mitigate the effects of the high-rise lifestyle, but such campaigns are expensive and their effects can only be judged empirically by their direct impact on climate modification, for which data on spatial aspects of the urban heat islands would be of obvious value. An evaluation of Landsat TM (Thematic Mapper) thermal data for microclimate monitoring in nine of Singapore's high-rise housing estates is described below.

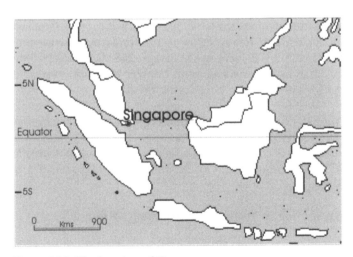

Figure 11.3 The location of Singapore.

11.3.1.1   Image processing to derive surface temperature (T_s)

Conversion of image DN (Digital Number) values to T_s was carried out on a pixel-by-pixel basis. First, radiance values were converted to T_b (black body equivalent temperature) using a quadratic conversion for Landsat 5 (Malaret *et al.* 1985). Second, emissivity correction was carried out according to land cover, of which the most important distinction in the Singapore environment is between *vegetated* and *unvegetated* areas. Unvegetated areas in the study area comprise high-rise apartment blocks and associated hard-surfaced walkways, roads and car parks while the remainder, comprising grassy surfaces bordered by ornamental shrubs and trees, are classed as vegetated. These are easily differentiated using Landsat TM visible wavebands, and a classified vegetation index image was produced for the study.

In this image, pixels representing vegetation were given emissivity values, derived from field measurement, of 0.95 and those representing non-vegetated areas a value of 0.92 (Nichol 1994). T_s was then obtained by ratioing this reclassified (emissivity) image with T_b using Equation (11.1). Thus T_s values for the study area between 30.7°C and 38.5°C were obtained:

$$T_s = T_b/[1 + (\lambda T_b/\alpha) \ln \varepsilon] \tag{11.1}$$

(Artis and Carnahan 1982), where λ = wavelength of emitted radiance, $\alpha = hc/K$ (1.438 ∗ 10 mK), $K =$ Stefan Bolzmann's constant (1.38 ∗ 10/K), $h =$ Planck's constant (6.26 ∗ 10 sec), $c =$ velocity of light (2.998 ∗ 10 m/sec), ε = emissivity.

The equation derives emissivity values from the reclassified NDVI (Normalized Difference Vegetation Index) band, having a 30-m pixel size, and ratios this band with the thermal band, which has a 120-m pixel size. This operation effectively decreases the pixel size of the thermal data from 120 m to the 30-m pixel size of the TM visible wavebands, while correcting for differential emissivity within the 120-m pixels according to vegetation status. Since vegetation is recognized as the main influence on day-time T_s (Carlson and Boland 1981) the method enhances the spatial resolution and spectral accuracy of the data. The correction is not equivalent to increasing the spatial resolution to 30 m since image brightness temperature represents an average over a ground surface of 120 m. However, there is an enhanced spatial and spectral accuracy within each 120-m pixel. The degree of enhancement depends on the amount of green biomass within each pixel and the extent to which T_s is biomass dependent. After the correction, single lines of trees along roadsides became readily discernible as cool corridors (Figures 11.4a and 11.4b).

In the humid tropics image-derived temperatures may be significantly lower than actual, due to absorption by atmospheric water vapour, and if

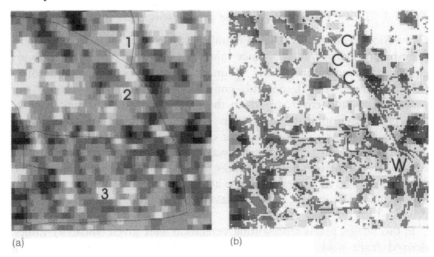

Figure 11.4 (a) Landsat thermal image of black body temperature (T_b) of three
housing estimates: 1 = Ang Moh Kio, 2 = Bishan, 3 = Toa Payoh.
The image is uncorrected for emissivity and the spatial resolution is
120 m. (b) Surface temperature image (T_s) of the same area corrected
for emissivity: pixel size = 30 m. The Cs indicate the cooling effect of a
row of trees and W the high temperature of a road without trees. The
darker the shading the lower the temperature.

absolute values are required a correction must be made. On two Landsat
TM images, May 1989 and March 1991, approximate atmospheric
corrections were made through calibration with a range of marine and
inland water temperatures, which vary by less than 1°C from year to year
(emissivity is assumed to be 1). These suggested that T_s values were
approximately 11°C and 15°C lower than actual ground temperature for
the May 1989 and March 1991 images, respectively, due to absorption of
radiation by the atmosphere. This is likely to be at least as accurate as the
use of radiosonde data, for which accuracies of ±2°C are achievable (Price
1979).

11.3.1.2   *Relationship between T_s and T_a in urban canyons of high-rise housing estates*

Surface and air temperature measurements were taken at half-hourly
intervals between 9.30 a.m. and 4.30 p.m. over a 5-week period during
April to May, for horizontal and vertical surfaces comprising street canyons,
at different elevations up to 43 m above ground (Figure 11.5). A strong
relationship was established between T_s and T_a for both vertical and
horizontal surfaces (Table 11.1).

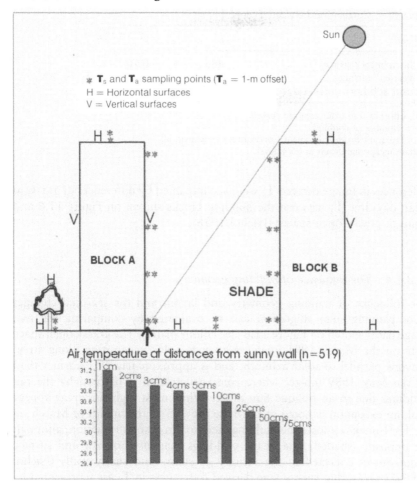

Figure 11.5 Sampling locations for T_a and T_s for horizontal and vertical surfaces within a street canyon (T_a measurements at 1-m offset from T_s), and air temperatures obtained at increasing distances from walls.

11.3.1.3 *The influence of high-rise buildings*

Figure 11.6 (see colour plates) is a road and building plan of Serangoon Central New Town in Singapore, superimposed on processed and classified thermal image data. The Figure shows a remarkable degree of similarity in T_s values between the two image dates, with open spaces around and between high-rise buildings being significantly warmer than the buildings themselves. High-rise blocks over 45-m high at (A, B and C) are cool on both image dates ($T_s = 34$–$35°C$). Three distinct and extensive hot spots (T_s above $36°C$) corresponding to grassy open spaces within the new town built-up area (D, E and F) were observed for both image dates. Such

Table 11.1 Correlation between T_s and T_a*

Surface	$n^\dagger$	$r^\ddagger$
All horizontal surfaces	486	0.70
All vertical surfaces	1,038	0.49
Vertical and horizontal surfaces	1,524	0.52

* T_a offset by 1 m from adjacent surface.
† n = Number of readings.
‡ r = Pearson's Product Moment Correlation Coefficient, all
relationships significant at 95% level.

differences in image-derived T_s were accompanied by differences of 1.5°C in mean day-time T_a between the building blocks shown on Figure 11.6 and adjacent grassy open spaces (Nichol 1998).

11.3.1.4   *The influence of building geometry*

The influence of building geometry and layout and the seasonal changes accompanying solar migration can be evaluated by comparing the two image dates shown on Figure 11.6 (see colour plates). The grassy open space at D on the March 1991 image is open to solar penetration along street canyons parallel to solar azimuth, and is approximately 2°C warmer than on the May 1989 image, where solar penetration is blocked by the tall buildings due to an oblique Sun angle. While most high-rise areas appear cool, an exception is noted at G where the 98° solar azimuth of March on the 1991 image exposes the building faces to direct solar radiation, although the opposite (shaded) side of the buildings remains cool. Air and surface temperatures collected in the vicinity of sunny and shady walls (Nichol 1996a) suggest that such image-derived differences in T_s are represented by mean day-time T_a differences of 1.5°C. The eastern portion of the open space at F is significantly cool compared with the western portion on both image dates. Significantly, at the image time 10.40 a.m., a Sun angle of 51° gives the adjacent buildings 60-m high a shortest shadow length of 50 m; thus almost the length of two image pixels remains in shadow.

11.3.1.5   *High-rise vs. low-rise buildings*

Three extensive areas on the periphery of Serangoon New Town (at H, I and J on Figure 11.6, colour plates) are warmer than areas corresponding to high-rise residential buildings. At H and I these warm patches correspond to older, medium density, low-rise residential estates that happen to be within the New Town administrative boundary. The warm area at J corresponds to a school, having low-rise buildings and extensive playing fields.

11.3.1.6 *Influence of vegetated areas*

Grassy open spaces (D, E and F, colour plates), in the north and south centre of the town, appear consistently warmer than built areas, but vary in the magnitude of this difference. This has been noted in other study sites such as Bishan New Town (Nichol 1996a) and is not always attributable to differences in surrounding building geometry. The relatively cooler T_s at grassy space E could be caused by the condition of the grassy vegetation due to a combination of climatic and human factors (e.g., no recent cutting and high soil moisture), thus promoting greater cooling by latent heat transfer. Figure 11.4b shows how effective the cooling mechanism of a single row of trees along road verges (°C) is, as well as the high temperature of roads without trees (W).

11.3.1.7 *Building height and volume*

In low-rise estates, as at H, I and J on Figure 11.6 (see colour plates), since a greater portion of the active surface is horizontal, more of it is heated, thus giving a warmer mean active surface during the day-time. The short building shadows comprise only a minor portion of a pixel even at a 30-m pixel size. Moreover, smaller buildings have lower thermal inertia; thus day-time heat is less effectively retained. They also form shallow canyons defining the depth of the urban canopy layer. Conversely, in high-rise areas the retention of day-time heat in large buildings and deep canyons would produce a marked night-time heat island. The effects of large building mass in high-rise estates (Tso *et al.* 1994) smooth the peaks in temperature fluctuations and cause a lag effect in heating and cooling of the surface and atmosphere. Thus at the time of the satellite overpass 10.40 a.m., large building masses, as at A, B and C on Figure 11.6 (colour plates), would be expected to be cooler than their surroundings (i.e., act as a heat "sink" relative to low-rise areas), but would exhibit a marked night-time heat island.

11.3.1.8 *The satellite view of the active surface*

Previous studies (Roth *et al.* 1989; Nichol 1994) suggest that since the satellite views proportionally more of the active surface in low-rise areas with shorter building sides, these areas would appear disproportionately hot relative to high-rise areas due to more horizontal surfaces "seen". However, since the present study indicates that horizontal surfaces are more representative of air temperature than vertical surfaces (Table 11.1) this difference of the satellite viewpoint may not be significant in the utilization of satellite-derived T_s as an indicator of T_a.

11.3.1.9   Planning implications

Given the close relationship between T_s and T_a (Section 11.3.1.2), Landsat TM thermal data are able to indicate hot and cool spots within the urban fabric of high-rise housing estates in Singapore, at the level of the city block, even to the extent of detecting shadows cast by high-rise residential buildings, which create cool areas within inner courtyards, and to the non-sunny side of housing blocks oriented at right angles to the Sun direction. Planning objectives in a humid tropical city such as Singapore differ from those in temperate cities in that heat loads always need to be minimized. Methods include consideration of the high Sun angle year-round, causing faster heating of urban canyons, thus a need to minimize solar penetration to street level. The following recommendations arise from the present study:

- Image data indicated a minimum difference in horizontal surface temperatures of 2°C on the images (4°C in the field data) between shaded and unshaded surfaces, which demonstrates the importance of avoiding building alignments at right angles to the dominant east–west solar angle.
- In a humid tropical city such as Singapore, low-rise, medium-density, private housing developments appear to constitute the warmest day-time living environments, indicating the need for more rigid environmental controls in these areas.
- The high Sun angle in low-latitude cities means that short-wave radiation may be lost by direct reflectance due to a high sky view factor. Additionally, the close spacing appears advantageous and ensures the inter-building space is not greater than the shadow length; thus shady canyons are maintained between blocks at most times during the day.
- The data indicate cool areas and, when overlaid on large-scale city maps or air photos showing vegetation distribution, can be used to evaluate the effectiveness of greening campaigns. Using 30-m pixels a tree-lined road appearing as a cool linear feature is potentially a corridor for the dispersal of warm, polluted air masses or may interrupt spatially extensive heat generating surfaces.
- The effectiveness of costly greening campaigns can be evaluated since not all tree-lined roads and grassy areas are equally cool.
- Overlay of the imagery onto maps or air photos in a GIS (Geographical Information System) may permit modelling of the impact of proposed land cover changes in terms of the topological arrangement of warm and cool polygons. Thus the optimum location of green corridors and buffer zones can be suggested.
- Satellite thermal data can contribute to the understanding of the type and spatial arrangement of different surface materials in urban micro-climatology, as well as indicate potential areas of energy interaction

with the overlying boundary layer for modelling of air pollution dispersal.

11.3.2   Case study of Kano, northern Nigeria: a semi-arid zone city

Kano in northern Nigeria is a mediaeval Islamic city situated at 12°N in the semi-arid Sudan savanna zone of West Africa (Figure 11.2). With a population probably approaching two million, it is the largest city in savanna Africa, and due to its expansion at different periods its mixture of settlement styles makes it a useful template for microclimate modelling of African cities. The mediaeval earth-walled city is the most densely populated and built up section and is little changed today. Figure 11.7 is an aerial photograph showing the ancient footpaths through the city wall, used by farmers transporting head or donkey loads to the central market. The inset shows the typical narrow alleyways and houses with walls and roofs constructed from sun-baked laterite comprising the built environment of the old city.

Kano temperatures are hottest during the late dry season (February to May), with air temperatures reaching 40°C, and 36°C in the shade. Figure 11.8a is a Landsat false colour composite showing the city and surrounding farmlands. The mid-dry season thermal images (Figures 11.1b and 11.8b) suggest that the city constitutes a heat sink relative to rural areas, persisting through the day from the 9-a.m. overpass time for Landsat, to 2.30 p.m. for AVHRR. However, it is possible that some satellite-derived heat sinks may be an artefact of lower emissivity values in urban areas giving lower image radiance values (T_b) than the actual surface temperature. For example, emissivity values of 0.92 and 0.98 are often quoted for concrete and forest, respectively (e.g., Jensen 2000). Since considerable temporal and spatial variability can occur in object emissivity over time and space, such standard values may not be accurate and Artis and Carnahan (1982) developed a method for estimating ε for land cover types within a scene, using the ratio between T_b values derived from two image thermal wavebands (Equation 11.2):

$$\varepsilon = \exp[(\alpha T_{b^1} - T_b)/T_b T_{b^1}(\lambda^1 - \lambda)] \qquad (11.2)$$

where λ = wavelength of emitted radiance, $\alpha = hc/K$ (1.438 ∗ 10 mK), K = Stefan Bolzmann's constant (1.38 ∗ 10/K), h = Planck's constant (6.26 ∗ 10 sec), c = velocity of light (2.998 ∗ 10 m/sec), ε = emissivity. [NB. λ may be taken as the central wavelength (e.g., for AVHRR Bands 4 and 5, 10.8 μm and 12 μm).]

The method is approximate since it assumes the emissivity of a pixel is the same for both wavelengths used. Emissivity values for land cover types on the Kano Landsat image were derived according to this method using the 10.3–11.3-μm and 11.5–12.5-μm bands of AVHRR. These values are

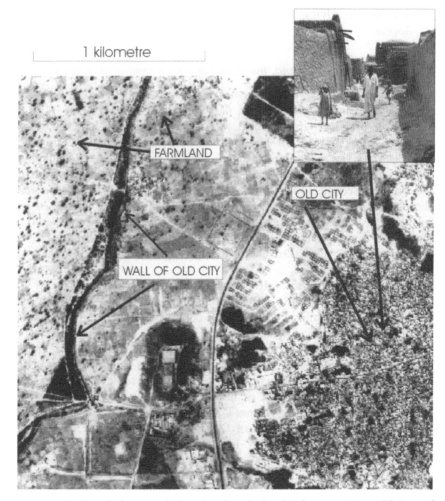

Figure 11.7 Aerial photograph showing the relationship between Kano old city and its rural hinterland.

shown in the bottom row of Table 11.2, though the image itself Figure 11.8b was not corrected. The ε values obtained agree well with empirically derived reflectivity values for land cover types in Nigeria (Oguntoyinbo 1970) where reflectivity is $1-\varepsilon$. After emissivity correction the city would still constitute a heat sink, but $\Delta T(u - r)$ would decrease by approximately 1°C, from −2°C to −1°C. The most obvious effect of the correction is the removal of the anomalously low image values for areas dominated by metal-roofed warehouses (industrial areas on Figure 11.8b). Due to their very low emissivity values these appear much cooler than they really are on this uncorrected image.

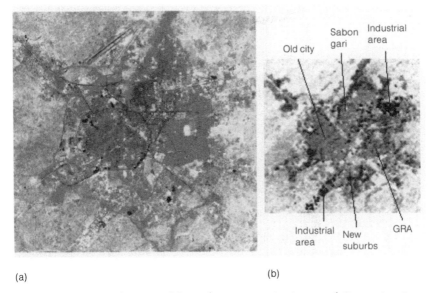

(a)   (b)

Figure 11.8 (a) Landsat TM false colour composite image of Kano city: image width = 1.5 km. (b) Landsat TM thermal waveband of the same area: arrow indicates wall of old city as on Figure 11.7.

The irrigated gardenscape of the Government Residential Area (GRA on Figure 11.8b) is approximately 3°C cooler than other parts of the urban area on this mid-dry season image, but surprisingly this area is the warmest part of the city on an April, late dry season image. The response of trees to very high temperatures by stomatal closure may explain this, and the observation confirms a tendency noted in Singapore that low-rise bungalow developments may comprise hot spots in tropical cities. The traditionally built old city of Kano appears cooler than newer suburban areas of the city on both image dates, and provides a $\Delta T(u - r)$ magnitude of approximately -2°C compared with rural areas; no doubt an incentive, over the centuries, for rural traders to reach the Kano market in good time. These observations are relevant, since tree planting in the GRA of Kano city may mitigate heating except in the hottest season; indeed it is noted that evapotranspiration from one large tree in the tropics has the same cooling effect as five room air conditioners. The use of traditional building styles and materials in Kano old city appears to confer a reduction in T_s of 2°C, and such building styles may remain the best way to moderate day-time air temperatures in Third World cities where air conditioning is not available, since to reduce air temperature by even 1°C uses as much energy as to raise it by 3°C (Wilhelm 2000).

Table 11.2 Mean T_b and T_s (°C) from training area data from Landsat TM image shown as Figure 11.8

	Rural	City racecourse	Old city	Sabongari	New suburbs	Government residential area	Industrial areas	All
T_b – no ε correction								
19 December 1986	23.1	22.8	20.6*	20.9*	20.5	20.3**	16.7	22
20 April 1985	39.0	38.1*	36	35.8	36*	36.5**	27.5	38
Emissivity***	0.92		0.91	0.90	0.90	0.93	0.86	
T_s – ε corrected								
19 December 1986	28.7		26.9	27.8	27.8	24.9	26.1	

* Denotes significant correlation with NDVI.
** Denotes significant negative correlation.
*** ε derived from AVHRR data (Artis and Carnahan 1982).

11.4   GIS and visualization

Two-dimensional models, such as those described in Section 11.3.1 where image data are superimposed on maps of the urban infrastructure, suffer from an incomplete representation of the active radiating surface of cities (Section 11.2.2). Developments in scientific computing, permitting visualization of structures in three dimensions (McCormick *et al.* 1987; Nichol 1998) can now be used to model the temperature of the complete radiating surface in urban areas, as opposed to the satellite's two-dimensional view, but require input of additional data (see Baltsavias and Gruen, Chapter 3 in this book). Anisotropy is particularly important in high-rise areas. For example, the total surface area measured in a Singapore Housing and Development Board estate is 1.7 times the planimetric area, if vertical building faces and tree canopies are included. Figure 11.9 (see colour plates) shows a three-dimensional model of T_s in Toa Payoh housing estate in Singapore, where the relationship between T_s, building geometry and solar azimuth can be visualized (Nichol 1994). The model was created as follows.

11.4.1   Two-dimensional procedures

Following conversion of image DN values to T_s in °C (Section 11.3.1.1) the data were converted to vector polygon format. The resulting polygons were overlaid with a street plan and building outlines. A statistical GIS function was then used to calculate the mean temperature for each building polygon, which was written to a record theme representing the polygon (building) temperatures as GIS database attributes. These were then classified into seven colour classes.

11.4.2   Three-dimensional procedures

In order to visualize the complete active surface, including both horizontal and vertical faces it was necessary to construct building surfaces as three-dimensional representations whose temperatures could be varied according to the effects of Sun angle and azimuth at the time of imaging. Thus the building polygons were extruded as three-dimensional wire frame models. Solid three-dimensional faces were added as graphic primitives by layer, according to four aspects. The GIS layer in this case is used as a surrogate for a three-dimensional database attribute (aspect) thereby permitting subsequent retrieval by this attribute. Each three-dimensional face could then be selected and modified using a Structured Query Language (SQL)-based heuristic filter, according to two attributes, colour (temperature class) and aspect. Temperatures were determined for the vertical facets in relation to the horizontal surfaces, according to values and thermal relationships derived from a previous study (Nichol 1996a). Thus the temperatures of vertical building faces were adjusted to compensate for their consistently cooler temperatures than the horizontal (imaged) surfaces observed at the same time of day and weather conditions of imaging (mid-morning on a sunny day).

11.4.3 Advantages of the three-dimensional model

The model permits visualization of the relative proportions of vertical building surfaces as well as their different aspects according to solar azimuth at the image time – shady (south and west) and sunny (north and east). Based on model data, this would constitute an image error at the time of imaging of +1.4°C of T_s compared with the temperature of the complete surface. The error ranges from +2°C for high-rise to +0.2°C for low-rise areas where anisotropy is less. It can thus be appreciated that, at the time of imaging, the satellite nadir viewpoint appears to be significantly inaccurate in high-rise areas due to the omission of vertical surfaces that are all much cooler than horizontal ones. The magnitude of error would decrease toward midday since sunny wall temperatures would then be similar to ground temperature. The inclusion of vegetation in the model (Figure 11.9, colour plates) is merely to enhance visualization since empirical assessment of the error contributed by unrepresentatively warm tree canopies is complex. In a study of Vancouver (Voogt and Oke 1996) where T_s from vertical and off-nadir thermal scanner imagery was supplemented by vehicle-based infrared transducers to obtain temperatures of vertical surfaces, it was discovered that off-nadir imagery in the direction of the coolest (most shaded) face was the most representative of the complete surface (T_s), within approximately 1°C of T_s.

11.5   Applications of remote sensing to global climate change modelling

Working Group 1 of IPCC (IPCC 2001) asserts that global average near-surface air temperature increased by 0.6 ± 0.2°C during the twentieth century. This value is about 0.15°C larger than the previous estimate due to the inclusion of various adjustments including the urban heat island effect. Air temperature deviations over cities are important in determining heat and radiation fluxes and various hydrological parameters for input to Global Climate Models (GCMs). Low-spatial-resolution sensors such as NASA's HCMM, Meteosat and NOAA's AVHRR, which have both visible and thermal sensing capabilities, may be used to derive parameters indicative of air temperature. For example, Price (1979), using HCMM, converted the satellite-derived T_s elevation of urban over rural areas to excess radiated power (W m^{-2}) for whole cities. Cities were defined as areas of $\Delta T_s(u - r) > 9$°C. Since a temperature elevation of 1°K at 300°K represents $+6\,W\,m^2$, the excess radiated power from New York City, for example, was 40,000 kW.

More recently, attempts have been to derive near-ground air temperature directly from image data. For example, Lakshmi *et al.* (2001) correlated AVHRR and observations from the TIROS Operational Vertical Sounder (TOVS) with air temperature, and Cresswell *et al.* (1999) derived air temperature from the solar zenith angle corresponding with Meteosat imagery. However, the difficulty of verifying the accuracy of the resulting air temperatures by relating point ground observations of air temperature with areally integrated pixel data from such low resolution sensors is acknowledged. Other global-scale estimates of urban heat islands have utilized vegetation indices. For example, Gallo *et al.* (1993) using day-time AVHRR weekly composite imagery for thirty-seven US cities obtained a somewhat higher correlation between $\Delta T(u - r)$ and NDVI than for radiant surface temperature, the NDVI producing a negative correlation with air temperature. The results are inconclusive since the image dates used to create the composite image were those with the maximum NDVI values. Such methods also depend on vegetation being the only influence on temperature, ignoring others such as soil moisture, wind and shadow. Furthermore, the NDVI is known to be a poor indicator of vegetation in arid and semi-arid regions, due to the darkening effect of deciduous plant canopies in the dry season (Moleele *et al.* 2001). The Landsat TM images of Kano region in Figure 11.8 show no correlation between NDVI and the thermal response.

A further compelling reason for the accurate quantification of the air temperature anomalies of urban heat islands is the recognition that the majority of long-term climate stations are located in increasingly urbanized areas, thus biasing long-term temperature records. This bias may be eliminated by satellite-based observations, given adequate historical depth and accurate interpretation of the signal.

11.6 Conclusion

Present and future satellite sensors offer a challenging environment for studying urban climates, both at microscale, where the term "surface temperature patterns" may be more relevant than "heat island" (Nichol 1996), and at macroscale, where heat islands can be viewed synoptically. Regardless of the level of detail, however, this chapter has emphasized the importance of accurate emissivity correction for accurate surface temperatures, as well as the understanding of the relationship between the latter and near-ground air temperature. Both these conditions are particular to tropical and Third World cities due to climatic, structural and building material differences between them and temperate zone cities. In temperate zone cities, for example, east–west street orientations are contraindicated due to the high Sun angle and thus the ability of the Sun to penetrate to street level in urban canyons at all times of day (e.g., Terjung and Louis 1973; cf. Oke 1984). However, in Singapore, at 1°12″ north, a more important consideration is the avoidance of direct solar illumination of building walls, especially if they are high-rise residential blocks; thus an east–west long axis of buildings is advocated and practised, giving street canyons an east–west orientation. The resulting heating of the east–west-oriented street canyons is then mitigated by strategic tree planting programmes. Atmospheric correction is also required if multi-images are involved.

It is urgent to investigate data from newer remote sensors such as: Landsat's ETM+ (Enhanced Thematic Mapper Plus) whose 10.4–12.6-µm waveband has a spatial resolution of 60 m, for detailed observations of the urban thermal environment, as well as data from hyperspectral systems; Advanced Spaceborne Thermal Emission and Reflection Radiometer (ASTER) on the Earth Observing System (EOS) Terra satellite with five thermal bands of 90-m spatial resolution and Moderate Resolution Imaging Spectrometer (MODIS) on the same platform with thirty-six wavebands including eight thermal bands and a spatial resolution of 1 km. The benefits for tropical and Third World cities could include healthier conditions, improved efficiency, less wastage of energy and water, and reduced air pollution, if climatological concerns are included in their design, and, for the scientific community, a vastly extended and more accurate means of monitoring the Earth's temperature.

11.7   References

Artis, D. A. and Carnahan, W. H., 1982, Survey of emissivity variability in thermography of urban areas. *Remote Sensing of Environment*, **12**, 313–29.

Chang, J. H., 1965, On the study of evapotranspiration and water balance. *Erdkunde*, **19**, 141–50.

Carlson, T. N. and Boland, F. E., 1978, Analysis of urban-rural canopy using a surface heat flux temperature model. *Journal of Applied Meteorology*, **17**, 998–1013.

Chandler, T. J., 1965, *The Climate of London* (London: Hutchinson).

Cresswell, M. P., Morse, A. P., Thomson, M. C. and Connor, S. J., 1999, Estimating surface air temperatures from Meteosat land surface temperatures, using an empirical solar zenith angle model. *International Journal of Remote Sensing*, **20**(6), 1125–32.

Gallo, K. P., McNab, A. L., Karl, T. R., Brown, J F., Hood, J. J. and Tarpley, J. D., 1993, The use of NOAA AVHRR data for assessment of the urban heat island effect. *Applied Meteorology*, **32**(5), 899–908.

Goldreich, V., 1985, The structure of the ground level heat island in a central business district. *Journal of Climatology and Applied Meteorology*, **2**, 1237–44.

IPCC, 2001, *Summary for Policymakers*, Report of Working Group 1 of IPCC. (www.ipcc.ch/pub/sarsum1.htm: Inter-governmental Panel on Climate Change).

Jauregui, E., 1984, Tropical urban climates: Review and assessment. Paper given at Urban Climatology Mexico Conference (Geneva: World Meteorological Organisation).

Jensen, J. R., 2000, *Remote Sensing of the Environment: An Earth Resource Perspective* (Englewood Cliffs, NJ: Prentice-Hall).

Lakshmi, V., Czajkowski, K., Dubayah, R. and Susskind, J., 2001, Land surface air temperature mapping using TOVS and AVHRR. *International Journal of Remote Sensing*, 22(4), 643–62.

McCormick, B. H., Defanti, T. A. and Brown, M. D., 1987, Visualisation in scientific computing. *Computer Graphics*, 21, 6.

Malaret, E., Bartolucci, L. A., Fabain Lozano, D., Anuta, P. E. and McGillen, C. D., 1985, Thematic Mapper data quality analysis. *Photogrammetric Engineering and Remote Sensing*, 51, 1407–16.

Moleele, N., Ringrose, S., Arnberg, W., Lunden, B. and Vanderpost, C., 2001, Assessment of vegetation indexes useful for browse (forage) prediction in semi-arid rangelands. *International Journal of Remote Sensing*, 22, 741–56.

Nichol, J. E., 1994, A GIS based approach to microclimate monitoring in Singapore's high-rise housing estates. *Photogrammetric Engineering and Remote Sensing*, 60, 1225–32.

Nichol, J. E., 1996a, High resolution surface temperature patterns related to urban morphology in a tropical city: A satellite-based study. *Journal of Applied Meteorology*, 35, 135–46.

Nichol, J. E., 1996b, Analysis of the urban thermal environment of Singapore using LANDSAT data. *Environment and Planning B*, 23, 733–47.

Nichol, J. E., 1998, Visualisation of urban surface temperatures derived from satellite images. *International Journal of Remote Sensing*, 19, 1639–49.

Oguntoyinbo, J. S., 1970, Reflection coefficient of natural vegetation, crops and urban surfaces in Nigeria. *Quarterly Journal of the Royal Meteorological Society*, 96, 430–41.

Oke, T. R., 1982, Overview of interactions between settlements and their environments. *WMO Meeting of Experts on Urban and Building Climatology*, WCP-37 (Geneva: World Meteorological Organisation), Appendix D, pp. 1–5.

Oke, T. R., 1984, Urban climatology and the tropical city: An introduction. *Proceedings of the Technical Conference of the World Meteorological Office, November, Mexico*, No. 625 (Mexico City: World Meteorological Organization), pp. 26–30.

Oke, T. R., 1987, *Boundary Layer Climates* (Methuen, NY: Routledge).

Price, J. C., 1979, Assessment of the urban heat island effect through the use of satellite data. *Monthly Weather Review*, 107, 1554–7.

Rao, 1972.

Roth, M., Oke, T. R. and Emery, W. J., 1989, Satellite derived urban heat islands from three coastal cities and the utilisation of such data in urban climatology. *International Journal of Remote Sensing*, 10, 1699–720.

Singapore Meteorological Service, 1982, A study of the urban climate of Singapore. In L. S. Chia (ed.) *The Biophysical Environment of Singapore and Its Neighbouring Countries* (Singapore: Geography Teachers Association of Singapore).

Terjung, W. H. and Louie, S. S-F., 1973, Solar radiation and urban heat islands. *Annals of Association of American Geographers*, 63, 181–207.

Tso, C. P., Toh, K. C. and Nichol, J. E., 1994, A case for the inclusion of the effect of buildings on the outdoor thermal environment in EIA studies. Paper given at Conference of the 12th ASEAN Federation of Engineering Organisations, November, Bandar Seri Begawan, Brunei.

UNEP, 2001, World Development Indicators (http://www.worldbank.org/data/wdi/home.html)

Voogt, J. A. and Oke, T. R., 1996, Complete urban surface temperatures. Paper given at 12th Conference on Biometeorology and Aerobiology (Atlanta, GA: American Meteorological Society), pp. 438–41.

Wilhelm, I. N., 2000, Keeping cool in the tropics without wasting energy. *The Courier*, **182**, 57–9.

Part III
Cities by night

12 LandScan

A global population database for estimating populations at risk

Jerome E. Dobson, Edward A. Bright,
Phillip R. Coleman and Budhendra L. Bhaduri

12.1 Introduction

Where are the world's 6,000,000,000 people? Obviously, they reside, work, travel and play on 510,000,000 km² of land, water and ice that comprise the Earth's surface. At that spatial resolution, we can state their *location* unequivocally, though the *count* itself is clouded in uncertainty. Even the most geographically challenged observer will note, however, that most people live on land, not in the ocean. Suddenly, 361,637,000 km² of sea surface is removed from the distribution, and population density jumps from 12 people per km² to 40 people per km². This simple, compelling act of removing the oceans from the total area is what cartographers call dasymetric interpolation.

With adequate resources, one might go on and on refining the spatial distribution and density calculation by removing uninhabited ice caps, deserts, mountain peaks, lakes and wetlands on the one hand and drawing people to cities, villages, houses, roads and waterways on the other. Fundamental to the task is the requirement to weigh some factors more heavily than others. As the number of variables increase and spatial resolution gets finer, the algorithm grows extremely complex, data volume expands geometrically and data availability becomes a bounding constraint. But residential population is only one part of the picture. People move about in regular cycles, such as the daily journey to work, or irregularly, as on vacation travel. Consequently, city centres experience huge daily fluctuations, and even the most remote locations occasionally receive visitors. Yet official census counts generally record only where inhabitants sleep most nights in the occasional census year. For instance, the US Census 2000 indicates that there are only fifty-five people in the large urban block that used to contain the World Trade Center. Again dasymetric interpolation, guided by spatial and behavioural models, can help locate ambient populations and monitor population mobility for day vs. night, seasonal migrations and special events. Remote sensing is an essential aid to locating settlements, transportation networks, agriculture and other indicators of human occupancy.

These same principles apply down to the most precise geometry of human location, and dasymetric interpolation is an invaluable aid to refining even the best available census data. Already, dasymetric interpolation and remote sensing have contributed to LandScan1998 (Dobson *et al.* 2000) and LandScan2000 at 30-arcsec (< 1 km) resolution, the most spatially precise global population distributions ever produced and the first to employ satellite imagery worldwide (Figure 12.1, see colour plates). This chapter describes their use in estimating population distributions and settlement densities at that resolution worldwide and at much finer resolution in the USA.

12.2   Global threats to local places

If ever there was doubt as to the need for high-precision estimates of populations at risk, it was dispelled on 11 September 2001. While the world watched in horror, the World Trade Center collapsed on an unknown number of people. Initial estimates ranged from 10,000 to 50,000 potential victims. A large medical corps was rushed into place, and despondence soon set in among emergency personnel when so few victims showed up for treatment. As it turned out, the number of injured was smaller than anticipated because more people escaped than might have been expected and few of the injured escaped alive. For several days no estimate of the missing was given. Then, for several weeks the estimate fluctuated by hundreds and later by thousands. The number of people at risk and the number who actually suffer injury or death are urgent and important pieces of information during any disaster. The penalty for overestimating is misallocation of resources, although not a serious problem in New York because the USA is blessed with abundant resources. In other parts of the world, however, the cost of such an error might have been substantial. That is certainly true when international relief organizations dispatch emergency medical teams to distant disaster areas. It is also true when poor countries themselves transfer emergency personnel from other care facilities, which are stretched thinly at the best of times. The error then compounds into unnecessary economic hardship and diminished medical care in other parts of the country. The penalty for underestimating is that more victims suffer and die. Triage is the hardest decision that physicians ever face. Its crucial variable is time, and the equation rests on the number of physicians and operating tables and quantities of medical supplies available at key instants soon after a disaster occurs.

These same principles apply to terrorist attacks, wars, earthquakes, floods, fires, explosions, chemical accidents and hurricanes. In every instance, lives depend on rapid deployment of personnel, medical equipment and emergency supplies from blankets to bandages. Those decisions, in turn, depend on reasonable estimates of populations at risk. Humanitarian crises and relief efforts thus generate immediate needs for local information anywhere on Earth. Long before the attacks on New York City, Washington, DC and rural Pennsylvania, Oak Ridge National Laboratory (ORNL) developed global population databases to address precisely the same kinds of issue that assaulted public conscientiousness on that day.

12.3   LandScan global population model

LandScan1998 was developed initially for use in emergency management by national and international organizations. It was adopted by the US Department of Defense (DoD) and the US Department of State, several agencies of the United Nations and some foreign nations as a world standard for estimating populations at risk. From April 1999 through March 2001, LandScan1998 was available on CD-ROM, and more than 250 organizations and individuals requested copies for applications as diverse as agriculture, forestry, health and telecommunications. LandScan2000 is an updated and enhanced version based on new census data and improved ancillary data. The new database can be downloaded via the Internet (http://sedac.ciesin.columbia.edu/plue/gpw/landscan/). The fundamental structure of the population model changed little between LandScan1998 and 2000, but the input data improved tremendously in spatial resolution and quality. The spatial resolution of the output data (30 arcsec) is unchanged, but the resulting pattern more precisely reflects local population distributions on a cell-by-cell basis (Figure 12.2).

LandScan global distributions represent ambient populations that integrate diurnal movements and collective travel habits into a single measure. This is suitable for emergency response when the event is unanticipated or the time of an anticipated event cannot be predicted. Fortuitously, it is easier

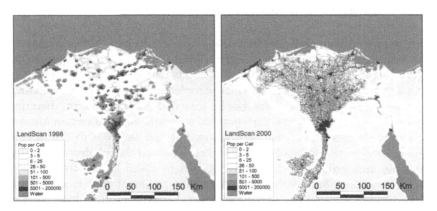

Figure 12.2 LandScan1998 compared with LandScan2000 in the Nile Delta of Egypt.

to accomplish with currently available global imagery and other geographic data. A regular grid of 30-arcsec latitude by 30-arcsec longitude is imposed on the polygonal geometry of all census areas worldwide. Vector census geometry is converted to raster geometry, and each cell is placed within its proper census unit approximating the finest available census geometry (e.g., block, county, province). Census counts serve as control totals for the sum of all cells associated with each polygon (block, county, province, nation, world). Each cell is assigned a dasymetric value based on its relative attraction for people. For example, any individual within a given census area may be hundreds or thousands of times more likely to occupy a cell with gentle slope, developed (urban or other intensive) land cover, bright and intense night-time lights and a road than in a cell with steep slope, forest or bare land cover, no night-time lights and no road. The relative attractiveness of each variable differs markedly from region to region – hence, the dasymetric value is determined heuristically for each of about 100 functional geographic regions worldwide.

The primary determinants of accuracy and precision in the LandScan model are: (i) size of census units, (ii) age and quality of census counts and (iii) precision of cartographic and other ancillary databases. The amount of ancillary data necessary to support such a model at the chosen resolution is massive. Indeed, the initial shipment of digital data sent from the DoD to ORNL literally consisted of a "ton of data" – more than 9,000 CD-ROMs occupying two pallets with a total shipping weight of 948 kg and containing about six terabytes of data. Several hundred updates and additions arrived over the following year. All the ancillary databases were processed and employed in the calculation. The imagery and scanned maps were used, wherever needed, to refine urban built-up areas and were employed extensively in verification and validation of the final population distribution.

12.4   Dasymetric interpolation

Dasymetric interpolation was first proposed in 1936 by John K. Wright of the American Geographical Society (Wright 1936). The technique is applicable to many geographical distributions, but Wright's initial application was population. He started with choropleth maps of population density on Cape Cod in Massachusetts, but he reasoned that the uniform densities produced by this traditional cartographic technique masked certain known or knowable elements of the distribution. His first step was to distinguish uninhabited from inhabited areas. He then weighted the inhabited areas by land use and settlement characteristics. Wright called his technique dasymetric mapping, but Chrisman (2002) later proposed that it should be considered more properly as a form of areal interpolation. Indeed, Tobler *et al.* (1995) called for "smart interpolation or co-Kriging" when they suggested that global population density estimates could be improved in accuracy and precision by similarly incorporating ancillary data such as location and size of towns and cities, roads, railroads, natural features and night-time lights. ORNL has used the technique for mapping population since about 1980 (Durfee and Coleman 1983). See Langford (Chapter 6 in this book) for more discussion on dasymetric mapping.

12.4.1 *Census counts and census geometries*

All census counts, including the official censuses of advanced nations like the USA, are stochastic estimates. Accuracy and precision are limited by time of day, frequency of repetition, resources, census takers' access to homes and understandings of personal work and travel habits, and sometimes outright manipulation to meet political objectives. For much of the world, the best available official census data are at subprovince level (i.e., two administrative divisions below national) ranging from current to less than a decade old. In extreme cases, the best available official counts are national totals taken many years ago.

To be used effectively in a population allocation model, all census counts must be associated with well-defined polygons representing the census areas for which each count is taken. The census geometry must be digital. Ideally, as in the USA, digital boundary files are provided to census takers and counts are associated with the files procedurally from the initial field survey through to the final reporting of results. The Topologically Integrated Geographic Encoding and Referencing (TIGER) files, developed by the US Census Bureau in conjunction with the US Geological Survey (USGS) have become the standard geometric reference not only for census data, but for digital street atlases and other high-precision applications as well. Most countries, however, do not operate in such an automated fashion. Often, it is necessary for others to scan or otherwise digitize hard copy maps. Fortunately, the global collection of digital census geometry files is substantial and growing. LandScan2000 uses all census geometry files for which associated census counts are available and, conversely, all census counts for which associated geometry files are available. The variable quality of census figures from country to country presents a major challenge to global population distribution efforts such as LandScan. Official census counts must be acquired from published sources and evaluated sceptically. Fortunately, for most countries the demographic literature is surprisingly rich, deficiencies are recognized by scholars and adjustments have been proposed in the literature. In addition to published articles and reports, the Internet has become an invaluable resource in locating and acquiring population data and understanding consensus among demographers.

Ultimately, LandScan must employ a single population total for each country, province, or subprovince. For LandScan2000 we relied on the International Programs Center (IPC) of the US Bureau of the Census to provide the best available census counts at the finest available census geometry for every country in the world. For at least three decades, IPC has been recognized for its expertise in acquiring, maintaining and interpreting population estimates for all nations (Leddy 1994). The LandScan team relied on IPC as a major source of census counts during the production of LandScan1998. That relationship was formalized for LandScan2000 through an agreement funded by the US DoD. Nominally, all counts are at subprovince level with demographic distinctions between males and females by age in 5-year intervals. Only for a few small countries are national counts not divided into province or subprovince counts.

The control total population for each area is mathematically determined by four input variables: (i) the most recent official census count, (ii) adjustments to official counts based on a consensus of opinion among demographers, (iii) proportional values used (in some countries) to apportion national totals to provinces and (iv) the annual growth rate used to project from most recent official census to the year 2000. Adjustments are made by IPC routinely when major disruptions, such as wars and natural disasters, occur and occasionally when the integrity of an official count is questioned by IPC and the demographic community at large.

12.4.2   Cartographic base

At three international geographical congresses from 1891 through 1913 Albrecht Penck (Wright 1952) proposed a concerted international effort to map the world at 1 : 1,000,000 scale. Cartographic standards were established, but the work was never completed. The greatest success was the millionth-scale *Map of Hispanic America*, consisting of 107 sheets, compiled by the American Geographical Society between 1920 to 1945. After World War II, Penck's dream was superseded by the Defense Mapping Agency's 1 : 250,000 topographic map series, which improved on precision but failed to meet Penck's thematic goals.

For global applications, most digital users remain at Penck's 1 : 1,000,000 scale, based on Vector Map (VMAP) Level 0 (formerly Digital Chart of the World) produced by the National Imagery and Mapping Agency (NIMA, formerly Defense Mapping Agency). It is readily available for the entire Earth and is a tremendous aid to global studies. Today, NIMA is producing VMAP-1 at 1 : 250,000 scale, and the improvement is astounding. VMAP-0, for instance, only shows major highways, while VMAP-1 shows secondary and tertiary roads and even many trails, and the number of "point" locations for populated places shown in VMAP-1 is at least an order of magnitude greater than the number in VMAP-0. Populated polygons in VMAP-0 are out of date and usually underrepresent city size, while VMAP-1 is good for identifying built-up areas. Coverage is complete for much of the world, but substantial areas remain to be completed in scattered regions. LandScan1998 employed VMAP-0. LandScan2000 employed VMAP-1 for every available tile, and that greatly improved road networks, populated places, boundaries and waterbodies. The improvement is so great that we dichotomously distinguish LandScan2000's accuracy between tiles with VMAP-1 and tiles without it. While LandScan2000 was being produced, VMAP-1 was restricted to official use only; now it is available to the public on NIMA's website.

12.4.3   Land surface slope

Digital elevation data at coarse resolution are readily available for the entire world, but fine resolution is restricted. Digital Terrain Elevation Data Level 0 (DTED-0) at 30-arcsec resolution are complete for the entire world and readily available to anyone through NIMA's website. DTED-1 at 3-arcsec resolution, however, is available for official use only. Coverage is complete for most of the world, but substantial areas remain to be completed in scattered regions. There is a great deal of excitement about the National Aeronautics and Space Administration's (NASA) Shuttle Radar Topography Mission (SRTM) data, but radar processing is complex and products will appear gradually. Ultimately, SRTM will cover most of the world at the equivalent of DTED-1 (3-arcsec) and DTED-2 (1-arcsec) resolution, but DTED-2 is unlikely to be released for quite some time. LandScan1998 employed DTED-0, while LandScan2000 employed DTED-1 for all tiles available. For 1998, we calculated average slope for each cell as a general indicator of slope suitability. For 2000 we calculated how much land within each cell is suitable for settlement. Each LandScan cell (30 arcsec) thus contains 100 elevation cells (3-arc sec) with slope calculated for each. The LandScan cell score then incorporates all 100 slope values into a single index representing both severity of slope and percentage of area.

12.4.4   Land cover

Currently, the best available land cover database worldwide is USGS's Global Land Cover Characteristics (GLCC) database derived from Advanced Very High Resolution Radiometry (AVHRR) satellite imagery at 1-km resolution (Loveland *et al.* 1991). Commensurate with its intended global use, GLCC is reasonably reliable for all land cover types except wetlands and developed lands, but there is considerable variation in accuracy from cell to cell. In test comparisons in the USA, most wetlands were recorded as water in GLCC. For areas tested elsewhere, GLCC's developed land cover category was a rasterized version of VMAP-0's "populated polygons", with attendant limitations.

For LandScan1998, we modified GLCC extensively. We geo-registered the data at 30-arcsec resolution in a common grid for the entire globe and devoted substantial effort to reconciling positional accuracy against other global databases, especially NIMA's World Vector Shoreline (WVS) at a 1 : 250,000 scale. Typically, this coastline differs somewhat from the related line representing the seaward boundary of administrative units, and both of these differ from the land/water boundary indicated on the GLCC gridded database. In the LandScan1998 land cover database, the WVS takes precedence, and water is assigned to all cells extending more than one-half cell beyond the WVS coastline. Wherever the GLCC land surface had to be expanded to reach the WVS shoreline, we inserted an "unclassified land" class. We replaced the GLCC "urban" class with two new classes – "developed" and "partly developed". The "developed" class is composed of GLCC's urban cells plus other cells identified through cartographic analysis of maps, images and other data sources. The "partly developed" class is derived from *Night-time Lights of the World* (Sutton 2002) and contains all cells with a frequency value of 90 per cent or greater. The "partly developed" class typically includes suburban areas, small towns, scattered industries, airports, etc. We distributed the modified land cover database along with LandScan1998 in the hope that these enhancements might be of use to others. For LandScan2000, we further improved the land cover database substantially using NIMA's VMAP-1, scanned maps and panchromatic imagery. The resulting land cover data cannot be released to the public until all the NIMA databases are released.

12.4.5   Night-time lights

Night-time lights are an invaluable global resource available from the National Geophysical Data Center (NGDC). Chris Elvidge of the Desert Research Institute, in residence at NGDC, produced the original frequency data (Elvidge *et al.* 1997; Elvidge *et al.*, Chapter 13 in this book) and the new *Stable Lights and Radiance Calibrated Lights of the World* (Sutton 2002).

The frequency data available for LandScan1998 were good for moderate settlement densities but tended to miss the diffused lights of small settlements and to saturate in large cities (Sutton 1997; Sutton *et al.* 1997). Radiance-calibrated images are derived from the Defense Meteorological Satellite Program (DMSP) Operational Linescan System (OLS) Night-time Imagery from 1996 and 1997 at 30-arcsec resolution. Their main advantage over the previously available frequency data is three separate gain settings to capture low-intensity lights in the countryside, saturated lights in cities and everything in-between. The new data available for LandScan2000 vastly improved our ability to detect small towns and villages and to discriminate densities within large cities. (See Chapter 13 in this book by Elvidge *et al.* for an introduction to night-time imagery and the OLS scanner.)

12.5   Applications

At the height of the Kosovo refugee crisis, the Departments of Defense and State requested an update of LandScan population distributions for Kosovo,

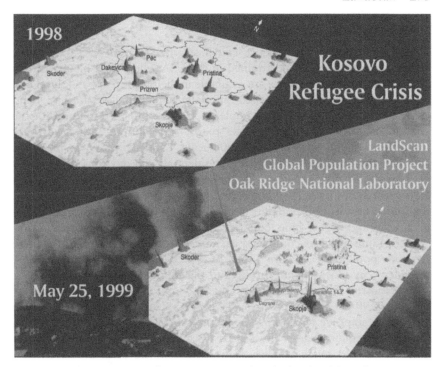

Figure 12.3 LandScan1998 for Kosovo updated at the height of the refugee crisis on 25 May 1999.

the rest of Yugoslavia and the surrounding countries. The update was accomplished using ARCVIEW GIS as a spatial accounting tool to reconcile spatially and categorically diverse estimates of: (i) known dead, (ii) known migrants to neighbouring countries in the region, (iii) known migrants out of the region by country of origin and destination and (iv) unaccounted. More than twenty-five information sources were synthesized and reconciled to produce a revised LandScan distribution with an effective date of 25 May 1999 (Figure 12.3). We compared this with the LandScan1998 baseline and demonstrated dramatic patterns of change: the depopulation of Priština, Peč and other cities; the appearance of huge refugee camps in Albania and Macedonia; and the growth of many small camps in central Kosovo.

During the vast Mozambique Flood of 2000, NIMA mapped floodwaters and intersected LandScan to estimate the number of victims. LandScan, combined with other data, such as the location of airfields, helped plan rescue and recovery efforts. The resulting maps prepared by Andrew Bauer of NIMA were posted on the UN's ReliefWeb site.

Currently LandScan2000 is being used to plan seed supplies for the spring planting in Afghanistan and nearby areas impacted by the ongoing refugee crisis.

12.6   LandScanUSA

The global LandScan experience convinced investigators that much finer spatial resolutions could be attained in the USA where high-quality census counts and census geometries are available in digital form at block level and ancillary data are available at exceptionally fine spatial resolutions. Land cover has been classified, for example, at 1-arcsec (30 m) resolution for the contiguous forty-eight states, while roads, railroads and waterways are available in digital form at 1 : 100,000 scale. LandScanUSA was proposed as a national database of night-time (residential) *and* day-time populations at 3-arcsec (90 m) resolution. By the beginning of 2002, the contiguous forty-eight states have been completed at 15-arcsec (500 m) resolution for night-time (residential) population only. A prototype study focusing on an area covering twenty-nine counties in south-eastern Texas and south-western Louisiana was completed at 3-arcsec resolution for night-time and day-time populations (Figure 12.4). At 3-arcsec resolution, LandScanUSA provides "city block" detail throughout the countryside as well as in cities. In the Houston Prototype, that represents an improved spatial resolution for more than 99 per cent of the area (Figure 12.5). Indeed, even the LandScan global cells improve resolution for 89 per cent of the area. In other words, most of the countryside is covered by census blocks larger than LandScan global cells and much smaller than LandScanUSA cells. Relatively few scattered census blocks, mostly near city centres, approach the "city block" size of LandScanUSA.

Night-time (residential) populations are distributed through dasymetric interpolation with census block counts serving as control totals for all cells associated with each block polygon. The algorithm employs variables similar to those of LandScan2000. Slope is based on the same 3-arcsec elevation data employed in LandScan2000, but other variables are at much finer spatial resolution than their global counterparts. Roads, railroads, waterbodies, parks, prisons and other landmarks are from TIGER at a 1 : 100,000 scale. Land cover is based on the National Land Cover Database, derived from Landsat Thematic Mapper (TM) imagery at 1-arc second (30 m) resolution by the federal interagency Multi-resolution Land Cover Characterization Task Force.

Daytime populations are based on similar procedures and data sources plus additional measures to account for workers, students and other day-time travellers. Block-to-block worker flows are obtained from the US Census. Schools, available on CD-ROM from the Environmental Systems Research Institute, are located by cell and student populations are allocated by age category based on the type of school (elementary, junior high, high school). Workers and students are subtracted from residential populations. Finally, estimates of people engaged in shopping and other day-time activities are subtracted from the remaining residential populations and allocated to roads and commercial land use areas.

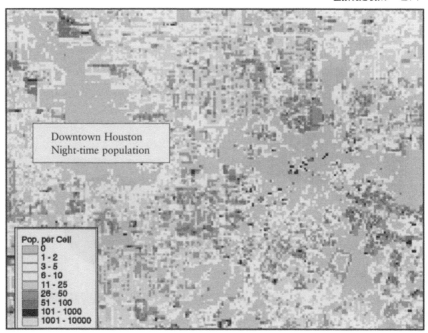

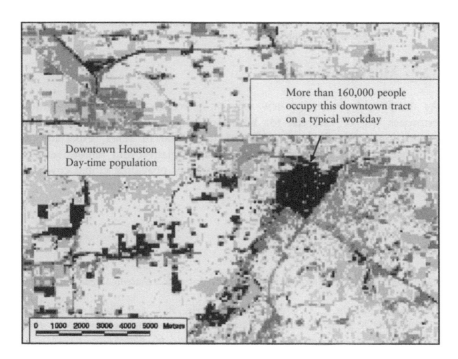

Figure 12.4 LandScanUSA night-time and day-time populations for Houston, Texas.

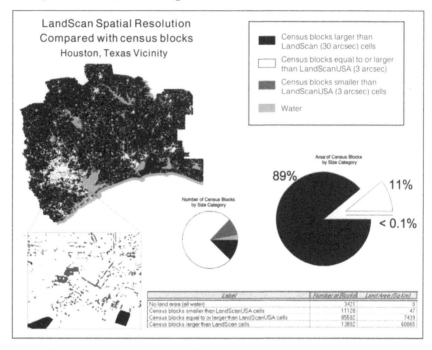

Figure 12.5 Houston Prototype study area, comparing the spatial resolution of
official US census blocks, LandScan2000 (global) and LandScanUSA.

12.7   Meeting future needs

As the World presses its new war on terrorism, improved population
estimates will be essential for homeland security, protection of civilian
populations around the world and humanitarian relief, especially in
response to refugee crises. Dasymetric interpolation and remote sensing
may prove to be the only reasonable means of providing global coverage
with improved accuracy, finer precision and frequent updates.

12.8   References

Chrisman, N., 2002, *Exploring Geographic Information Systems* (New York: John
Wiley & Sons).
Dobson, J. E., Bright, E. A., Coleman, P. R., Durfee, R. C. and Worley, B. A., 2000,
LandScan: A global population database for estimating populations at risk.
Photogrammetric Engineering and Remote Sensing, **66**, 849–57.
Durfee, R. C. and Coleman, P. R., 1983, *Population Distribution Analyses for
Nuclear Power Plant Siting*, Technical Report No. NUREG/CR-3056 (ORNL/
CSD/TM-197) Washington, DC: US Nuclear Regulatory Commission).

Elvidge, C. D., Baugh, K. E., Kihn, E. A., Kroehl, H. W. and Davis, E. R., 1997, Mapping city lights with night-time data from the DMSP Operational Linescan System. *Photogrammetric Engineering and Remote Sensing*, **63**, 727–34.

Leddy, R., 1994, Small area populations for the United States. Paper given at Conference of the Annual Meeting of the Association of American Geographers, San Francisco, CA (http://infoserver.ciesen.org/datasets/cir/jdysedit.html).

Loveland, T., Merchant, J., Ohlen, D. and Brown, J., 1991, Development of a land cover characteristics database for the conterminous US. *Photogrammetric Engineering and Remote Sensing*, **57**, 1453–63.

Sutton, P., 1997, Modeling population density with night-time satellite imagery and GIS. *Computers, Environment, and Urban Systems*, **21**, 227–44.

Sutton, P., 2002, The lights come out at night: night-time satellite imagery helps scientists track human activities through time. *GeoWorld*, **15**, No. 9

Sutton, P., Roberts, D., Elvidge, C. and Meij, H., 1997, A comparison of night-time satellite imagery and population density for the continental United States. *Photogrammetric Engineering and Remote Sensing*, **63**, 1303–13.

Tobler, W. R., Deichmann, U., Gottsegen, J. and Maloy, K., 1995, *The Global Demography Project*, Technical Report No. 95-6 (Santa Barbara, CA: National Center for Geographic Information and Analysis, UCSB).

Wright, J. K., 1936, A method of mapping densities of population with Cape Cod as an example. *Geographical Review*, **26**, 103–10.

Wright, J. K., 1952, Geography in the making: The American Geographical Society, 1851–1951. (New York, NY: American Geographical Society).

13 Overview of DMSP OLS and scope of applications

Christopher D. Elvidge, V. Ruth Hobson,
Ingrid L. Nelson, Jeffrey M. Safran,
Benjamin T. Tuttle, John B. Dietz and
Kimberly E. Baugh

13.1 Introduction

Cities, like cats, will reveal themselves at night.
Rupert Brooke, *Letters from America*, 1916

Much global change research is dedicated to discerning and documenting the impact of human activities on natural systems. Human population numbers were an estimated 750 million in the mid-1700s, rose to six billion in 1999 and could hit twelve billion before 2100 if current pace in growth continues. Human activities, which are known to be cumulatively altering the global environment, are responsible for greenhouse gas emissions from fossil fuel consumption, biomass burning, air and water pollution, and land cover/land use change. Far from being evenly distributed across the land surface, to a great extent human activities with environmental consequences are concentrated in or near human population centres. Having a capability for frequent global observation of a widespread and distinctly human activity that varies in intensity could substantially improve understanding of the scale of human enterprise and modelling its impact on the environment.

One approach to modelling the spatial distribution of human activities is to use population density as an indicator for the phenomenon of interest (e.g., percentage coverage of impermeable surfaces). There are two primary disadvantages to using population density as an indicator for human activities with environmental consequences. At a global level, population density is not well characterized. Previously available global population density grids (e.g., Tobler *et al.* 1995) were too coarse for many environmental applications. In areas where high spatial resolution population density data sets are available, the environmental applicability of the data suffer due to the fact that population density is defined as a residential parameter. Consequently, transportation corridors along with public, commercial and industrial zones are assumed to have very low population densities. In reality, these areas have much higher densities during the typical 8–12 hours of daylight than residential zones, which are occupied more during the night. Thus, the use of population density as an indicator for percentage of land area covered by impermeable surfaces (roads, roofs, parking lots, etc.) would result in a substantial skewing of the results toward residential areas. Satellite sensors capable of acquiring frequent global coverages would seem to have the potential for producing and updating global maps of city and town locations. Unfortunately, urban areas, villages and towns are difficult to identify with coarse spatial resolution satellite data from sensors such as the National Oceanic and Atmospheric Administration's (NOAA) Advanced Very High Resolution Radiometer (AVHRR) and the National Aeronautics and Space Administration's (NASA) Moderate Resolution Imaging Spectrometer (MODIS) due to the spectral heterogeneity of the various zones within cities and spectral confusion with other land cover types. While spatial features such as road networks and many buildings can be identified using high spatial resolution data produced by Landsat, SPOT (Système Pour l'Observation de la Terre) and other satellites, the prospects of generating and updating a global map showing where people are concentrated is daunting, even from a data acquisition standpoint.

Since the 1970s, the US Air Force Defense Meteorological Satellite Program (DMSP) has operated satellite sensors capable of detecting VNIR (Visible and Near InfraRed) emissions from cities and towns. The DMSP Operational Linescan System (OLS) acquires global day-time and night-time imagery of the Earth in the visible (VIS) and Thermal InfraRed (TIR) spectral bands. The night-time VIS bandpass straddles the VNIR portion of the spectrum (0.5 μm–0.9 μm). The VIS band signal is intensified at night using a photomultiplier tube (PMT), making it possible to detect faint VNIR emission sources. The PMT system was implemented for the detection of clouds at night. An unanticipated consequence of the night-time light intensification is the detection of city lights, gas flares and fires. Figure 13.1 shows city lights in Europe on 17 August 2001.

The potential use of night-time OLS data for the observation of city lights and other VNIR emission sources was first noted in the 1970s by Croft (1973, 1978, 1979). However, during the first 20 years of the DMSP no digital archive was maintained and only film strips were available to the scientific community. This limited the scientific value of the OLS observations, though the potential value of the data for the study of cities and human activity was clearly noted (e.g., Welch 1980; Forster 1983). Sullivan (1989) produced a 10-km spatial resolution global image of OLS-observed VNIR emission sources using filmstrips archived at the National Snow and Ice Data Center. In 1992, the US Department of Defense (DoD) and NOAA established a digital archive for the DMSP program at the NOAA National Geophysical Data Center. To date, three styles of night-time lights data of human settlements have been developed from digital OLS data. In each case, large numbers of night-time orbital segments are used to generate cloud-free composites of light detections, which are then filtered to remove noise,

Figure 13.1 Visible and thermal band DMSP OLS imagery of Europe acquired 17
August 2001. Note the city lights, glare, sunlit data and the higher noise
level present in the outer quarter panels of the visible band image.

ephemeral lights from, say, fire and lights from persistent sources other than
human settlements. These products are made using data collected during the
dark halves of lunar cycles to avoid the detection of moonlit clouds and
moonlight reflected from bodies of water. The three products are: (i) stable
lights – which identify the location, uncalibrated brightness and frequency of
detection for human settlements, (ii) radiance-calibrated lights and (iii)
night-time lights change data – indicating changes in the lights over time.

This chapter provides an overview of the DMSP OLS, a description of the
algorithms and procedures used to make the three types of products, and a
review of the applications that have been developed or proposed for DMSP
night-time lights observations of human settlements. Subsequent chapters in
this book will examine the application of DMSP OLS to population
estimates (Sutton, Chapter 14), GDP estimates and CO_2 measurements
(Doll, Chapter 15) and crime deterrence (Weeks, Chapter 16).

13.2 The DMSP OLS

The OLS is an oscillating scan radiometer designed for cloud imaging
with two spectral bands (visible and thermal) and a swathe of ~3,000 km.
The visible band straddles the VNIR portion of the spectrum from 0.5 μm
to 0.9 μm. OLS visible band data have 6-bit quantization, with digital
numbers (DNs) ranging from 0 to 63. The thermal band has 8-bit

quantization and a broad bandpass from 10 μm to 12 μm. The wide swathe widths provide for global coverage four times a day, at dawn, day, dusk and night. DMSP platforms are stabilized using four gyroscopes (three axis stabilization), and platform orientation is adjusted using a star mapper, Earth limb sensor and a solar detector. The visible band PMT system was implemented to facilitate the detection of clouds at night using moonlight. With sunlight eliminated, the light intensification results in a unique data set in which city lights, gas flares, lightning illuminated clouds and fires can be observed.

An optical instrument's gain can be thought of as the amplification of the incoming signal, from the front end of the telescope to the output of the DN data stream. The full system contains both gains and losses; however, the overall system amplifies the original signal. The OLS has analog pre-amplifiers and a post-amplifier with fixed gains as well as VDGA (Variable Digital Gain Amplifier) gain. The gain of the PMT also contributes to the overall system gain for night scenes. Primary control of the low-light imaging gain is achieved through ground command of the VDGA, which has values from 0 to 63 dB (dB is the log of the amplification).

Prior to launch, the OLS is calibrated under conditions that simulate the space environment. During the calibration the OLS views light sources having known irradiance and DNs are recorded over the full range of VDGA. The calibration data equate telescope input luminance to DNs at specific VDGA gain settings. Figure 13.2 shows the relationship between

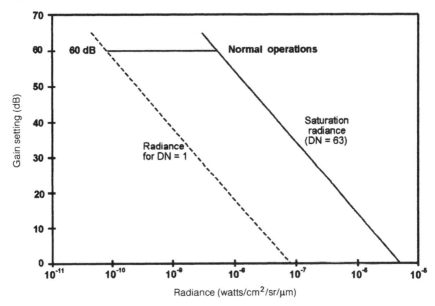

Figure 13.2 Pre-flight calibration results showing PMT gain setting vs. saturation radiance (solid diagonal line) for the DMSP F-12 OLS. The dashed line is the radiance for 1 DN.

VDGA gain settings and the observed range of radiances, based on the preflight sensor calibration for the OLS on DMSP satellite F-12. This calibration data makes it possible to relate OLS DNs back to laboratory-observed radiances known in terms of watts/cm^2/sr/μm. The solid diagonal line indicates the saturation radiance (where DN = 63). Observed pixels with radiances greater than this yield DN values of 63. The dashed diagonal line indicates the radiance for a DN of one, the lowest detectable radiance. At a VDGA gain setting of 63, the full system gain, including the fixed gain amplifiers, is 136 dB. This is a light intensification of $10^{13.6}$ times the original input signal, permitting measurement of radiances down to 10^{-10} watts/cm^2/sr/μm.

The visible band gains are computed on board based on scene source illumination predicted from solar elevation and lunar phase and elevation. The base gain is modified every 0.4 msec by an on-board, along-scan gain algorithm. In addition, a BRDF (Bidirectional Reflectance Distribution Function) algorithm further adjusts the gain in the scan segment where the illumination angle approaches the observation angle, resulting in enhanced specular reflectance. The objective of the two types of along-scan gain changes is to produce visually consistent imagery of clouds at all scan angles for use by Air Force meteorologists. During the darkest twelve nights of the lunar cycle illumination is too low to detect clouds in the visible band. Under these conditions the effects of the along-scan gain and BRDF algorithms are minimized and the gain is programmed to rise to its maximum monthly level. The high visible gain settings used when lunar illumination is low provides for the best detection of faint light sources present at the Earth's surface. The drawback of these high gain settings is that the visible band data of city centres are typically saturated. Data acquired under a full Moon when the gain is turned to a lower level show many fewer lights detected and less saturation for lights in city centres.

There are two spatial resolution modes in which data can be acquired. The full resolution data, having nominal Ground Sample Distance (GSD) of 0.5 km, is referred to as "fine". On-board averaging of five by five blocks of fine data produces "smoothed" data with a nominal 2.7-km GSD. Figure 13.3 shows simultaneously acquired smoothed vs. fine resolution night-time visible band images of the Nile Delta of Egypt. Primary lighting features can be seen in the smoothed data. The fine data provide some additional spatial detail, including the detection of small lights that could be confused with noise in the smoothed, and in some cases fine data reveal details on the internal structures of urban lighting patterns that are lost in the smoothed data. The OLS uses several methods to constrain the enlargement of pixel dimensions, which normally occur as a result of an across-track scanning (Lieske 1981). The OLS features a sinusoidal scan motion, which maintains a nearly constant GSD of 0.56 km between pixel centres in adjacent scan lines. In addition, the PMT electron beam is magnetically shifted during the outer quarter of each scan, reducing the size of the

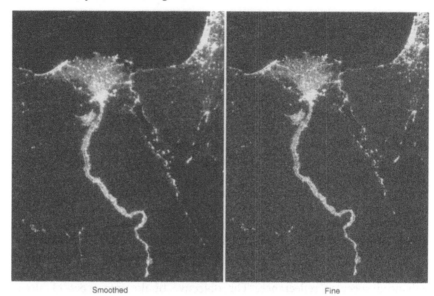

Smoothed   Fine

Figure 13.3 Simultaneously acquired smoothed vs. fine resolution night-time visible band imagery of the Nile Delta from 17 January 2000.

detector image on the ground surface. Note that the OLS gain is auto-matically increased for data acquired in the outer quarters of each scan. This gain increase causes noise and lights to have higher DN values in the outer quarter panels. The effect is quite noticeable for OLS data acquired at high gain settings (see Figure 13.1). When the PMT gain is turned down to avoid saturation in city centres the brightness difference between lights in the centre and edge of the scan are negligible. The OLS fine resolution night-time visible band Instantaneous Field of View (IFOV) data starts at 2.2 km at the nadir and expands to 4.3 km at 766 km out from the nadir. After the PMT electron beam is switched the IFOV is reduced to 3.0 km and expands to 5.4 km at the far edges of the scan. Thus the IFOV is substantially larger than the GSD in both the along-track and along-scan directions.

13.3   Stable lights product generation

The first global stable lights data set was produced by the National Geophysical Data Center (NGDC) using six months of night-time OLS data from the dark half of lunar cycles acquired between 1 October 1994 and 31 March 1995. This time period was selected to avoid the April through September solar contamination present in the night-time visible band data of mid to high latitudes in the northern hemisphere. For data acquired near the winter solstice, night-time lights can be observed at latitudes as far north as 70–75 degrees. In examining this initial product we found that six months of data from one DMSP satellite does not provide a sufficient number of cloud-free observations to identify all the stable lights present on the Earth's surface. Part of the problem is that some land areas end up with small numbers of cloud-free observations. In areas with large numbers of fires six months of data leave ambiguities in the discrimination of stable lights and fires. South-East Asia had both these problems in the 1994–95 stable lights product. The time period of the product covered the core of the burning season in Indochina and the rainy season in Indonesia. To address these insufficiencies, NGDC recently updated the stable lights of South-East Asia using a full twelve months of low-Moon data from the year 2000. We will use the South-East Asia stable lights update to illustrate how a stable lights product of human settlements is constructed using a time series of OLS observations.

The basic procedures used to generate the stable lights have been described by Elvidge *et al.* (1997a). Fire product processing can be grouped into three divisions: (i) processing done on OLS suborbits, (ii) processing done on geolocated suborbits and (iii) processing done on the composites.

13.3.1 *Processing on OLS suborbits*

The OLS suborbits are processed to identify and label the data types present in the visible band. Labelled data types include lights, glare, sunlit data, missing data, bad scan lines and noise. The pixel labels are written to a flag image, which directly overlays the OLS image and is subsequently processed along with the OLS image:

- *Missing data* Missing scan lines are identified based on the valid scan flag located in the scan line header.
- *Bad scan lines* In some cases the OLS imagery has bad scan lines present. These are lines with extended run lengths of pixels with spurious high and low DN values. These lines should not be included in night-time lights products. NGDC has developed an automatic algorithm for identifying and removing bad scan lines. The algorithm examines each scanline for the presence of at least one set of ten adjacent pixels with detected lights and no lights present immediately above or below.
- *Sunlit data* Night-time visible band data acquired at high latitudes near the summer solstice may be contaminated by sunlight. The OLS data stream includes solar elevations for the Earth's surface at the nadir of each scan line. Scan lines with solar elevations exceeding −14 degrees are flagged as sunlit.
- *Glare* One adverse effect of the light intensification is that the OLS is quite sensitive to scattered sunlight. Under certain geometric conditions, sunlight enters the OLS telescope while the sensor is in low light imaging

mode, resulting in visible band detector saturation (Figure 13.1), a condition referred to as glare. The exact shape and orbital position of the glare changes through the year. NGDC uses an algorithm to identify and label glare. This is based on the detection of blocks 100 by 100 pixels with saturated values in the visible band (DN = 63). Once such a block of data is encountered, all adjacent pixels having DN values greater than 40 are flagged as glare.

- *Lights*   Given that brightness variations occur within and between orbits, it is not possible to set a single DN threshold for identifying VNIR emission sources. We have developed an algorithm for automatic detection of VNIR emission sources (lights) in night-time OLS data using thresholds established based on the local background (Elvidge *et al.* 1997a). The algorithm uses a pair of pixel boxes. An outer box of 200 by 200 pixels in size is used to generate the background statistics for light detection. The light detection threshold is applied to an inner box of 50 by 50 pixels centred in the outer box. The algorithm then selects a dark background pixel set by analysing the visible band histogram generated from the pixels found in the outer box. This histogram features a prominent spike near the origin. In the next stage, the algorithm identifies the spike as the background pixel set, for which a mean and standard deviation are calculated. A light detection threshold is then set at mean plus four standard deviations. The resulting threshold is applied to the pixels in the inner box. The light detection algorithm proceeds by tiling the results from adjacent inner boxes. Transitions in the light detection threshold are smoothed by the inclusion of large numbers of common background pixels for adjacent inner boxes.

- *Noise*   In order to detect the maximum number of lights present at the Earth's surface the light detection filter allows a certain amount of the random noise present in the visible band data to be identified as lights. A very high proportion of this background noise is subsequently removed by an algorithm that identifies and removes single pixel light detections. Given the substantial overlap in the footprint of adjacent night-time visible band pixels even point sources of light present on the Earth's surface tend to be detected in more than one OLS pixel.

- *Geolocation*   The DMSP night-time lights products are generated in 30-arcsec grids, which correspond roughly to a square kilometre at the equator. The geolocation algorithm operates in the forward mode, projecting the centrepoint of pixels onto the Earth's surface. The geolocation algorithm estimates the latitude and longitude of pixel centres based on the geodetic subtrack of the satellite orbit, satellite altitude, OLS scan angle equations, an Earth sea level model and digital terrain data. The algorithm uses an oblate ellipsoid model of sea level and a terrain correction-based digital elevation from the Global Land One-kilometer Base Elevation (GLOBE) Project (http://www.ngdc.noaa.

gov/seg/topo/globe.shtml). The DN values of the visible, thermal and flag band data are used to fill the 30-arcsec grid cells containing the OLS pixel centres creating a "sparse grid". The remaining grid cells are filled using nearest neighbour resampling.

13.3.2   *Processing on geolocated suborbits*

The processing steps performed on geolocated suborbits include cloud detection and assembly of a set of composite images:

- *Cloud identification*   Given the low level of lunar illumination present in the night-time data, it is not possible to use the visible band to assist in the identification of clouds. Cloud identification relies on thermal band imagery, where clouds are generally cooler than the Earth's surface. There are several exceptions to this that complicate OLS cloud detection. Fog tends to be warm, with very little temperature contrast with the Earth's surface. In addition, there are latitudinal differences in cloud and Earth surface temperatures. During winter months at high latitudes the Earth's surface may be colder than local clouds. High mountain areas may be colder than local clouds. Another complication observed in Indonesia is that islands may be colder than the surrounding warm seas, making it easy to misidentify land as clouds. There are also latitudinal differences in cloud temperatures and the temperature contrast between clouds and the Earth's surface. Given these general complications to cloud detection and the fact that the OLS only has a single thermal band, we have been unable to fully automate OLS cloud detection. Our current method is semi-automated, employing a Graphical User Interface (GUI) designed to assist an analyst in the selection of a thermal threshold for cloud detection. In order to enhance the contrast between clouds and the Earth's surface a difference image is generated by subtracting a surface temperature grid from the geolocated OLS thermal band image. We use the global one-degree surface temperature grids generated by the NOAA National Center for Environmental Prediction (NCEP). The analyst reviews the displayed visible band, thermal band, the difference image and a histogram of the difference image, simultaneously. The histogram features a prominent complex of spikes near the origin, corresponding to cloud-free areas. Clouds that are colder than the Earth's surface show up in the long tail of the histogram. When the analyst selects a threshold for cloud detection from the histogram, the corresponding image pixels are coloured yellow, allowing the analyst to review the results. The analyst can then interactively adjust the threshold and review the results. When satisfied with the threshold selection, the analyst commands the GUI to record the threshold and is then prompted to select the next image. A separate program is used to apply the cloud detection threshold and record the results in the geolocated flag image.

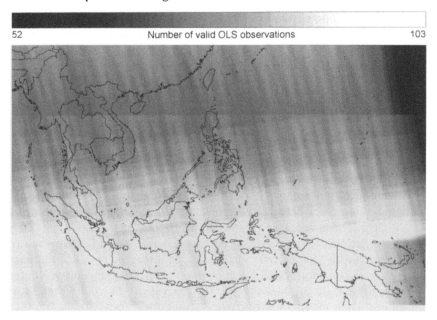

Figure 13.4 Composite image of the total number of valid OLS observations for the year 2000 stable lights of South-East Asia.

- *Identification of snow cover*   Snow cover causes lights to be brighter and to cover more area than observations made without snow cover. Daily one-degree NCEP snow cover grids can be used to flag lights that may have snow cover present.
- *Compositing*   A series of composite products are generated using the geolocated visible and flag images. This includes composites tallying the total number of usable OLS observations, the total number of cloud- and snow-free observations, the total number of cloud- and snow-free light detections and the average DN of the cloud- and snow-free light detections. These are shown for South-East Asia for the year 2000 stable lights product in Figures 13.4–13.7.

Figure 13.4 shows the composite image tallying the total number of valid OLS observations. This tally excludes missing scan lines, sunlit data and glare, recording the total number of usable observations that have been processed for the composite. The product is used to generate a companion image recording the number of cloud-free observations, and is also used to confirm that there are no gaps in the geographic distribution of valid OLS observations. Figure 13.5 shows the total number of cloud-free observations, which is generated by subtracting the grid cells flagged as clouds from the composite recording the total number of valid OLS observations. This composite is used later to normalize the detection frequency of lights and to

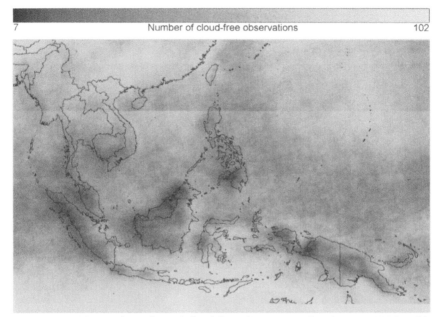

| 7 | Number of cloud-free observations | 102 |

Figure 13.5 The total number of cloud-free observations for the year 2000 stable lights of South-East Asia.

confirm that all areas in the composite have cloud-free observations. Figure 13.6 shows the total number of light detections within the set of cloud-free observations covering the Bangkok, Thailand region.

13.3.3 Processing on composite images

There are three objectives for processing the composite images: (i) removal of noise, (ii) separation of ephemeral and persistent (stable) lights and (iii) removal of persistent lights from sources other than human settlements (e.g., gas flares). As an initial step in noise removal procedure, the light counts image (Figure 13.6) is processed to remove lights detected only once. The primary separation of ephemeral and stable lights is performed using a persistence threshold of 10 per cent. This involves generation of a percentage frequency image, calculated by dividing the cloud-free light detection counts (Figure 13.6) by the total number of cloud-free observations (Figure 13.5) and multiplying by 100. Following application of the 10 per cent detection frequency threshold, the results are displayed with the cloud-free light detection counts as a colour composite and shoreline vectors are added. This image is visually inspected and manually edited to remove clusters of fire detections that passed through the 10 per cent threshold test. After manual editing, a land–sea mask is applied to remove lights from fishing boats and offshore oil and gas platforms. As a final step in the construction

0   Number of cloud-free light detections   35

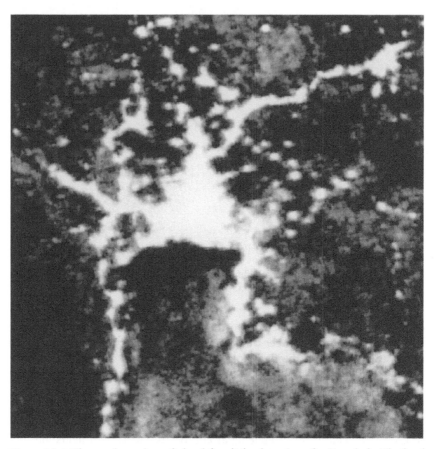

Figure 13.6 **The total number of cloud-free light detections for Bangkok, Thailand**
from the year 2000 stable lights.

of a stable lights product for human settlements, the resulting image is
further edited to remove lights from gas flares that have no associated city or
town marked in populated place vectors from the Digital Chart of the
World. The final product Figure 13.7 shows the average DN of the stable
lights in the Bangkok, Thailand region.

13.4   Radiance-calibrated night-time lights

NGDC produced a global radiance-calibrated night-time lights product
using 28 nights of OLS data acquired at reduced gain settings during

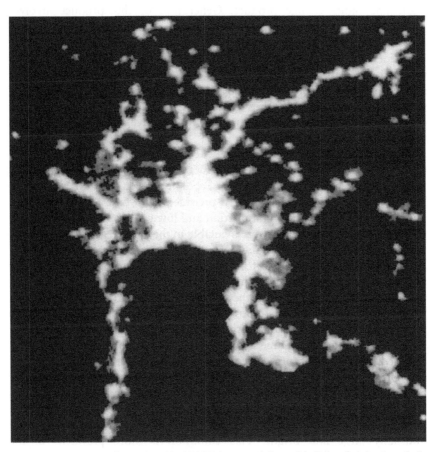

0   Per cent frequency   100

Figure 13.7 Per cent frequency (0–100%) image of the stable lights for the Bangkok, Thailand region from the year 2000.

March of 1996 and January–February of 1997 (Elvidge *et al.* 1999a). During these special data acquisitions the on-board ASGC (Along Scan Gain Control) and BRDF functions are turned off. It was discovered early on that it was not possible to cover the full range of brightness levels observable in cities and towns using a single gain setting. As a result OLS low-gain data acquisitions are split between visible band data acquired at low gain settings (e.g., 20 dB) to avoid saturation in the bright cores of urban centres and at a higher gain setting (e.g., 50 dB) to detect the dimmest lights detectable from the OLS. Low-gain data for generation of an improved, global radiance-calibrated, night-time lights product was acquired for 1999, 2000 and 2001.

13.5   Change detection lights

The most recent style of night-time lights to be developed by NGDC is the night-time lights change product. Generation of this product involves differencing of night-time lights products made from two or more time periods using OLS data acquired under similar conditions to processing, using identical algorithms and processing steps. The purpose of this product type is to identify areas where lighting conditions present on the Earth's surface have changed. This includes (i) the appearance of new lighting sources, (ii) the disappearance of lighting sources, (iii) expansion and contraction of the extent of lighting and (iv) positive and negative changes in the brightness of lights.

We have completed a night-time lights change product for the USA (forty-eight states) covering the 1992–93 to 2000 time period. The data used are from the dark half of lunar cycles in September, October and November. This time period was used because it provided night-time OLS data with a minimal amount of solar contamination and low probability of snow cover. Given that no low-gain data were available in the early years of the archive, we used data acquired under the normal operating conditions of the OLS. As a result, the bright cores of urban centres were saturated in both time periods, with no change in lighting detectable. The basic steps in generating the night-time lights product of the USA (1992–93 vs. 2000) share many of the processing steps outlined above for the stable lights product, with the following exceptions. First, as with the higher gain data used to make the radiance-calibrated night-time lights, only the centre halves of orbital scans are used to avoid brightness differences in the composite average DNs that arise from use of differential numbers of lights from the outer quarter panels. Second, the night-time visible band images for both time periods were visually inspected for "overglow" conditions, where the lights in portions of the upper Midwest and east coast of the USA are occasionally unusually large and bright. This condition is more prevalent in the November through March time period and may be due to snow cover or atmospheric scattering associated with winter air pollution. Lights in overglow areas were removed from the individual suborbits using manually drawn masks.

13.6   Comparison of three types of night-time lights products

Figure 13.8 (see colour plates) shows a comparison of the three types of night-time lights products for the city of Atlanta, GA. The image in Figure 13.8a is the percentage frequency of light detection from the 1994–95 stable lights product. Note that the city shows a uniform field of nearly 100 per cent detection frequency, only falling off at the extreme edges. Figure 13.8b shows the radiance calibrated lights from 1999 to 2000, revealing internal brightness structures of Atlanta. Figure 13.8c shows the night-time lights change image from 1992–93 to 2000, with areas that were saturated in both time periods as black, and areas that were substantially brighter in 2000 as red. The light grey areas had faint light detection in both time periods, but no major increase in the brightness of lights. Note the large rim of red surrounding Atlanta, corresponding to the rapid expansion of the city during the 1990s.

13.7 Conclusion

Nocturnal lighting could be regarded as one of the defining features of concentrated human activity. The authors have developed a methodology for producing geo-referenced images indicating location and extent of human settlements based on satellite observations of visible light emission sources present at the Earth's surface at night. The digital method for mapping night-time lights with OLS data utilize a large number of orbits to overcome the obscuring effects of clouds and to separate the observed lights into four primary categories: human settlements, gas flares, fishing boats and ephemeral lights (mostly fires). The lights in the resulting composite images are known to overestimate the actual size of human settlements. This area overestimation is due to a combination of factors, such as the large OLS pixel size, the OLS capability to detect subpixel light sources and geo-location errors. Surface effects, such as the presence of snow cover and the reflection of onshore lighting by near-shore waters, also contribute to the spread of light that can be detected by the satellite. These effects, present in data from single observations, are accumulated during the time series analysis. Imhoff *et al.* (1997a) developed thresholding techniques to map urban areas accurately. However, these techniques eliminate lights from small towns due to their low frequency of detection. A number of environmental monitoring activities are being developed using night-time OLS data. All these applications benefit from having periodic long-term composite images, such as the one presented in this chapter. Below is a brief description of the environmental applications for which OLS data are being developed or for which significant potential appears to exist:

- Sutton *et al.* (1997) examined the potential use of stable light data to apportion population spatially. Subsequently, DMSP night-time lights data of human settlements were combined with other data (e.g., topography, hydrography and land cover) to generate high spatial resolution population density grids (Dobson *et al.* 2000). Preliminary studies indicate that these provide more detailed and more accurate depictions of population density than previously available global population density data sets (Sutton, Chapter 14 in this book).

- Meteorological records extending back decades from ground stations in urban settings are of dubious value for analysing climate change due to the growth of urban heat islands as the urban areas expanded. This problem is serious given that many of the stations that began recording data in rural environments have been encroached upon or enveloped in urban sprawl. Gallo *et al.* (1995) and Owen *et al.*(1998) used night-time lights to assess urban heat island impacts on meteorological records. By plotting the location of recording stations on top of night-time light data, they have been able to identify which records are most heavily influenced by the urban heat island effect. Their ultimate objective is to remove the urban heat island influence on these records so that they may be pooled with records from observing stations in rural settings.

- The amount of urban sprawl in the past 20–30 years is staggering, particularly in the USA. From 1970 to 1990 the population of greater Chicago grew by 4 per cent, but the developed area grew by 50 per cent. For cities and towns in the USA, there is typically a 6–12 per cent increase in developed area for each 1 per cent of population growth. It is anticipated that the pattern of sprawl development in the USA will continue in the coming decades. The night-time light data provide perhaps the best continental- to global-scale depictions of urban sprawl. Imhoff *et al.* (1997a, b) used the stable lights to estimate the extent of land areas withdrawn from agricultural production. The World Resources Institute (Revenga *et al.* 1998) used night-time lights to assess the level of development present in major watersheds of the world. We anticipate that the night-time lights could be used to analyse the impacts of urban sprawl on the environment. For instance, they could be used to model the spatial extent of constructed material (buildings, roads, parking lots). This information could in turn be used to model the impacts of sprawl development on terrestrial carbon dynamics and alteration of the hydrologic properties of the land surface due to the construction of impervious surfaces. The night-time light data could also be used to assess the impact of urban sprawl on biodiversity. A recent study (Salmon *et al.* 2000) confirms the aversion of sea turtles to nesting on heavily lit beachfronts based on DMSP night-time lights.

- Several studies have noted the high correlation between the area, or brightness, of human settlements observed with DMSP data and energy-related trace gas emissions (Elvidge *et al.* 1997b, c; Doll *et al.* 2000; Doll, Chapter 15 in this book). The results suggest that the spatial distribution of carbon dioxide (CO_2) emissions from distributed fossil fuel consumption could be modelled based on night-time light data. By distributed sources, we refer to the consumption of fossil fuel by vehicles and in buildings of various types. Night-time lights would not provide an accurate depiction of emissions from major point sources, such as electric power generation facilities. The high spatial resolution depiction of the spatial distribution of CO_2 emissions from fossil fuel consumption has two primary applications. One, improved inverse modelling to estimate the magnitude of regional land–atmosphere and ocean–atmosphere fluxes, and, two, evaluating the validity of national trace gas emission estimates as reported under the UN Framework Convention on Climate Change and the Kyoto Protocol (as suggested by Saxon *et al.* 1997).

- DMSP night-time light data have also been used to detect power outages following hurricanes, earthquakes and other disasters (Elvidge *et al.* 1999b). The availability of night observations facilitates the use of DMSP data for measuring the reliability of electric power generation and delivery systems. A related application would be to use the data to track and monitor the effectiveness of rural electrification programs.

- Artificial lighting has reduced the visibility of the stars and planets to millions of people worldwide. Degradation in astronomical viewing conditions has forced many observatories to relocate, and substantially limits the value of many of the pioneering telescope facilities. Cinzano *et al.* (2001) used radiance-calibrated night-time lights to produce an atlas of artificial sky brightness and found that in many developed countries more than two-thirds of the population are no longer able to see the Milky Way from their homes.

The incredible low-light imaging capabilities provided by the DMSP program are expected to continue until at least the year 2010. The NOAA–DoD converged system of meteorological sensors (National Polar Orbiting Environmental Satellite System (NPOESS)), scheduled for deployment toward the end of this decade, will preserve the low-light sensing capability initiated with the OLS. Thus, the mapping of light sources present at the Earth's surface using night-time satellite sensor data can be expected to be a continuing source of information for the coming decades.

13.8 Acknowledgements

The authors gratefully acknowledge the US Air Force DMSP Program Office and the Air Force Weather Agency for providing the data used to construct the night-time lights products. The year 2000 stable lights of South-East Asia were generated for the Program to address ASEAN Regional Transboundary Smoke with support from the US State Department.

13.9   References

Cinzano, P., Falchi, F. and Elvidge, C. D., 2001, The first world atlas of the artificial night sky brightness. *Monthly Notices of the Royal Astronomical Society*, **328**, 689–707.

Croft, T. A., 1973, Burning waste gas in oil fields. *Nature*, **245**, 375–6.

Croft, T. A., 1978, Night-time images of the Earth from space. *Scientific American*, **239**, 68–79.

Croft, T. A., 1979, *The Brightness of Lights on Earth at Night, Digitally Recorded by DMSP Satellite*, Stanford Research Institute Final Report (Palo Alto, CA: US Geological Survey).

Dobson, J. E., Bright, E. A., Coleman, P. R., Durfee, R. C. and Worley, B. A., 2000, A global population database for estimating population at risk. *Photogrammetric Engineering and Remote Sensing*, **66**, 849–57.

Doll, C. N. H., Muller, J-P. and Elvidge, C. D., 2000, Night-time imagery as a tool for global mapping of socio-economic parameters and greenhouse gas emissions. *Ambio*, **29**, 157–62.

Elvidge, C. D., Baugh, K. E., Kihn, E. A., Kroehl, H. W. and Davis, E. R., 1997a, Mapping of city lights using DMSP Operational Linescan System data. *Photogrammetric Engineering and Remote Sensing*, **63**, 727–34.

Elvidge, C. D., Baugh, K. E., Kihn, E. A., Kroehl, H. W., Davis, E. R. and Davis, C., 1997b, Relation between satellite observed visible–near infrared emissions, population, and energy consumption. *International Journal of Remote Sensing*, **18**, 1373–9.

Elvidge, C. D., Baugh, K. E., Hobson, V. H., Kihn, E. A., Kroehl, H. W., Davis, E. R. and Cocero, D., 1997c, Satellite inventory of human settlements using nocturnal radiation emissions: A contribution for the global toolchest. *Global Change Biology*, **3**, 387–95.

Elvidge, C. D., Baugh, K. E., Dietz, J. B., Bland, T., Sutton, P. C. and Kroehl, H. W., 1999a, Radiance calibration of DMSP-OLS low-light imaging data of human settlements. *Remote Sensing of Environment*, **68**, 77–88.

Elvidge, C. D., Baugh, K. E., Hobson, V. R., Kihn, E. A. and Kroehl, H. W., 1999b, Detection of fires and power outages using DMSP-OLS data. In R. S. Lunetta and C. D. Elvidge (eds) *Remote Sensing Change Detection: Environmental Monitoring Methods and Applications* (Ann Arbor, MI: Ann Arbor Press), pp. 123–35.

Forster, J. L., 1983, Observations of the Earth using night-time visible imagery. *International Journal of Remote Sensing*, **4**, 785–91.

Gallo, K. P., Tarpley, J. D., McNab, A. L. and Karl, T. R., 1995, Assessment of urban heat islands – A satellite perspective. *Atmospheric Research*, **37**, 37–43.

Imhoff, M. L., Lawrence, W. T., Stutzer, D. C. and Elvidge, C. D., 1997a, A technique for using composite DMSP/OLS "city lights" satellite data to accurately map urban areas. *Remote Sensing of Environment*, **61**, 361–70.

Imhoff, M. L., Lawrence, W. T., Elvidge, C., Paul, T., Levine, E., Prevalsky, M. and Brown, V., 1997b, Using night-time DMSP/OLS images of city lights to estimate the impact of urban land use on soil resources in the US. *Remote Sensing of Environment*, **59**, 105–17.

Lieske, R. W., 1981, DMSP primary sensor data acquisition. *Proceedings of the International Telemetering Conference*, **17**, 1013–20.

Owen, T. W., Gallo, K. P., Elvidge, C. D. and Baugh, K. E., 1998, Using DMSP-OLS light frequency data to categorize urban environments associated with US climate observing stations. *International Journal of Remote Sensing*, 19, 3451–6.

Revenga, C., Murray, S., Abramovitz, J. and Hammond, A., 1998, *Watersheds of the World: Ecological Value and Vulnerability* (Washington, DC: World Resources Institute).

Salmon, M., Witherington, B. E. and Elvidge, C. D., 2000, Artificial lighting and the recovery of sea turtles. In N. Pilcher and G. Ismail (eds) *Sea Turtles of the Indo-Pacific: Research Management and Conservation* (London: Academic Press), pp. 25–34.

Saxon, E. C., Parris, T. and Elvidge, C. D., 1997, *Satellite Surveillance of National CO$_2$ Emissions from Fossil Fuels*, Development Discussion Paper No. 608. (Cambridge, MA: Harvard Institute for International Development, Harvard University).

Sullivan, W. T. III, 1989, A 10-km resolution image of the entire night-time Earth based on cloud-free satellite photographs in the 400–1100 nm band. *International Journal of Remote Sensing*, 10, 1–5.

Sutton, P., Roberts, D., Elvidge, C. and Meij, H., 1997, A comparison of night-time satellite imagery and population density for the continental United States. *Photogrammetric Engineering and Remote Sensing*, 63, 1303–13.

Tobler, W., Deichmann, U., Gottsegen, J. and Maloy, K., 1995, *The Global Demography Project*, Technical Report 95–6 (Santa Barbara, CA: National Center for Geographic Information and Analysis, University of California).

Welch, R., 1980, Monitoring urban population and energy utilization patterns from satellite data. *Remote Sensing of Environment*, 9, 1–9.

14 Estimation of human population parameters using night-time satellite imagery

Paul C. Sutton

14.1 Introduction

Following on from Chapter 13 by Elvidge *et al.* on the technical specifications of night-time sensor technology, the last three chapters of this book will explore relationships between night-time data and population counts (Sutton, this chapter), levels of GDP (Gross Domestic Product) and CO_2 emissions (Doll, Chapter 15) and crime deterrence (Weeks, Chapter 16). The strength of these relationships will, by inference, determine the potential of night-time data for forecasting population growth, economic activity, energy consumption and the creation of safer cities.

The first of these applications describes methods used to estimate two population parameters from the DMSP (Defense Meteorological Satellite Program) OLS (Operational Linescan Systems) night-time imagery. The first parameter is simply the total population of a city or urban cluster as identified by the imagery (referred to as aggregate estimation), and the second parameter estimated is the intra-urban population density within these urban clusters (referred to as disaggregate estimation). Two separate approaches for estimating the total population of cities are described in Sections 14.1.1 and 14.1.2, and two distinct approaches to estimating intra-urban population density are described in Section 14.2.1 and 14.2.2.

14.1 Aggregate methods: estimating the total population of a city

The total populations of cities throughout the world are estimated from their areal extent as identified by adjacent pixels of certain threshold value in the DMSP OLS imagery. The 1-km^2 pixels of the DMSP imagery are grouped on the basis of adjacency and then counted. The count of pixels contained in the urban cluster is considered to be a proxy measure of the areal extent of that urban cluster. The log–log relationship between the areal extent of an urban cluster and its corresponding population is the basis of this estimation (Stewart and Warntz 1958; Nordbeck 1971; Tobler 1969).

In Section 14.1.1 the high-gain "maximum observed" value of the DMSP data set was compared with a complete grid of population density for the continental USA. The population density data set was derived from the 1990 US Census (Meij 1995) (a summary of this work can be found in Sutton 1997). Because the quality of the US Census is so high it was possible to obtain a population estimate for *every* lit area in the continental USA. In addition, these measures of population would not be complicated by the variability of metropolitan statistical area definitions throughout the USA.

In Section 14.1.2 the high-gain "percentage observation" DMSP data set is compared with a point coverage of cities of the world with known populations. It was anticipated and demonstrated that the slope and intercept parameters for the ln(cluster area) vs. ln(cluster population) would vary from country to country. The slope and intercept parameters for this regression were obtained for every country of the world. Typically a country would have a small number of urban clusters that captured points with known populations. These points were then used to obtain slope and intercept parameters and then these parameters were used to estimate the population of all the lit clusters in the country. The sum of these estimates is considered to be an estimate of the urban population of the nation, and the total population of the nation can be calculated using known values for the per cent of population in urban areas for that nation. Consequently an estimate of the population of every country of the world was produced; the sum of which is an estimate of the global population.

14.1.1   Estimating the population of urban clusters (cities) of the continental USA using "high-gain" DMSP OLS imagery

An in-depth comparison of two images/data sets that cover the continental USA is described here (Figure 14.1). The first image is a 1-km^2 resolution grid of the population density derived from the 1990 US decennial Census. The grid was derived from the block group layer of the Bureau of the Census Topologically Integrated Geographic Encoding and Referencing (TIGER), and proportionally allocated to 1-km^2 cells. This data set was developed by the Socioeconomic Data and Applications Center (SEDAC) at the Center for International Earth Science Information Network (CIESIN) (Meij 1995). The second data set was produced by Elvidge *et al.* at the NOAA National Geophysical Data Center (Elvidge *et al.* 1995). This data set is a composite image of 231 orbits of the night-time passes of DMSP OLS platform over the continental USA.

14.1.1.1   Distribution of values in population density and DMSP OLS data sets

The DMSP OLS data have 6-bit quantization providing a dynamic range between 0 and 63. Once the stable lights image was geo-referenced, the land

Figure 14.1 Urban clusters identified by DMSP OLS imagery (top) and population density of USA, derived from 1990 US Census (bottom).

values were incremented by one unit to distinguish land pixels from ocean pixels. Of the pixels that are part of the land of the continental USA, 89 per cent have a value of one, 8 per cent have a value of 64 (saturated) and 3 per cent have intermediate values from 2 to 63 (Table 14.1). Table 14.1 describes the distribution of values of the overlaid pixels of the two data sets. The first column gives the values for the DMSP OLS pixels (1–64); the second column describes their relative frequency. The remaining columns provide descriptive statistics regarding the values of the population density

Table 14.1 DMSP OLS pixel values, statistics and corresponding population densities for conterminous USA

DMSP pixel value	# of pixels at this value	Mean of population pixels at value	Max of population pixels at value	Range of population pixels at value	Sum of population pixels at value	Std. dev. population pixels at value	Median population pixels at value	% of total total population at value	% of DMSP pixels
1	6,830,018	5.98	4,258	4,258	41,128,748	18.4	1	17.42	89.34
12	9	3.56	9	9	32	3.2	4	0.00	0.00
13	26	4.96	22	22	129	5.3	6	0.00	0.00
14	63	5.16	40	40	325	9.3	0	0.00	0.00
15	164	3.35	77	77	550	7.8	0	0.00	0.00
16	230	5.90	76	76	1,356	12.9	1	0.00	0.00
17	337	4.33	83	83	1,459	9.4	1	0.00	0.00
18	441	5.74	107	107	2,531	14.4	1	0.00	0.01
19	637	8.78	176	176	5,591	20.9	2	0.00	0.01
20	804	9.65	327	327	7,758	23.3	2	0.00	0.01
21	1,063	7.27	173	173	7,730	14.9	1	0.00	0.01
22	1,283	9.44	223	223	12,112	18.0	3	0.01	0.02
23	1,502	9.79	278	278	14,708	21.4	3	0.01	0.02
24	1,855	11.57	333	333	21,462	24.3	3	0.01	0.02
25	2,035	11.96	289	289	24,330	24.3	3	0.01	0.03
26	2,152	13.63	393	393	29,338	29.0	5	0.01	0.03
27	2,349	12.88	384	384	30,243	26.3	4	0.01	0.03
28	2,587	13.35	1,124	1,124	34,539	30.6	5	0.01	0.03
29	2,936	16.15	657	657	47,408	35.1	5	0.02	0.04
30	3,128	15.00	1,332	1,332	46,933	36.6	5	0.02	0.04
31	3,194	16.41	581	581	52,423	30.1	6	0.02	0.04
32	3,494	17.08	471	471	59,685	31.0	6	0.03	0.05
33	3,555	15.82	512	512	56,240	28.2	6	0.02	0.05
34	3,883	18.22	666	666	70,740	39.4	6	0.03	0.05

Age									
35	4,017	18.90	2,014	2,014	75,933	46.7	7	0.03	0.05
36	4,165	19.25	1,892	1,892	80,180	47.8	7	0.03	0.05
37	4,386	18.12	643	643	79,470	33.2	7	0.03	0.06
38	4,683	20.99	916	916	98,296	40.0	8	0.04	0.06
39	4,919	21.36	821	821	105,090	40.3	7	0.04	0.06
40	5,143	23.09	738	738	118,726	44.3	9	0.05	0.07
41	5,264	21.34	1,111	1,111	112,350	43.6	9	0.05	0.07
42	5,353	22.64	721	721	121,213	39.5	10	0.05	0.07
43	5,380	23.02	721	721	123,858	40.6	10	0.05	0.07
44	5,772	25.67	702	702	148,185	45.6	11	0.06	0.07
45	5,862	25.41	1,157	1,157	148,971	43.9	12	0.06	0.08
46	6,264	24.11	893	893	151,006	42.3	10	0.06	0.08
47	6,434	25.35	1,333	1,333	163,083	44.9	11	0.07	0.08
48	6,685	29.07	1,940	1,940	194,353	56.1	13	0.08	0.09
49	6,848	28.39	979	979	194,442	49.8	12	0.08	0.09
50	7,154	29.49	985	985	210,957	50.2	14	0.09	0.09
51	7,110	30.61	1,418	1,418	217,644	51.9	14	0.09	0.09
52	7,505	30.46	1,079	1,079	228,602	55.7	14	0.10	0.10
53	7,649	31.49	1,469	1,469	240,890	54.1	15	0.10	0.10
54	7,501	32.44	1,440	1,440	243,355	55.4	15	0.10	0.10
55	7,674	32.10	1,282	1,282	246,343	53.8	15	0.10	0.10
56	8,070	34.08	965	965	274,993	60.4	15	0.12	0.10
57	8,418	33.21	2,009	2,009	279,570	62.9	16	0.12	0.11
58	8,376	36.36	3,631	3,631	304,585	78.6	18	0.13	0.11
59	8,502	36.94	1,167	1,167	314,030	61.8	18	0.13	0.11
60	9,331	37.82	2,087	2,087	352,898	62.5	20	0.15	0.12
61	9,236	37.70	1,247	1,247	348,188	59.8	20	0.15	0.12
62	9,501	40.73	2,000	2,000	386,976	73.8	19	0.16	0.12
63	9,597	41.34	2,220	2,220	396,750	72.6	20	0.17	0.12
64	—	—	—	—	—	—	—	—	—
Total	586,561	321.36	53,465	53,465	188,497,243	798.3	7	79.69	7.62

pixels that overlap with the DMSP OLS pixels of that value. The first and last lines of this table are the most interesting. The first line represents those square kilometres that register "one" on the DMSP sensor (these pixels were really zeros but one was added to all pixels that were on the land to distinguish them from zero pixels on the ocean), these pixels constitute almost 90 per cent of the continental USA and only 17 per cent of the human population. The remaining 10 per cent of pixels (those with non-zero DMSP values) coincide with over 80 per cent of the human population. This may be one of the most valuable pieces of information with respect to how the "high-gain" DMSP OLS data can act as a proxy for population parameters. It should be noted that the frequency of all the intermediate values monotonically increases from low values to high. Clearly many of the pixels in urban areas are saturated at 64, and this presents many problems for identifying a quantitative correlation between light saturation and population density (particularly in areas of high population density). It also presents difficulties for transforming either variable to improve the correlation.

The population density image has a distinctly different distribution. Like the DMSP OLS image, the most common values are the low values of zero and one person per square km (41 per cent and 57 per cent of the pixels, respectively); however, the frequency of higher values decrease monotonically in a manner suggesting exponential decay. The values range from a minimum of zero to a maximum of over 50,000 persons/km^2.

14.1.1.2   *Measures of correlation at disaggregate and aggregate levels*

The first and most direct comparison of these two images was a simple correlation between the raw value of the population density pixel and the DMSP pixel value. This cross-correlation resulted in an $R^2 = 0.07$. One explanation for the low correlation is simply the fact that the DMSP saturates in most areas of high population density. The distribution of the DN (Digital Numbers) values of the two images clearly shows why direct correlations will be problematic. Transforming the population density data by taking its natural logarithm is one means of improving the correlation. The correlation between the DMSP image and the ln(population density) raises the correlation to $R^2 = 0.16$. This result remains unsatisfactory, and there is no obvious physical justification for performing such a transformation.

The "high-gain" DMSP OLS imagery locates over 80 per cent of the continental US population on only 10 per cent of the land. The third column of the table is a measure of the average population density for pixels at the corresponding value of the DMSP OLS pixel. There is a trend toward increasing population density with increasing DN values for the DMSP OLS imagery; however, this trend makes a jump at the last DMSP value (i.e., from 63 to 64). The mean population density value increases from around

10 persons per square kilometre at a DN of 1, to 41 persons per kilometre at a DN of 63, but it jumps to 321 at the next increment of the DMSP value. This is a result of the saturation of the DMSP sensor in what are heavily urbanized areas. Further exploration with both the low-gain DMSP OLS data set and the per cent detection version has shown stronger direct relationships with some population parameters on a pixel-by-pixel basis.

14.1.1.3 Estimating city population: the log(area) vs. log(population) relationship

Another manipulation of the data was performed in which all the saturated pixels in the DMSP OLS stable lights image were grouped in such a way that all adjacent saturated pixels were clustered into independent urban "clusters". These urban clusters were overlaid over the population density data set to produce a set of over 5,000 paired data points in which the first value was the area of saturation or urban cluster and the second value was the actual population that lived within that area. It should be noted that *every* lit area in the DMSP image was associated with a corresponding population from the population density image.

A simple linear regression between these two variables produced an R^2 of 0.62 (Figure 14.2). A quick glance at this plot might suggest that there is a problem with heteroscedasticity or unequal variance. This appearance may merely be a result of the paucity of data points at higher values or it may indicate that the relationship is even stronger for larger urban areas (i.e., has lower variance for large values). The ln(area) vs. ln(population) relationship can be dramatically improved by weighting the points by their respective populations. Without weighting, a small town in Nevada will influence the line as much as the New York City cluster. Weighting improves the R^2 to 0.98. This increased R^2 results from larger cities at the "less populated" end of the scatterplot having greater influence on the estimation of the regression parameters. Without weighting, ten points representing extremely small portions of the population will have ten times more influence on the regression than a single point that has more than a thousand times the population of those ten points combined. The use of a weighted regression eliminates disproportionate representation.

Saturated areas of the imagery can be grouped into urban clusters. The linear relationship between the natural log of these areas and their corresponding populations produce slope and intercept parameters. The following research demonstrates that these parameters will vary from one region of the world to the next. If these parameters can be either identified for other parts of the world or shown to be related to simple national aggregate statistics such as per cent of population living in urban areas, GDP per capita and/or energy consumption, the methods described are useful. This potential is explored in Section 14.1.2.

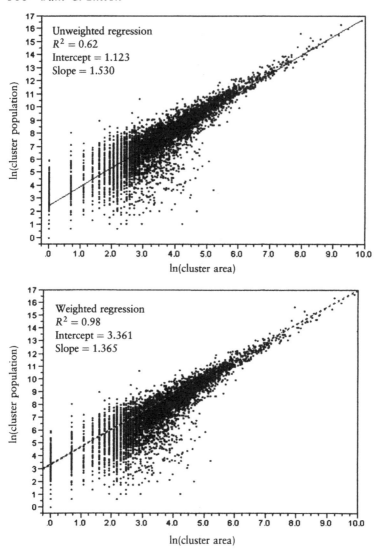

Figure 14.2 Regression of ln(area) vs. ln(population) for all urban clusters identified by DMSP OLS imagery over the continental USA.

14.1.2 *Estimating the population of the world's cities (urban clusters) using "high-gain" per cent observation DMSP OLS data*

In Section 14.1.2 the methods described in Section 14.1.1 are applied to all the countries of the world. The one difference being that a composite "per cent observation" DMSP OLS global data set is used instead of the "high-gain" data set. This data set is compared with a dataset of over 1,500 cities with known populations. This analysis produces an estimate of the population of every lit area of the world identified by the DMSP imagery. These estimates are obtained with nationally varying slope and intercept parameters. The following is a more detailed description of these efforts and the resulting aggregate estimates of both the global human population and the population of all the world's nations.

14.1.2.1 Night-time satellite imagery data

Section 14.1.2 describes efforts at estimating the population of every country of the world using the stable night-time lights data set produced from hundreds of orbits from the DMSP OLS (Elvidge *et al.* 1995). Night-time satellite imagery acquired by DMSP OLS is being used to map human settlements globally (Elvidge *et al.* 1997a; Dobson *et al.*, Chapter 12 in this book), map urban extent nationally (Imhoff *et al.* 1997a) and has been strongly correlated to population and energy consumption at nationally aggregated levels (Elvidge *et al.* 1997b; Doll, Chapter 15 in this book). The pixels of this data set are not a measurement of light intensity, but a record of the percentage of light observations in a given location. Consequently, the values of the pixels range from 0 to 100. The global hypertemporal mosaic image shown in previous chapters is classified into four categories: city lights, gas flares from oil processing, lantern fishing and forest fires. The city lights image has pixel values that vary from 0 to 100. A value of 100 means that light (of any intensity) was observed in 100% of the many orbits used to create the image (the number of orbits used to determine this value varies spatially due to clouds, glare, etc.). This analysis uses only those pixels classified as city lights.

14.1.2.2 Area vs. Population relationship for cities of known population

The ln(cluster area) vs. ln(cluster population) relationship used to estimate the population of the urban areas of the world is described in Section 4.1.1. For this image there were 22,920 urban "clusters". The image was super-imposed on a data set of 1,597 cities with known populations resulting in 1,383 points of analysis. This reduced number of 1,383 is due to the effects of conurbation. Each point is referred to as an "urban cluster" with known areal extent and population. Conurbation of cities is often reproduced in the night-time satellite image when several cities fall into one "lit" area. When this occurred the populations of those cities were summed for that urban cluster. A scatterplot (Figure 14.3, see colour plates) of the ln(area) vs. ln(cluster population) of these 1,383 clusters provides a sense of the strength of this relationship at five levels of aggregation: globally (all the points), for high-income countries (green points; GDP/capita >$5,000), for medium-income countries (blue points; $1,000 \le$ GDP/capita $\le$ $5,000), for low-income countries (red points; GDP/capita < $1,000) and for the urban clusters of Venezuela (black points).

14.1.2.3   Exclusion of Chinese cities and the circularity problem

The scatterplot of Figure 14.3 does not include points representing the cities of China. These points were excluded from the parameterization process for two reasons: (1) the population figures for the cities of China were unreliable, and (2) by excluding the Chinese cities an estimate of the urban and national population of China could be derived in an independent manner. There is an element of circularity in this method of estimating the population of urban clusters by using known populations. By removing the Chinese cities and applying the results of this study to them, the issue of circularity is addressed and the robustness of the method is demonstrated.

14.1.2.4   Setting light thresholds for various nations of the world

The fundamental objective of using the night-time satellite imagery in the context of this research is as a proxy measure of the areal extent of urban centres. Thresholding is a process where the value of the pixels that define urban extent is set on a national basis. A threshold of 80 for a particular nation would use only those pixels in which light was observed 80–100 per cent of the time. For example, consider a pixel on the outskirts of Mumbai, India. If any light was observed in 25 out of 50 orbits of the satellite that pixel would receive a value of 50. Varying levels of industrialization throughout the world suggest that a single threshold on this data set would be inappropriate. Setting image thresholds at the national level is a trade-off between two conflicting objectives. Low thresholds capture more area and smaller urban clusters that would otherwise not be identified at high thresholds; this is generally needed in the less developed countries. Higher thresholds prevent the "melding together" or conurbation of urban clusters into mega-clusters that do not lend themselves to accurate population estimates. Earlier work using this data set by Imhoff *et al.* (1997) has shown that a threshold of 82 per cent was a good measure of urban extent within the USA. In an attempt to capture all the urban population of the world thresholds of 40 per cent, 80 per cent, and 90 per cent were used. It is hoped that the thresholding process will be greatly improved via the use of the low-gain data set described earlier (Elvidge *et al.* 1998) and other national studies.

14.1.2.5   Estimating the urban and national population of a nation

The urban and national population of every nation of the world was estimated using these methods. The urban clusters with known populations were used to identify a slope and intercept parameter for each nation from a linear regression between the ln(area) and the ln(cluster population) (Sutton *et al.* 2001). The regression was weighted by the known populations of their spatially matching clusters. The slope and intercept obtained were then used to estimate the population of every urban cluster in the country identified by the imagery. The sum of the populations of these urban clusters is the estimate of the urban population of the country. The total population of the country is calculated using published values for the percentage of the nation's population that lives in villages and cities of over 2,000 people (Haub and Yanagishita 1997).

A large number of countries did not have a sufficient number of cities with known population to perform this regression. In these cases, pooled regression statistics were used. A regression was run on the clusters of all the low-income countries, medium-income countries and high-income countries (based on their GDP/capita). If a country had no cities of known population the parameters from the appropriate pooled regression were used. However, if the country had a city of known population, then the pooled slope was used and the city with a known population defined the intercept.

14.1.2.6 *Globally aggregated estimates*

This research is consistent with existing geographic theory, applies it to estimating national and global populations, and provides an empirical definition and estimate of urban land cover. The strong linear correlation between the natural logarithms of a city's population and its areal extent as measured by night-time satellite imagery validates the theories of Nordbeck and Tobler. The increased strength of the relationship that results when the cities are classified by nation or by GDP/capita is consistent with theory in the areas of urban and transportation geography. Empirical estimates of these parameters for all the nations of the world produce a global population estimate of 6.295 billion, which is 7 per cent greater than the accepted estimate of 5.867 billion. The independent estimate of China's population is 1.512 billion, which is 22 per cent greater than the accepted estimate of 1.243 billion. The methods used implicitly measure the areal extent of urban land cover for every nation of the world. An independent estimate of urban land cover (USGS/IGBP [US Geological Survey/International Geosphere – Biosphere Programme]) was obtained (Belward 1996; Loveland *et al.* 1991) and compared with the DMSP OLS derived estimate. In virtually all cases the urban extent as measured by the night-time satellite imagery exceeds those of the USGS/IGBP global land cover data set. The global estimate of urban land cover was almost 1 per cent of the Earth's land surface. Estimates of the total populations of countries aggregated for (1) all the low-income countries, (2) all the medium-income countries, (3) all the high-income countries and (4) the whole planet are reported (Table 14.2). The percentage error between the aggregate estimates and known aggregate populations is less than 8 per cent. Errors in the estimates grow as national income drops; however, this may be due to the fact that the percentage of population in urban areas generally increases as income drops and this has a multiplicative effect on the error in national population estimates.

Table 14.2 Aggregate estimates of population and extent of urban land cover

Population estimation (N = # of clusters with known population)	Slope	Standard error of slope	Intercept	Standard error of intercept	R^2	Actual population	Estimated population	% Error
Total world (N = 1,404)	0.85	0.016	9.107	0.102	0.68	5,867,566,272	6,224,626,469	6.1
High income (N = 471)	1.065	0.027	7.064	0.201	0.77	1,074,842,200	1,105,827,550	2.9
Medium income (N = 575)*	1.011	0.022	8.174	0.141	0.78	1,127,715,000	1,166,132,502	3.4
Low income (N = 347)*	1.189	0.031	7.425	0.19	0.81	3,665,009,072	3,952,666,417	7.8

Urbanization estimation		Total area (km²)	Urban area by DMSP	Urban area by IGBP	% Urban cover by DMSP	% Urban cover by IGBP
Total world		143,393,924	1,158,977	257,085	0.81	0.18
High income	(GDP/Capita > $5,000)	40,314,690	646,722	132,710	1.60	0.33
Medium income	($1,000 ≤ GDP/Capita ≤ $5,000)	47,557,079	298,300	89,641	0.63	0.19
Low income*	(GDP/Capita < $1,000)	55,522,155	258,012	34,734	0.46	0.06

*Threshold = 40.

A frequently asked question regarding the utility of these methods is: "*If we know the urban populations of the cities, which we use to calibrate this imagery, why do we need to use the imagery to determine the population of the cities?*" This fundamental question deserves several responses. First of all, the ln(area) – ln(population) relationship is interesting in its own right (as determined by Nordbeck and Tobler), but it is also interesting that this relationship is even stronger when night-time satellite imagery is used to measure the areal extent of urban areas. Second, the night-time satellite imagery used in tandem with simple GIS operations demonstrates how the ln(area) – ln(population) relationship varies systematically with such a basic number as national GDP. Third, by using the ln(area) – ln(population) relationship in tandem with nationally aggregate statistics such as GDP (future research may identify additional variables), independent estimates of the populations of urban areas can be made with these methods. Clearly these methods are no substitute for a thorough population census; however, these methods can be used to: (i) supplement censuses of population in countries that lack the resources to conduct exhaustive population censuses, (ii) serve as alternative measures of population censuses in areas with dubious population measures (e.g., if China's population figures are as unreliable as their fish harvest measures these methods may prove informative) and (iii) inform studies that investigate how urban extent varies with population of urban areas globally.

Table 14.2 summarizes the night-time satellite and USGS/IGBP-based measures of urban land cover at these aggregate scales. The DMSP OLS-based measure estimates 0.81 per cent of the Earth's land to be urbanized, a number that is 4.5 times larger than the IGBP figure of 0.18 per cent. The DMSP OLS-derived night-time imagery is being considered by the USGS as an improved means of delineating urban extent for its global land cover maps (Loveland 1997). At the aggregate level, the estimates of the slope and intercept parameters for the low-, medium- and high-income countries are statistically significantly different. However, the significance of these differences weaken or disappear at national levels; Pakistan and Japan's parameters are not significantly different from one another despite the large difference in their GDP per capita.

14.1.2.7 *Nationally aggregated estimates*

Information on the national statistics, identified parameters and estimated populations for the sixty most populous countries of the world all warrant consideration. The percentage error of these estimates is more variable than the aggregate measures based on national income. For twenty of the sixty most populous nations a single large city was used to estimate the intercept of the regression. A table of the parameters, populations, errors and thresholds for all the countries is shown in Table 14.3.

Table 14.3 National statistics, identified parameters and estimated populations for the sixty most populous nations of the world

Country	% population urban	GNP capita	Income class	Slope	Standard error of slope	Intercept	Standard error of intercept	R^2	Threshold	# Clusters known population	# of clusters from DMSP	Total national population	DMSP estimate of national population	Absolute error (1,000's)	% Error	Minimum det. population
China	30	3,640	Low	1.185	0.031	7.224	0.190	n/a	40	None used	944	1,242,500,000	1,512,611,819	270,112	22	1,372
India	26	340	Low	1.339	0.028	6.317	0.178	0.93	40	179	1952	969,729,000	1,026,965,385	57,236	6	554
USA	75	26,980	High	1.252	0.049	4.851	0.390	0.84	80	116	4215	267,661,000	285,641,333	17,980	7	128
Indonesia	31	980	Low	0.997	0.055	8.165	0.369	0.94	40	21	476	204,323,000	221,735,484	17,412	9	3,516
Brazil	76	3,640	Medium	1.186	0.048	7.446	0.317	0.87	80	85	944	160,343,000	190,086,842	29,744	19	1,713
Russia	73	2,240	Medium	1.202	0.051	6.986	0.300	0.84	80	105	796	147,264,000	140,560,274	-6,704	-5	1,081
Pakistan	28	460	Low	1.302	0.113	5.955	0.743	0.90	40	18	396	137,752,000	100,315,307	-37,437	-27	386
Japan	78	39,640	High	1.389	0.069	5.166	0.548	0.81	80	62	487	126,054,000	135,212,821	9,159	7	175
Bangladesh	16	240	Low	1.369	0.053	6.180	0.336	0.99	40	10	100	122,219,000	91,530,000	-30,689	-25	483
Nigeria	16	260	Low	1.185	0.038	5.937	*Ibadan*	n/a	40	1	115	122,000,000	204,711,250	82,711	68	8,111
Mexico	71	3,320	Medium	1.508	0.042	4.653	0.303	0.97	80	55	635	95,724,000	95,507,606	-216	0	105
Germany	85	27,510	High	0.858	0.080	8.120	0.523	0.76	40	39	602	82,022,000	57,798,542	-24,223	-30	3,361
Vietnam	20	240	Low	0.926	0.161	9.111	0.980	0.89	40	6	66	75,123,000	89,219,000	14,596	19	9,054
Phillipines	47	1,050	Medium	1.090	0.071	8.680	0.470	0.89	80	13	40	73,419,000	84,700,851	11,282	15	10,467
Iran	58	n/a	High	1.003	0.041	7.862	0.949	0.95	90	34	405	67,540,000	87,544,483	20,004	30	2,597
Egypt	44	790	Low	1.617	0.457	5.131	2.970	0.86	90	4	108	64,792,000	82,422,273	17,630	27	169
Turkey	63	2,780	Medium	0.917	0.052	9.198	0.307	0.91	80	37	245	63,674,000	88,942,857	25,269	40	9,877
Thailand	19	2,740	Medium	1.011	0.022	7.707	*Bangkok*	n/a	80	1	127	60,088,000	75,687,368	15,599	26	2,224
UK	90	18,410	High	1.005	0.072	7.673	0.465	0.81	80	48	274	58,800,000	40,789,111	-18,011	-31	2,150
Ethiopia	15	100	Low	1.185	0.038	7.518	*Addis Abbaba*	n/a	40	1	28	58,733,000	34,251,200	-24,482	-42	10,775
France	74	24,990	High	1.270	0.075	5.787	0.525	0.91	80	30	386	58,633,000	45,240,676	-13,392	-23	326
Italy	67	19,020	High	0.877	0.068	8.265	0.418	0.82	80	41	345	57,429,000	53,629,701	-3,799	-7	3,885
Ukraine	68	1,630	Medium	1.105	0.098	6.976	0.603	0.86	40	24	409	50,719,000	49,354,860	-1,364	-3	1,071
Zaire	29	120	Low	1.185	0.038	6.292	*Lubumbashi*	n/a	40	1	27	47,440,000	28,264,862	-19,175	-40	8,587
Myanmar	25	n/a	Low	1.185	0.038	7.654	*Rangoon*	n/a	40	1	55	46,822,000	47,718,000	896	2	9,490
South Korea	74	9,700	High	1.085	1.061	8.143	0.412	0.96	80	16	99	45,850,000	44,867,027	-983	-2	3,439
South Africa	57	3,160	Medium	1.011	0.022	7.138	*Johannesburg*	0.89	80	9	254	42,465,000	31,577,393	-10,888	-26	1,259

Spain	64	13,580	High	1.185	0.079	5.975	0.532	0.90	80	26	433	39,330,000	34,920,313	-4,410	-11	393
Poland	62	2,790	Medium	0.912	0.049	7.650	0.304	0.93	40	26	352	38,648,000	35,625,771	-3,022	-8	2,101
Colombia	70	1,910	Medium	0.871	0.087	8.484	0.516	0.85	80	21	252	37,418,000	34,367,000	-3,051	-8	4,837
Argentina	87	8,030	High	1.247	0.037	6.362	0.267	0.98	80	19	277	35,558,000	31,491,839	-4,066	-11	579
Canada	77	19,380	High	1.286	0.087	5.058	0.657	0.91	80	25	919	30,142,000	39,815,325	9,673	32	157
Algeria	50	1,600	Medium	1.273	0.041	6.806	0.825	0.87	80	8	183	29,830,000	34,106,072	4,276	14	903
Tanzania	21	120	Low	0.867	0.137	9.025	0.706	0.91	40	6	31	29,461,000	27,316,429	-2,145	-7	8,308
Kenya	27	280	Low	1.185	0.038	6.746	*Nairobi*	n/a	40	1	48	28,803,000	24,234,667	-4,568	-16	6,088
Morocco	51	1,110	Medium	1.230	0.176	7.276	0.980	0.86	80	10	99	28,217,000	28,674,118	457	2	1,445
Sudan	27	n/a	Low	1.185	0.038	5.788	*Khartoum*	n/a	40	1	59	27,899,000	34,473,293	6,574	24	5,693
Peru	70	2,310	Medium	1.298	0.068	6.715	0.439	0.97	80	14	83	24,362,000	20,826,857	-3,535	-15	825
North Korea	61	n/a	Low	1.185	0.038	8.357	*Pyongyang*	n/a	40	1	16	24,317,000	7,921,574	-16,395	-67	20,171
Uzbekistan	38	970	Low	1.487	0.127	4.070	0.838	0.94	40	10	148	23,672,000	20,264,842	-3,407	-14	59
Nepal	10	200	Low	1.185	0.038	7.063	*Kathmandu*	n/a	40	1	22	22,641,000	27,385,300	4,744	21	5,162
Venezuela	85	3,020	Medium	1.196	0.143	6.251	0.950	0.84	80	15	205	22,576,000	23,608,824	1,033	5	519
Romania	55	1,480	Medium	1.039	0.089	8.397	0.430	0.92	80	14	59	22,535,000	15,525,185	-7,010	-31	4,434
Afghanistan	18	500	Low	1.185	0.038	4.605	0.230	0.51	40	0	13	22,132,000	22,963,667	832	4	15,245
Taiwan	75	n/a	High	1.065	0.027	7.460	*Tai-Pei*	n/a	40	1	30	21,535,000	32,999,736	11,465	53	1,737
Iraq	70	n/a	Low	1.041	0.041	7.346	0.651	n/a	40	1	147	21,177,000	21,488,916	312	1	1,550
Malaysia	51	3,890	Medium	0.610	0.093	9.982	0.559	0.90	80	7	73	21,018,000	26,904,314	5,886	28	21,634
Uganda	11	240	Low	1.185	0.038	7.252	*Kampala*	n/a	40	1	11	20,605,000	22,130,636	1,526	7	7,824
Saudi Arabia	80	7,040	High	1.065	0.027	6.719	*Jiddah-Mecca*	n/a	80	1	280	19,494,000	32,602,000	13,108	67	828
Australia	85	20,090	High	1.186	0.179	5.989	1.295	0.84	80	10	248	18,700,000	19,760,941	1,061	6	399
Sri Lanka	22	700	Low	1.185	0.038	5.235	*Colombo*	n/a	40	1	37	18,665,000	12,984,318	-5,681	-30	2,827
Mozambique	28	80	Low	0.985	0.317	8.871	1.693	0.90	40	3	21	18,355,000	20,481,929	2,127	12	7,122
Ghana	36	390	Low	1.185	0.038	5.960	*Accra*	n/a	40	1	41	18,102,000	17,891,778	-210	-1	5,437
Kazakhstan	56	1,330	Medium	0.814	0.186	8.802	0.995	0.76	80	9	80	16,433,000	16,558,929	126	1	6,648
Netherlands	61	24,000	High	0.612	0.045	10.258	0.274	0.97	80	8	48	15,598,000	16,267,148	669	4	28,510
Yemen	25	260	Low	1.185	0.038	6.620	*Sanaa*	n/a	40	1	66	15,214,000	14,313,992	-900	-6	2,276
Ivory Coast	46	660	Low	1.185	0.038	7.218	*Abidjan*	n/a	40	1	38	14,986,000	17,579,978	2,594	17	10,604
Syria	51	1,120	Medium	0.838	0.097	8.154	*Damascus*	n/a	80	1	53	14,951,000	7,309,078	-7,642	-51	3,477
Chile	85	4,860	Medium	1.011	0.022	8.585	*Santiago*	0.66	80	1	72	14,800,000	18,849,529	4,050	27	5,351

Note: If standard error of intercept column has a *City Name* in it that city was used to define the intercept.

The intercept of the regression for each country provides a theoretical minimum detectable population for a 1-km^2 urban area. For the same threshold of 80, these values were 1,169 for high-income countries, 3,548 for medium-income countries and 7,251 for low-income countries. (A lower threshold of 40 was used for the lower income countries in which the minimum detectable population was 1,201.) These results agree with a simple theory of more light emission per capita as an increasing function of income. Many wealthy countries such as Japan, Canada and the USA had low minimum detectable populations as expected (1-km^2 lit areas had predicted populations of 175, 157 and 128, respectively). High minimum detectable populations tended to occur for the less developed countries: North Korea, Ethiopia, Afghanistan and the Ivory Coast (1-km^2 lit areas had predicted populations of 20,171, 10,775, 15,245 and 10,604, respectively). However, not all nations fit these trends. For example, Uzbekistan, which has a relatively low GDP/capita, had the lowest minimum detectable population (57) of these 60 nations; while the Netherlands recorded the highest minimum detectable population with a value of 28,510. Some of these anomalies can be attributed to small sample sizes or the unusual population densities of single large or heavily urbanized cities in the nation in question; nonetheless, GDP/capita alone does not explain the variation.

The minimum detectable population has straightforward implications on national estimates of total urban population. This model tends to under-estimate the total national populations for less developed countries including Pakistan, Bangladesh, Ethiopia, Zaire and North Korea. This would be expected if the minimum detectable population is much larger than the Population Reference Bureau's definition of urban area. Some of the largest percentage overestimates occurred in the more developed countries, such as Australia, Canada and Taiwan. However, these trends had many notable exceptions including substantial underestimates of Germany and the UK, and overestimates of Afghanistan and Nigeria. Nonetheless, this model's estimates of the national populations were within 25 per cent of accepted values for more than 65 per cent of the sixty most populous nations. These results are impressive considering the fact that the model is derived from a single night-time image mosaic.

14.1.2.8   *Between-nation variation of the slope and intercept parameters*

At the aggregate level (i.e., the whole world) a strong correlation is observed between the areal extent of an urban cluster (as identified by night-time satellite imagery) and its corresponding population. Disaggregation of this information to three nationally aggregated income levels greatly increases the strength of this correlation. General trends across income levels are observed with respect to both the slope and intercept of this relationship. However, the effect of the intercept is more pronounced and suggests that a small city of a given area in a low-income country will have a higher population (and population density) than a city of the same areal extent in a wealthy country. The cause of this is likely to be the availability of modern transportation: cities in which cars, buses and rail are not as available will be more compact and have higher population densities. A time-series study of the same ln(area) vs. ln(population) of the city of Ann Arbor shows a dramatic change in the relationship after World War II; this happens to correspond with the widespread use of cars in that city (Tobler 1975). The variation in the intercept parameter was more significant than variation in the slope between nations and income classes. This agrees with the earlier findings of Stewart and Warntz (1958) in a similar analysis of cities in Europe.

Aggregation averages the errors in the over- and underestimates of both the cluster and national population estimates; this is similar to how signal averaging eliminates noise. However, if aggregation were simply analogous to signal averaging then the scatterplot of the ln(area) vs. ln(population) for the cities of known population in a particular country would not be expected to have any stronger a correlation than a random selection of cities from around the world. Nonetheless, this is not what was observed. The increased strength of the population vs. area relationship via aggregation by GDP per capita has been demonstrated. The fact that a much stronger relationship exists for the urban clusters of any given nation hints that the spatial attributes of the data in question can prove to be useful in explaining the relationship between the areal extent and population of an urban area.

14.1.2.9 *Discussion of error*

Explicitly defining the error in this study is difficult if not impossible. There is no "official" estimate of the world's population and our knowledge of the world's total human population has at most two significant figures and most likely only one (i.e., billions represent the only significant figure in estimates of global population) (Cohen 1995). Estimates of the population of the cities and countries of the world are just that: *estimates*. The errors described here are comparisons between DMSP OLS-based estimates and existing estimates. Existing estimates of city and national populations vary dramatically in quality and availability. Future research in this area will require an accurate population figure for a large number of cities throughout the world.

A simple model based on a potentially crude set of city population values has been parameterized. The circularity introduced by using known populations of cities will introduce bias in the results if there is any bias in the city population estimates. It is possible that many city population estimates are underestimates for several reasons. First, many city population figures represent the population within the city boundary rather than the metropolitan area population. Second, large conurbations often include many small cities (typically less than 100,000) that were not on the list of cities with known population, thus biasing the estimates of many of the conurbation clusters toward the low end. Finally, virtually every city for which extra information was sought had a population substantially higher than the published figures.

For example, conventional wisdom in Mexico (and among many population experts) suggests that the population of Mexico City is somewhere between 21 and 24 million. The official figure used was 18.5 million. This has a substantial influence on the estimate of the national population of Mexico. The estimate of Mexico's population would have been significantly higher if a larger estimate for the population of Mexico City had been used. The case for Riyadh, Saudi Arabia was so dramatic that the published population of Riyadh was not used and the conurbation of Jiddah and Mecca defined the regression parameters for Saudi Arabia. The Jiddah–Mecca conurbation and Riyadh had areal extents of 2,213 km^2 and 2,317 km^2, respectively. Yet the population numbers for these two clusters were 3.2 and 1.3 million, respectively. The resulting estimate of Riyadh's population was 3.2 million and agreed closely with expert opinion on the population of Riyadh (3.3 million) (Al-Sahhaf 1998). It is believed that there is a bias toward underestimation of the "known" city populations, and therefore it is believed that it is very unlikely that the aggregate estimates are overestimates. If the numbers used for the percentage of population in urban areas on a national basis are accurate, it is very likely that the global population estimate of 6.3 billion is reasonable or perhaps even an *under*estimate. In any case, it is a sobering thought to contemplate the implications of 400 million more people on the planet. The number is sufficiently different from the conventional 5.9 billion to have significant impact on commonly calculated numbers such as global grain production per capita, per capita availability of fresh water and per capita consumption of seafood.

14.2   Disaggregate methods: estimating intra-urban population density

The following describes two methods in which the disaggregate population density within these urban clusters can be estimated. The third method (Section 14.2.1) discussed is based on utilizing existing urban geographic theory about population density decay as a function of distance from the city centre. A new twist in the analysis is provided in that the distance functions are defined from the edge of the urban area rather than the centre of the city. The second method is explored as a result of the limitations of precision of the first method. The second method explores the variation of population density within an urban area as a function of observed light intensity at night. This is a substantial departure from existing theories of urban geography. Admittedly, it is an approach that has only recently been feasible as a result of improved technology (remote sensing and GIS [Geographical Information Systems]) for making and analysing empirical observations. However, the empirical success of night-time light intensity as a measure of temporally averaged population density suggests that the method has great potential.

These studies use the USA as a case study for identifying appropriate spatial and temporal scales for modelling that can be generalized to other parts of the world. The data used is DMSP OLS data and the actual population density data derived from the 1990 US Census described in Section 14.1.1. It is hoped that this exploration of the night-time satellite imagery and population density will expose some systematic general properties of the relationship between night-time satellite imagery and population density within urban centres. The quality and high spatial resolution of the US census data validates the ln(area) – ln(population) relationship while the large sample size of the ln(area) vs. ln(population) for many cities of the world demonstrates how this relationship varies with nationally aggregate statistics. Future research will undoubtedly identify how these relationships can be built upon to improve means by which remote sensing can enhance our understanding of human population distributions.

Section 14.2.1 uses the high-gain "maximum observed DN" DMSP OLS data set to define the urban clusters of the continental USA. A model for allocating population density to the 1-km^2 pixels within these urban clusters was developed that uses the distance of each pixel from the edge of its cluster as a primary determinant of population density. Those pixels that are deeper in the cluster are given higher population density values based on several different population density decay functions. This is essentially identical to using the negative exponential functions identified by Clark. However, distance is defined from the edge of the cluster rather than from the centre of the city. This is summarized in Sutton (1997).

In Section 4.2.2 a model for allocating population density to the 1-km^2 pixels is developed that uses the light intensity of each pixel as derived from the "low-gain" version of the DMSP OLS imagery. Higher light intensity values result in higher population density estimates. This method was explored because of the lack of precision of the distance-based models. This is an exploration of a potentially new empirical paradigm through which a better theoretical and empirical understanding of the urban environment may be apprehended.

14.2.1 Estimating population density using distance from the edge of urban cluster (e.g., reversing the direction of Clark's studies)

The data sets used to perform these analyses were the same continental coverages of the USA at a resolution of 1 km^2 as described in Section 14.1.1. The second data set was an image or "grid" of the continental US population density derived from 1990 census data at the block group administrative boundary level. The methods described here explain various means of disaggregating the earlier results back to the 1-km^2 pixel resolution.

14.2.1.1   Theoretical models

The methods adopted involve taking advantage of the spatial nature of the DMSP data. The DMSP image ultimately used was a simple binary image with saturated values and dark values. The saturated pixels of the DMSP OLS image were grouped into urban clusters based on their adjacency (Rook's rule) to other saturated pixels. Each saturated pixel in the DMSP OLS image was classified based on a number uniquely identifying the cluster to which it belonged and a number representing the distance of that pixel to the edge of the cluster to which it belonged. This gave a unique number to each pixel that identified the cluster of all the "lighted" pixels that were both saturated in the DMSP image and adjacent to one another. This resulted in over 5,000 distinct clusters. One additional manipulation was incorporated to account for coastal cities. This was done to account for the fact that the densest part of cities such as Chicago, Los Angeles, etc. is often right on the coast. Consequently the pixels of the ocean, lakes, and Mexican and Canadian borders were treated as if they were urban. This resulted in coastal and border cities having the highest distances to the edge on the centre of their coastal or border contacts. Figure 14.4 shows the distribution of distance values and data structure for three clusters of different shape but equal area including a coastline cluster.

The data structure of Grids in Arc/INFO includes a table called the Value Attribute Table (VAT) that contains two items, a *Value* and a *Count*. The value represents the value assigned to the pixel or cell. The count is the number of pixels that have that particular value. In order to produce a grid that incorporated both the distance-to-edge information and cluster identification, the following manipulation was performed. First, the grid that had values for all the unique clusters (which had a VAT whose count value represented the area of each urban cluster in square kilometres) was multiplied by a thousand. This would result in Cluster Number 1 becoming Cluster Number 1,000. The distance-to-edge grid was multiplied by 10 and truncated to integers. This resulted in each pixel having values that represented their distance to dark continental US land in tenths of kilometres. At this point, however, a grid cell on the edge of the New York cluster is the same as the pixel on the edge of any other urban cluster including the Los Angeles, Phoenix (AZ) and Tucumcari (NM) clusters. The values of all these pixels are less than 100 km; consequently, adding the cluster ID grid to the distance-to-edge grid results in a grid where each pixel has a value in which the numbers from 0 to 999 represent the pixel's distance to the edge of its cluster, and the numbers from one thousand upward represent a number that uniquely identifies the cluster in which that pixel occurs (e.g., the pixels

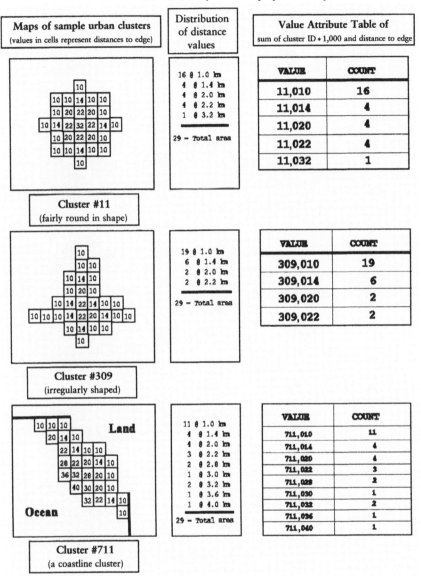

Figure 14.4 Three sample urban clusters of equal area on left with Value Attribute Table data structure on the right.

on the edge of Cluster Number 1 have values of 1,010 (10 being the smallest value). The VAT table produced for this grid conveniently supplies the number of unique occurrences of each value. Thus the VAT table for this grid has values that contain both the distance-to-edge information and count

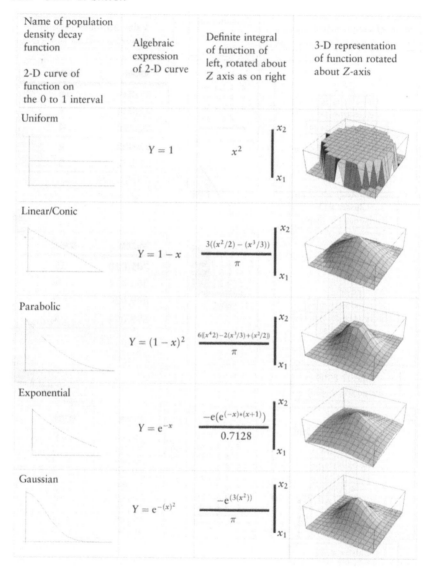

Name of population density decay function 2-D curve of function on the 0 to 1 interval	Algebraic expression of 2-D curve	Definite integral of function of left, rotated about Z axis as on right	3-D representation of function rotated about Z-axis
Uniform	$Y = 1$	x^2	
Linear/Conic	$Y = 1 - x$	$\dfrac{3((x^2/2) - (x^3/3))}{\pi}$	
Parabolic	$Y = (1 - x)^2$	$\dfrac{6((x^4/2) - 2(x^3/3) + (x^2/2))}{\pi}$	
Exponential	$Y = e^{-x}$	$\dfrac{-e(e^{(-x)*(x+1)})}{0.7128}$	
Gaussian	$Y = e^{-(x)^2}$	$\dfrac{-e^{(3(x^2))}}{\pi}$	

Figure 14.5 Graphic and algebraic representations of the five urban population density decay functions used in this chapter.

information that indicates the number of pixels that are at X distance from the edge of Cluster Y (Figure 14.5).

The next stage of this analysis involves the manipulation of the VAT for the grid just described. The manipulation involves adding new columns to the table, which are used to model the population density within each cluster according to any integrable function describing the decay of population density from an urban centre. However, this model turns the distance function around and uses distance from the edge of an urban centre instead. Nonetheless, the decay functions used will be the traditional functions proposed describing population density as a function of distance from the urban centre. The question then becomes: how does one apply the traditional models based on a circular city to the irregularly shaped clusters identified from the DMSP night-time satellite imagery?

The way this was done was via manipulation of the *Value* and *Count* items of the VAT for this grid. A simple program was written that calculated the following: (1) cluster ID, (2) cluster area, (3) radius of equivalent circle, (4) total population of cluster, (5) upper limit of integration for equivalent circle at distance D, (6) lower limit of integration for equivalent circle at distance D and (7) population density estimate. Most of these values can be found for the clusters shown in Figure 14.4 in the table at the bottom of Figure 14.6. This table has population density estimates for all the urban decay functions described in Figure 14.5. Cluster ID was obtained simply by dividing by 1,000 and truncating. Cluster area was obtained by summing the count values by the cluster ID. Radius of equivalent circle was obtained by simply solving the equation:

$$\text{Cluster area} = \pi * (R)^2 \tag{14.1}$$

for R. Total population of cluster was obtained by using the formula obtained from previous work describing the linear relationship between the natural log of the population of urban areas and the natural log of the area of those clusters: total population of cluster $= \exp[1.365 + 3.361$ (cluster area)] (Sutton *et al.* 1997). The R^2 for the log–log relationship is 0.98. The limits of integration warrant further explanation. It is important to know that the VAT file is sorted in ascending order. The limits of integration are calculated in the following manner. The urban population density functions all begin at a distance of 0 and end at a distance of 1, with the highest or central population densities at 0. Figure 14.3 (colour section) and Figure 14.4 describe the urban decay functions used. The limits of integration are merely the radii of circles nested in a circle of radius one. An example is shown in Figures 14.3 and 14.4 and in more graphic detail in Figure 14.6. Suppose a cluster has 100 1-km^2 pixels in it. Thirty-five of these pixels are on the outer edge of this cluster with a distance value of 1 km from the edge. The upper limit of integration for these pixels would be 1.0 and the lower limit of integration for these pixels would be the radius of a circle that contained 65 per cent of the area of the unit circle (65 per cent comes from 100 total – the 35 in question). The unit circle has an area of π, thus by solving the equation:

$$0.65 * \pi = \pi * (R)^2 \tag{14.2}$$

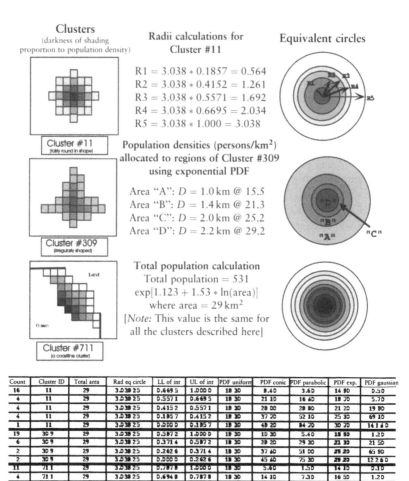

Figure 14.6 The three urban clusters with their proportional equivalent circle and a table denoting the various estimates of population density within them based on the five population density decay functions described in Figure 14.5.

for R, the lower limit of integration is obtained (e.g., 0.806). These radii are solved cumulatively within each cluster ID resulting in steadily decreasing limits of integration as distance from the edge increases to its maximum and the last lower limit of integration is 0.

These limits of integration are then used in the definite integral associated with various urban population density decay functions shown in Figure 14.5. The value of this definite integral is the proportion of the total population that live in the pixels that are that distance from the edge of the cluster. This fraction is multiplied by the estimated total population of the cluster and divided by the number of pixels at that distance to produce an estimate of the population density of those pixels. It may be important to note here that all the pixels at a distance x from the edge of a particular cluster will be assigned the same population density value. The integrals solved are the urban population density decay functions rotated about the y-axis. For a graphic description of this method of estimating population density see Figure 14.6.

Several different functions were used to model the population density within each urban cluster (Figure 14.5). The simplest was a uniform model that simply assigned each pixel the same average population density on a per-cluster basis. The second was a simple linear function that decreased from 1 to 0 over the range 0 to 1. The third was a parabolic function that decreased from 1 to 0 over the same 0 to 1 range. The fourth was an exponential decay that ranges from 1 to $\exp(-1)$ as distance varies from 0 to 1. The fifth was the standard Gaussian distribution for which the limits of integration were bumped up to 3 to include the first three standard deviations. An appropriate multiplicative constant was used on each of the definite integrals to ensure that they all integrated to unity over the limits of integration: 0 to 2π and 0 to 1. These models were all developed on a theoretical basis with no "training" from the empirical data. Empirical models derived from averages of the empirical data are also easy to produce. An exploration of the empirical models is described in the next section.

14.2.1.2 Empirical models

In addition to applying these theoretical models to estimate the population density inside these urban clusters it was also possible to determine empirical, population density decay functions from the actual population density data. This was accomplished in the following manner. The grid or image that contained the distance-to-edge information in all the saturated pixels was laid over the actual population density grid. An average population density was calculated for each distance value. In addition a standard deviation was determined for each distance value. The sample size for these plots decreases with increasing distance. The leftmost point for both these plots is based on calculating the average population density and standard deviation of the population density for all the pixels that are on the edge of any cluster. As the distance to the edge of the cluster increases, the sample size (e.g., number of pixels used to calculate these values) drops significantly. As one moves to the right in these plots to greater distances one is moving into the hearts of these urban clusters. A simple linear fit on this curve produces an R^2 of 0.87. This linear fit was used as an empirical model for the population density.

14.2.1.3   Results for both theoretical and empirical models

All the curves show population density increasing with distance in ways reminiscent of the theoretical exponential decays of traditional urban geographic theory. However, the empirical models outperformed all the theoretical mathematical models described in Figure 14.5. None of the models including the theoretical model had R^2 values greater than 0.30. This shows that, in order to improve correlations, future models must include information that accounts for the variability in population density that occurs at constant distances to the edge of these clusters.

The method of modelling population density presented here requires very little input information: a binary DMSP image and a relationship between area and population for *urban* clusters. The model is then developed using the spatial information in the image and applying various urban population density decay functions to the *urban* clusters. These theoretical models explain almost as much variation as empirically derived, population density decay functions that depend on distance alone. These rather simple beginnings may prove to be a good starting point for the development of a more accurate method of modelling population density that uses additional independent sources of information such as the greenness index of AVHRR (Advanced Very High Resolution Radiometry), digital elevation data and the low-gain version of the DMSP imagery (which shows much more variation within urban clusters).

The concept of population density is an abstract one. Consequently attempting to model it raises many questions. The ground truth data used in this chapter were derived from a vector data set of the block group polygons of the 1990 US Census. Assumptions were made to convert this data set to a grid. One assumption is the uniformity of population density within the block group polygons. Using this dataset as the ground truth may not be entirely appropriate. Alternative ground truthing methods may show that the correlations obtained here are actually underestimates. The census is also a measure of the night-time population location. Despite the fact that the DMSP imagery is also produced at night it may be providing clues as to where the population is in the day also. The downtown areas of many cities show up as very bright despite the fact that they often have lower population densities. These population densities are much higher during the working hours of the day. How much does the population density of an urban centre vary on a temporal basis? Clearly it is difficult to measure the error inherent in a model of this nature because the truth is an elusive quantity that varies temporally and spatially in complex ways that are difficult if not impossible to measure for appreciably sized areas. These issues suggest that the most appropriate models will be at a coarser resolution than the 1-km scale focused on in this chapter. The model described in this chapter accounts for 25 per cent of the variation in the population density of the urban areas in the continental USA from information contained in a binary image derived from DMSP OLS imagery and some relatively simple spatial analysis. Section 14.2.2 explores the potential of the low-gain DMSP OLS data for improving these disaggregate population density estimations.

14.2.2   Within-cluster population density estimation using "low-gain" DMSP light intensity (a new empirical paradigm?)

In this section the actual intensity or brightness of light is used as a direct measure of population density within the urban clusters that have been discussed up to this point. The "low-gain" DMSP OLS imagery is used as a proxy measurement of urban population density. The model estimates population density in the urban areas defined by stable night-time light. These estimates can be made unbiased at any spatial scale by incorporating two additional pieces of information: (1) the total population of the region and (2) the percentage of the population that lives in urban areas. These are typically available for most nations of the world and often available at subnational levels also. These figures are not needed to use this model; however, if the figures are valid they can reduce regional biases.

The night-time imagery used in this case was the "low-gain" DMSP OLS imagery. In all cases the comparisons performed are for the urban areas of these images only. The estimate of urban population density at each non-zero pixel of the night-time satellite image is based on the following model:

$$
\begin{pmatrix} \text{Estimated} \\ \text{Population} \\ \text{Density} \end{pmatrix} = \begin{pmatrix} \text{Estimated} \\ \text{Cluster} \\ \text{Population} \end{pmatrix} \times \frac{\text{DMSP Low-Gain Pixel Value}}{\begin{array}{c} \text{Sum of all DMSP Low-Gain} \\ \text{Pixel Values in Clusters} \end{array}}
$$

or

$$
E_j = E_k * (\mathrm{lgv}_i / \Sigma(\mathrm{lgv}_i)) \qquad i \in M_k \tag{14.3}
$$

where lgv_i = low-gain value of pixel i, k = index of cluster, j and i = index of pixels, E_k = estimated total population of cluster k, E_j = estimated population density of pixel i and M_k = pixel membership of cluster k.

The estimated cluster population is derived using the parameters of the weighted regression, which shows the log–log relationship between cluster area and cluster population. One attribute of the population density data set that presents problems is its high degree of spatial variability. A one-pixel misregistration of a perfect model to the actual data set will only result in an R^2 of 0.74. These issues are addressed indirectly via spatial aggregation and smoothing. Spatial aggregation reduces the variability of pixel values toward their mean. Smoothing retains the spatial resolution of the data set but reduces the variability of the pixels locally (Holloway 1958). In this study the ground truth data set was smoothed using a 5 by 5 mean filter and an 11 by 11 mean filter. The justification for mean smoothing is the fact that census data record where people are at night. A mean spatial filter is a means of performing a smoothing that in reality is created through time as people move from home to office to store to school.

The model of population density derived from night-time satellite imagery was evaluated on the 1-km² DMSP OLS low-gain image, and on aggregations of the image to pixels with 5- and 10-km sides. The larger pixel images were simply mean aggregations of the finer resolution image. The resulting "predictions" of population density were then compared with both the smoothed and unsmoothed "ground truth" population density. This model has two primary sources of error when predicting the population density of a pixel. The first is error associated with the cluster as a whole. In a sense this is a regional bias on a per-cluster basis. The second source of error manifests from spatial misallocations within the cluster, as happens in well-lit areas that have low population densities according to the census data. The relative contribution of these two sources of error can be measured. This is accomplished by calculating the Mean Absolute Deviation (MAD) of the total error grid. The MAD of the error grid was 214 persons/km². This is the total error of the model. To estimate the relative contributions of the first and second types of error another grid was created. This grid uses the areally averaged error of each cluster and adds it to the error grid resulting in an image that is what the error grid would look like if each urban cluster's population was estimated perfectly. This grid had a MAD of 198 persons/km². This shows that only 7 per cent $(100 * (214 - 198)/214)$ of the error is of the first kind (e.g., errors in estimating the cluster populations). Most of the error (93 per cent) is of the second kind, which results from spatial mismatches of population density within the clusters.

An example is provided by the Minneapolis–Saint Paul (MN) cluster (Figure 14.7). The images in Figure 14.7 include unsmoothed population density derived from the block group polygons, the 5×5 and 11×11 mean filtered versions of those data, the model derived from weighted, least squares, low-gain DMSP OLS imagery, and the 5-min by 5-min global demography project. The average error for the Minneapolis–Saint Paul cluster is -170 persons/km² (representing an overestimate of the population of this cluster by 812,900 people and divided by an area of 4,782 km²: $170 = 812,900/4782$).

Additional sources of error are illustrated by the Los Angeles urban cluster (Figure 14.8). This has been adjusted to have a mean of zero and the resulting error classification is based on standard deviations. This error has been subtracted from each pixel to provide an unbiased image of the error that results from misallocating population density within the cluster. Many of the errors are clearly interpretable. For example, the population density of urban centres are typically underestimated. These areas tend to be well lit in

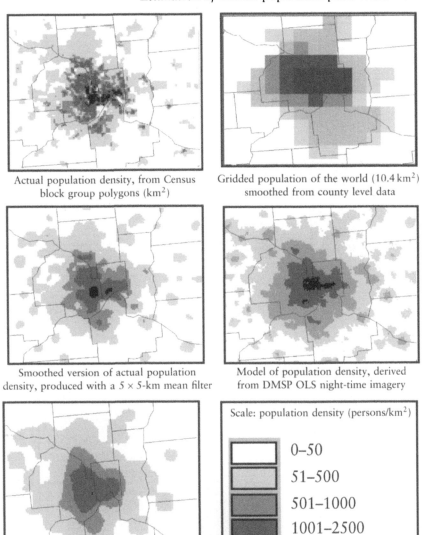

Actual population density, from Census block group polygons (km²)

Gridded population of the world (10.4 km²) smoothed from county level data

Smoothed version of actual population density, produced with a 5 × 5-km mean filter

Model of population density, derived from DMSP OLS night-time imagery

Scale: population density (persons/km²)

0–50

51–500

501–1000

1001–2500

2501–8,000

Smoothed version of actual population density, produced with a 11 × 11-km mean filter

Figure 14.7 Representations of the population density of Minneapolis–Saint Paul from various versions of the data sets described in this chapter.

the image; however, the linear proportion of population allocated to these areas is insufficient to account for the true population density. These errors could be mitigated with the use of a non-linear allocation of population density. The two large airports in the cluster are significant overestimates.

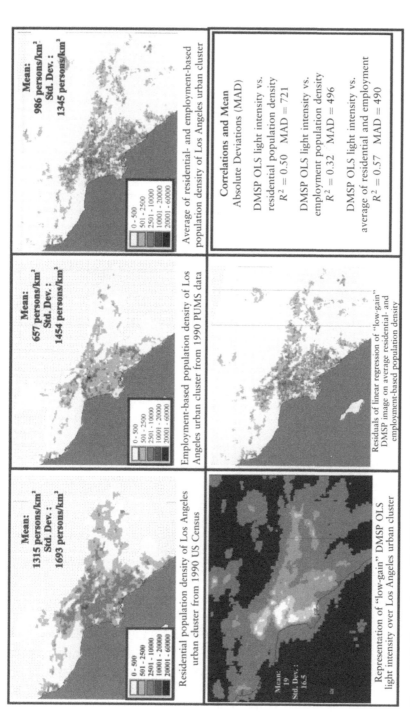

Figure 14.8 Comparisons of DMSP OLS-based images of night-time light intensity over Los Angeles with residential- and employment-based measures of population density.

The area around the Los Angeles International Airport (LAX) shows up as a large error of overestimation of population density. LAX is one of the largest employers in the city and tens of thousands of people fly in, out or through the airport on a daily basis; yet, the area has a relatively low population density according to the data derived from the census.

14.3 Conclusions

There are many benefits provided by a greater understanding of the relationship between remotely sensed imagery and ground-based measures of socio-economic and demographic information. Many countries of the world lack the financial and/or institutional resources to conduct useful censuses. Models derived from readily available satellite imagery that have been validated in parts of the world where good ground-based information is available could serve as reasonable proxy measures in countries that lack such information. In addition, if a country has some limited resources with which to conduct an incomplete census of its population, existing imagery for that country could be used to help design statistical sampling strategies for a limited census. These sampling strategies could be designed to maximize the effectiveness and accuracy of proxy measures derived from satellite imagery. These methods could also be used to provide inter-censal estimates. This symbiosis allows for and benefits from independence of ground- and imagery-based measures.

These proxy measures of socio-economic and demographic information provided by remotely sensed imagery are easily incorporated into myriad environmental analyses such as global warming, deforestation, loss of biodiversity, etc. The finer spatial and temporal resolution of these proxy measures of population will be useful in many studies for planning, mitigation and response to natural and anthropogenic disasters. If these kinds of model are developed and maintained over time the time series data could be used for validation of urban growth models, planning for habitat preservation and informing "smart growth" initiatives. Establishing models that estimate demographic information from remotely sensed imagery also has the potential to increase cooperation between nations and between those nations and the UN.

14.4 Acknowledgements

Parts of this chapter have been reproduced with permission of the American Society for Photogrammetry and Remote Sensing for use of material from the paper by Sutton, P., Roberts, D., Elvidge, C. and Meij, H., 1997, "A comparison of night-time satellite imagery and population density for the continental United States". *Remote Sensing*, **63**, 1303–13.

Parts of this chapter have been reproduced with permission of Taylor & Francis for use of material from the paper by Sutton, P. C., Roberts, D., Elvidge, C. and Baugh, K., 2001, "Census from heaven: An estimate of the global human population using night-time light satellite imagery". *International Journal of Remote Sensing*, 22, 3061–76. http://www.tandf.co.uk

14.5 References

Al-Sahhaf, N., 1998, Personal communication on population of Riyadh, Saudi Arabia.

Belward, A. S., 1996, *The IGBP-DIS Global 1 km Land Cover Data Set (DISCover) – Proposal and Implementation Plans* (Toulouse, France: IGBP-DIS).

Cohen, J. E., 1995, *How Many People Can The Earth Support?* (London: W. W. Norton).

Elvidge, C. D., Baugh, K. E. and Davis, E. A., 1995, Mapping city lights with night-time data from the DMSP Operational Linescan System, *Photogrammetric Engineering and Remote Sensing*, 63, 727–34.

Elvidge, C. D., Baugh, K. E., Kihn, E. A., Kroehl, H. W. and Davis, E. R., 1997a, Mapping city lights with night-time data from the DMSP Operational Linescan System. *Photogrammetric Engineering and Remote Sensing*, 63, 727–34.

Elvidge, C. D., Baugh, K. E., Kihn, E. A., Kroehl, H. W., Davis, E. R. and Davis, C., 1997b, Relation between satellite observed visible – near infrared emissions, population, and energy consumption. *International Journal of Remote Sensing*, 18, 1373–9.

Elvidge, C. D., Baugh, K. E., Hobson, V. H., Kihn, E. A., Kroehl, H. W., Davis, E. R. and Cocero, D., 1997c, Satellite inventory of human settlements using nocturnal radiation emissions: A contribution for the global toolchest. *Global Change Biology*, 3, 387–95.

Elvidge, C. D., Baugh, K. E., Dietz, J. B., Bland, T., Sutton, P. C. and Kroehl, H. W., 1999, Radiance calibration of DMSP-OLS low-light imaging data of human settlements. *Remote Sensing of Environment*, 68, 77–88.

Elvidge, C. D., Baugh, K. E., Dietz, J. B., Bland, T., Sutton, P. C. and Kroehl, H. W., 1998, Radiance calibration of DMSP-OLS low-light imaging data of human settlements, *Remote Sensing of Environment*, 68, 77–88.

Haub, C. and Yanagishita, M., 1997, *World Population Data Sheet* (Washington DC: Population Reference Bureau).

Imhoff, M. L., Lawrence, W. T., Stutzer, D. C. and Elvidge, C. D., 1997a, A technique for using composite DMSP/OLS "city lights" satellite data to accurately map urban areas. *Remote Sensing of Environment*, 61, 361–70.

Imhoff, M. L., Lawrence, W. T., Elvidge, C., Paul, T., Levine, E., Prevalsky, M. and Brown, V., 1997b, Using night-time DMSP/OLS images of city lights to estimate the impact of urban land use on soil resources in the US. *Remote Sensing of Environment*, 59, 105–17.

Loveland, T., 1997, DMSP OLS imagery to be used for urban land cover classification.

Loveland, T., Merchant, J., Ohlen, D. and Brown, J., 1991, Development of a land cover characteristics database for the conterminous US. *Photogrammetric Engineering and Remote Sensing*, 57, 1453–63.

Meij, H., 1995, *Integrated Datasets for the USA* (New York: Center for International Earth Science Information Network).

Nordbeck, S., 1971, Urban allometric growth, *Geografiska Annaler*, 533, 54–67.

Stewart, J. and Warntz, W., 1958, Physics of population distribution. *Journal of Regional Science*, 1, 99–123.

Sutton, P., 1997, Modeling population density with night-time satellite imagery and GIS. *Computers, Environment, and Urban Systems*, 21, 227–44.

Sutton, P. C., Roberts, D., Elvidge, C. and Baugh, K., 2001, Census from heaven: An estimate of the global human population using night-time light satellite imagery. *International Journal of Remote Sensing*, 22, 3061–76.

Tobler, W., 1969, Satellite confirmation of settlement size coefficients. *Area*, 1, 30–4.

Tobler, W., 1975, *City Sizes, Morphology, and Interaction* (Laxenburg, Austria: International Institute for Applied Systems Analysis).

15 Estimating non-population activities from night-time satellite imagery

Christopher N. H. Doll

15.1 Introduction

The expansion of global change research has opened a Pandora's box of data requirements of increasingly diverse parameters. Initially, environmental monitoring was confined to the condition of the biosphere with little regard to the overarching agents of change. As we improve our understanding of the interactions of the various processes involved, so too does our identification of the key factors lying at the heart of global change. The ever-broadening remit of global change demands an increasing number of disciplines to be employed to fully describe it.

The global economy is an example of such a key factor. There is a growing realization that it is a major driving force of global change as it influences, first, urbanization and, second, levels of power consumption. Taken together, these two activities spawn serious issues facing human populations, such as increased pollution and its associated health effects, loss of agricultural land and pressure on water resources. Economic activity is not only a driving force of environmental stress, but also a determinant of the adaptability and ameliorative capacity of a society to deal with its impacts. A joint study by the Environment Programme (UNEP) and Development Programme (UNDP) of the United Nations highlighted the economy as one of the top ten high-priority data themes for environmental assessment (UNEP/UNDP 1994). Additionally, the issues of land use change, hydrology, air and water quality are also regarded with the same importance. Many of the major data requirements can be mapped by satellites, increasingly more so with the launch of new sensors and satellite missions such as the National Aeronautics and Space Administration's (NASA) Earth Observing System (Kaufman *et al.* 1998a). This programme will seriously augment current remotely sensed data sets by establishing a database of Earth science products for the land, ocean and atmosphere that is expected to last until at least 2015.

The increasing use of satellite sensor data for the creation of systematic environmental data products stimulates the requirement to provide data at a constant resolution. This is something that is not often available for socio-economic data since they are often reported at regional, national, sub-national or other irregular spatial units. The question therefore presents itself that assuming night-time lights are representative of human activity, how well do they correlate with facets of human activity such as economic activity and energy consumption? By association they may also serve as indirect tracers of greenhouse gas production (Southwell 1997).

Night-time satellite imagery has provided researchers with a means to push back the boundaries of what is conventionally thought to be the limits of environmental remote sensing. As a data source, it has the potential to offer information on human activity unlike anything that can be derived from other forms of remotely sensed imagery. This chapter is concerned with how night-time satellite imagery as captured from the Defense Meteorological Satellite Programme's Operational Linescan System (DMSP OLS) sensor described in previous chapters can model and map various facets of human activity.

15.2   Global mapping of GDP and CO_2 emissions from DMSP OLS imagery

The potential for night-time light imagery to describe economic activity and energy usage has been noted in early papers written on the subject (Croft 1978; Welch 1980; Sullivan III 1989). Elvidge *et al.* (1997) examined correlations of lit area and Gross Domestic Product (GDP) and lit area and power consumption for twenty countries in the Americas. Doll *et al.* (2000) developed this further by establishing global regression relationships between lit area in a country and GDP, and CO_2 emissions. Global 1-degree resolution maps were created from these relationships.

15.2.1   *Mapping methodologies*

Producing maps at 1-degree resolution from night-time satellite imagery requires generalization of the existing data, which are at a much finer resolution (30 arcsec in this case). The night-time light data were summarized into 1-degree units by calculating the "percentage lit" value for each new cell. Global maps and models invariably use geographic projection to represent the Earth's surface as it is one of the most convenient for displaying global data at a variety of resolutions and allows easy indexing of any position through the use of latitude/longitude coordinates. This presents a problem when using an empirically derived relationship based on area to distribute data, since it is not an equal area projection. Converging lines of longitude result in a square degree at the equator being larger than the same unit at higher latitudes. Although most of the data of interest are within 60° north or south of the equator, a degree of longitude is still half the length at 60° than it is at the equator.

Table 15.1 Dimensions of the latitudinal zones used to create maps

Latitudinal zone	Dimension at 30" (E/W × N/S) (m)	Area at 1 degree (km^2)
60°+ (northern hemisphere)	400 × 929	5,351.040
50°–60°	531.25 × 927.5	7,095.375
40°–50°	655 × 926.1	8,734.975
30°–40°	755 × 924.6	10,052.251
10°–30°	850 × 922.5	11,291.400
−10°–10°	920 × 921.5	12,208.032

In order to map onto this system, mean cell areas were calculated for ten latitudinal zones (Table 15.1). The summarized night-time light data could then be calculated directly to a lit area figure for each cell. The relationships between lit area and GDP, and CO_2 emissions calculated are given below:

$$CO_2 = 10^{(1.2508*(\log area)-0.7291)} \tag{15.1}$$

$$GDP = 10^{(0.9735*(\log area)+0.7124)} \tag{15.2}$$

The global product was generated by applying the relationship to each latitudinal band and merging the subgrids.

The CO_2 emissions map was compared with one previously prepared at 1-degree resolution by the Carbon Dioxide Information and Analysis Center (CDIAC) (for method see Andres *et al.* 1996 and ORNL/CDIAC for data location). Their map used a global, gridded, population density map developed by Li (1996) to distribute the emission figures. Human population centres and the location of night-time lights are also very highly correlated, which helps to explain the excellent correspondence of emission locations between the two maps. The two histograms also resemble each other, both exhibiting a bimodal distribution with the largest class in each case belonging to the lowest range of emissions (0–25 kilotonnes of carbon [ktC]) and a second peak in the 100–250-ktC class. This may be interpreted as the division between developing and developed countries since maximum emissions in developing countries rarely exceed this figure.

A difference map CDIAC OLS (Figure 15.1, see colour plates) was used to identify those areas over- and underestimated by CDIAC. The modal class of this histogram centred on the zero value, but generally the night-time light map underestimated emission values. These differences were greatest in those regions that had the highest values. More interesting, however, are those areas that have been overestimated in the DMSP OLS results by applying the global relationship. These areas include parts of Canada, Scandinavia and southern Brazil. This could be indicative of energy generation from non-fossil fuels. The Itaipu Dam in Paraguay for instance also supplies energy to the neighbouring southern states of Brazil, and Sweden has particularly low relative levels of fossil fuel consumption. The emissions total of the night-time light map is 1.53 billion tonnes of carbon, around 25 per cent of the CDIAC map total. The widespread under-estimation of emissions suggests that other factors need to be considered in the model. Certainly issues such as the contribution of fossil fuels to national energy consumption, and its relationship to lighting practices would appear to be appropriate places to start to identify the biasing factors between countries which affect global mapping of CO_2. A more detailed discussion of the assumptions and extra data sources required to refine greenhouse gas emissions mapping from radiance-calibrated night-time imagery is given in Section 15.3.4.

Validation of the GDP map is more speculative as there is no other map of its kind. In terms of a global figure of economic activity, the map sums to 22.1 trillion international dollars (Intl.$), some 80 per cent of the total for the input data. The international dollar is the currency unit for the purchasing power parity GDP statistic, which attempts to account for the purchasing power of different currencies thus making international compar-isons more meaningful. There is a much wider range of values in the GDP map due to the higher intercept of the allometric relationship. Even though it has a lower gradient than the CO_2 relationship, the value of the constant is an important feature in log–log space where the gradients are similar as it influences the magnitude of the retrieved anti-log. Features such as the Trans-Siberian Railway are apparent as well as hubs of economic activity within a country such as Moscow and New Delhi. The developed nations on the other hand exhibit less variation between high and low values. Successive night-time light data sets will allow for analysis of the change in relationship parameters, as well as tracking the development of burgeon-ing economies such as China (see Figure 15.2, colour plates).

Using global relationships to disaggregate data from the country level, national totals, from which the relationships were derived, are not preserved in the maps due to the best fit nature of a regression line. An alternative to this approach was presented by Li *et al.* (1996) who used a political map of the world segmented over a 1-degree grid. The study aimed at mapping concentrations of hexachlorocyclohexane (HCH) an organochlorine insec-ticide, which was widely employed to improve agricultural yield, protect livestock and eliminate vector-transmitted disease. The figure for usage of this compound was also quoted at the country level. A NASA/GISS (NASA's Goddard Institute for Space Studies) map of cultivation intensity at 1° by 1° resolution (Matthews 1983) was used to disaggregate the country level totals. Five intensity levels were identified from these data. Since the disaggregation data were already at the output resolution of the map, no generalization was required. However, instead of deriving a global correla-tion between cultivation intensity and HCH concentrations, the segmented 1-degree grid was used to distribute the national total into the grid cells lying within a national boundary. Where a grid cell was intersected by one or more political units, the relative contribution of the cell for each unit was calculated and then assigned the commensurate proportion of the intensity value according to the following formula:

$$V_{ij} = \frac{C_i R_{ij}}{\sum_i C_i R_{ij}} V_j \tag{15.3}$$

where for a cell i and political unit j: C is the cultivation intensity value, R_{ij} is the percentage of total area j in cell i and V_j is the total value of the parameter to be mapped (HCH). In this case, C and V_j would be replaced by radiance and GDP/CO_2, respectively. This method assumes that the quantity is evenly distributed within a cell. This is unlikely to be the case for night-time imagery as lights tend to be clustered. The problem of splitting allocated proportions in a cell by intersecting it with a line or polygon coverage is a fundamental problem in geographic information science. The error involved in allocating data to cells is inversely proportional to the resolution of the cells in the grid.

15.3 The potential of radiance-calibrated data

The maps created in the previous section (Figures 15.1 and 15.2, both in colour plate section) were the results of applying an empirical relationship based on area. A radiance-calibrated data set offers an expanded range of possible correlations since the radiance profile within a lit area can be used as weights. Doll and Muller (1999a) used radiance-calibrated data to examine how they could be used to improve its population predictions from the traditional allometric lit area relationships identified by Tobler (1969). In this case, the radiance-calibrated data offered the refinement of modelling intra-urban population density and was found to offer much more robust estimates for the twelve countries in their study. Radiance-calibrated data can be used in much the same way for estimating other parameters. GDP, CO_2 emissions and power consumption have all been shown to be highly correlated with lit area figures at the country level (Doll *et al.* 2000; Elvidge *et al.* 1997). When estimating a parameter like power consumption, a radiance value may be a more useful attribute to take as a proxy than the lit area. This section examines the relationship of "cumulative radiance" to these parameters, and also the nature of relationships at different spatial scales. Cumulative radiance is here defined as the summed radiance value over a number of pixels.

15.3.1 *Subnational spatial classifications*

The wealth of statistical data for the UK, produced by the Office of National Statistics (ONS), provides a way of assessing how well an approach like this performs at a smaller subnational scale. Nomenclature of Units for Territorial Statistics (NUTS) is a hierarchical classification of spatial units that provides a breakdown of the EU's territory for reporting regional statistics that are comparable across the Union. Starting at the NUTS-0 (country) level, countries are divided up into broad regions (NUTS-1) and then individual counties or groupings thereof (NUTS-2). There are eleven NUTS-1 regions in the UK and around three NUTS-2 areas in each NUTS-1 region. Many social and economic data sets can be requested and downloaded from their website (ONS 2002).

15.3.2   *Radiance and GDP: an analysis for the UK*

The established area–GDP relationship developed at the national level appears to be invalid at the subnational scale. Table 15.2 shows the correlation between night-time light parameters and GDP at different spatial units. The cumulative radiance correlates better but the correlation statistic is still low. The Greater London region has an anomalously high GDP for both its cumulative radiance and its area. Although London accounts for just under 2 per cent of the total lit area of the UK, it contributes 15.9 per cent to national GDP (Table 15.3). Conversely, the East Midlands and Scotland have a greater proportion of lit area compared with their proportion of GDP contribution. In this sense sole reliance on an areal relationship between light and economic activity at the subnational regional level is unrealistic, the R^2 value of 0.04 testifying to this. However, when the Greater London region is removed from the data, a positive correlation is apparent.

If one hypothesizes that increasing radiance is linearly related to GDP, then one way to disaggregate the statistic is to divide the GDP by the total

Table 15.2 Correlation coefficients for fourteen European countries (NUTS-0), and NUTS-1/2 regions for the UK between lit area and radiance with GDP (excluding Northern Ireland)

Parameter	R^2 *(linear–linear)*	R^2 *(log–log)*
NUTS-0 Europe lit area	0.84	0.72
NUTS-0 Europe radiance	0.74	0.66
NUTS-1 lit area	0.04	0.01
NUTS-1 total radiance	0.30	0.50
NUTS-1 primary (lit area)	0.66	0.90
NUTS-1 primary (radiance)	0.19	0.33
NUTS-1 secondary (lit area)	0.27	0.18
NUTS-1 secondary (radiance)	0.78	0.71
NUTS-1 tertiary (lit area)	0.01	0.002
NUTS-1 tertiary (radiance)	0.20	0.42
NUTS-2 lit area	0.22	0.14
NUTS-2 radiance	0.61	0.72

Table 15.3 NUTS-1 statistics of lit area, cumulative radiance and GDP for the UK

NUTS-1 region	Lit area (km^2)	Cumulative radiance ($\times 10^{-10}$ W cm^2 μm^{-1} sr^{-1})	GDP (£ millions)
Scotland	11,880	743,420	58,578
North-east	3,403	327,523	24,321
North-west	7,989	618,702	72,475
Yorkshire and the Humber	9,614	686,588	53,002
East Midlands	11,106	544,521	47,289
Wales	5,508	253,741	27,912
West Midlands	7,231	499,110	58,053
East of England	12,695	642,582	72,229
South-east	14,320	678,347	107,630
South-west	8,821	368,894	53,453
Greater London	1,586	505,366	108,645

radiance to establish a "GDP per unit radiance" coefficient and apply this to lit pixels. By doing this, pixels of a given radiance will have the same GDP value and the country total is preserved. Total radiance and GDP at NUTS-1 are not well correlated in normal or logarithmic space ($R^2 = 0.3$ and 0.5, respectively). Comparing coefficients across NUTS-1 regions, there is a difference of a factor of 3 between the lowest (north-east) and the highest (Greater London). A single coefficient was calculated for the UK and applied to the eleven NUTS-1 regions. The resulting graph (Figure 15.3) shows that Greater London and the south-east are not well represented by this approach, both regions being significantly underestimated. In fact there is a general north/south divide with the greatest disparities of over- and underestimation being those regions farthest apart!

The distribution of radiance values over the UK is heavily biased toward the lower radiance values. Forty-one per cent of the land area of England, Scotland and Wales has a light source detectable from space. Of this, 37 per cent of the lit area occupies the five lowest radiance values. The connection between ground level light sources and radiance values in the product is not well established and thresholding out these very low radiance values only marginally improves the strength of the correlation.

Uncertainties exist concerning the manifestation of certain sectors of the economy by the presence of night-time lights; therefore it might be helpful to examine the nature of the light GDP relationship with respect to individual sectors of the GDP. The effect of the relative contributions from the agricultural (primary), industrial (secondary) and service (tertiary) sectors to GDP is also not well understood. The ONS produce NUTS-1 data for the UK split into the various categories of the economy. These can be grouped into the three principal sectors mentioned above. The UK economy is principally service-based, with this accounting for an average of 67 per cent of the GDP across the regions. The agricultural contribution is very

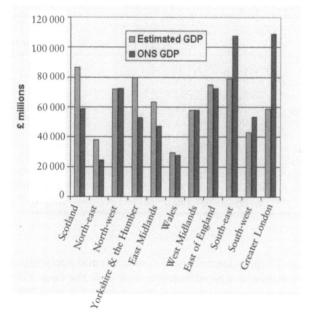

Figure 15.3 Estimation of GDP at NUTS-1 level using a single GDP/radiance coefficient.

low, only accounting for a maximum of 3 per cent (south-west). Nonetheless, the primary sector correlates very well with lit area, but poorly when compared with cumulative radiance. However, the opposite effect is observed when analysing the industrial sector. Here, cumulative radiance is shown to be the better descriptor. The tertiary sector was not well correlated to either measure from night-time light imagery despite being the largest sector of the UK economy.

The GDP to lit area relationship at the NUTS-2 scale is completely uncorrelated, but the cumulative radiance relationship is still good. It appears that the lit area relationships break down much more quickly than the radiance relationships. The strength of the cumulative radiance relationship appears to dip at NUTS-1. The cumulative radiance relationship at NUTS-2 is better than at NUTS-1. This could be due to less aggregation of different economic sectors in NUTS-2. NUTS-1 regions occupy larger areas and therefore have more heterogeneous sector contributions. However, aggregating light and GDP to NUTS-0 (country level GDP from the World Resources Institute) appears to efface these regional disparities and the relationship becomes stronger again. The individual sector figures suggest that NUTS-1 is the scale where confusion between light and GDP is the highest. This is likely to be an example of the modifiable areal unit problem (Openshaw 1984). As such it is not clear whether this trend would be repeated for other countries, or even different zoning schemes for NUTS-1 sectors in the UK. For instance, a better correlation may be achieved by constructing NUTS-1 regions by grouping NUTS-2 areas by dominant GDP sector. The correlation coefficients listed in Table 15.2 for NUTS-2 excludes London, which was not split into its inner and outer London NUTS-2 regions for this study.

15.3.2.1 *Possible solutions*

Considering the relative proportions of lit area and radiance to GDP contribution, it is apparent that the Greater London region skews any relationship, which one may seek to draw, because of its anomalously high economic contribution. The relationship between area, radiance and GDP is more complex at NUTS-1 than at the international level. The differences in the regional economy cannot be adequately modelled by a relationship relying on just one parameter. Greater London for instance has some very high radiance pixels over a small area. It would appear that using cumulative radiance as a proxy does not adequately describe the subnational relationship of light to GDP. Figure 15.4 demonstrates that a region can have a larger lit area and greater cumulative radiance than another region, but contribute less GDP. Table 15.2 gives these figures for NUTS-1 in the UK. The splitting of GDP by sector yielded some interesting results and may be helpful when planning a methodology for estimating economic activity in rural areas or countries with a highly agrarian economy.

The use of a single GDP/radiance coefficient assumes, first, an equal GDP generating capacity per unit radiance and, second, no consideration is given to its status with respect to neighbouring pixels. A more sophisticated model could be weighted toward individual pixels of a high radiance. Furthermore, clusters could be defined using some thresholding technique. A large cluster would have greater weighting than a smaller one. In this way, two areas of equal cumulative radiance can have different GDP values depending on the arrangement of pixels with their spatial unit. The Greater London region is a cluster in itself and would therefore be heavily weighted toward a higher GDP. The East Midlands in contrast is composed of a series of smaller clusters. The influence of clustering may help explain why the correlation improves at NUTS-2 level. As spatial units get smaller, there is less chance for larger clusters to develop.

15.3.3 *Radiance and power consumption*

The relationship between power consumption and lit area, and radiance was tested for different countries in Europe. The relationship could not be tested at subnational scales due to a lack of suitable data at this level. The relationships were found to be very strong in log–log space. Areal relation-

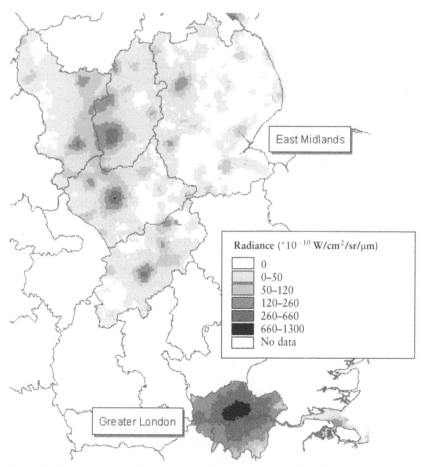

Figure 15.4 Comparison of Greater London with East Midlands. Though the East Midlands has more lit area and total radiance (Table 15.3), it only generates 43.5% of London's GDP.

ships performed marginally better than radiance ones ($R^2 = 0.91$ for area and $R^2 = 0.84$ for radiance). The law of allometric growth is observed in a host of geographic features from river morphology to volcanoes and also densely populated urban areas. Originally developed by biologists, it states that the rate of relative growth of an organ is a constant fraction of the rate of relative growth of the total organism (Nordbeck 1965). Taking 'y' to be the organ and 'x' to be the organism, the law of allometric growth can be expressed as $y = ax^b$. Taking the logarithm of both sides will linearize the equation thus:

$$\log y = b \log x + \log a$$

Welch (1980) previously investigated urban energy utilization patterns in the USA. A selection of US cities was mapped using the early DMSP OLS sensor, from which three-dimensional surfaces were computed of the illuminated urban area. These volumes were found to correlate highly ($R = 0.89$) with energy consumption (kWh), the regression model again taking the P = aAb allometric form. This suggests that the relationship of radiance to power consumption may be more robust to scale than for GDP. To map power consumption accurately at the subnational level, one has to make an assumption about what proportion of energy is being used for street and industrial/commercial lighting. Often this is not uniform across a country and also dependent on the location of heavy manufacturing and other industrial processes. The assumptions one is required to make are very similar to mapping of a related parameter: CO_2 emissions.

15.3.4 *Radiance and CO_2 emissions*

The discussion of critical factors concerning the mapping of GDP and power consumption can be extended to the monitoring of fossil fuel trace gas emissions. A first attempt at global mapping of this environmentally important parameter was discussed in Section 15.2 (see also Doll *et al.* 2000). The study confirmed that night-time lights appear to be a highly appropriate data set to use to disaggregate country level data. It was mentioned in the previous section that the poor agreement of the satellite-derived CO_2 map with the CDIAC estimates was possibly due to the mode of energy generation (e.g., nuclear, wind, solar) and how the energy produced was utilized. One has to consider what are the "invisible" energy uses, and how important they are in assessments. The disparate emission magnitudes between the two maps underline the requirement for a far more sophisticated model incorporating many more factors than just the relationship between lit area and emissions. Elvidge *et al.* (2000) described how this might be accomplished using a radiance-calibrated night-time lights data set. In the study Elvidge *et al.* (2000) presented a graph of cumulative brightness vs. energy-related carbon emissions for the forty-eight conterminous states of the USA, which showed that the two parameters are strongly positively correlated. Building on this result, they outlined the extensive ancillary data required to make sensible assumptions about distributing the data. It is likely that countries will also have to be grouped into those that are heavily reliant on fossil fuels, and therefore should have closer agreement with official estimates. It is also likely that GDP will be the key determinant of how to group these countries, as Sutton *et al.* (2001) has shown for population estimation using night-time imagery. Mapping of CO_2 emissions is potentially one of the most valuable applications of using night-time data, if it could be demonstrated to produce reliable and consistent results.

Terrestrial night-time lights can be attributed to three sources: biomass burning, gas flaring and human settlements. Their identification from satellite imagery is facilitated due to their distinct spatial and temporal features. Biomass burning can be identified in forest areas by lights that have a short temporal duration. Aside from their location, gas flares have a limited spatial extent and are extremely bright. Elvidge *et al.* (2001) used DMSP OLS data to assess the areal extent of biomass burning in Roraima, Brazil, while other studies have sought to establish relationships between the thermal properties of the fire, the rate of biomass consumption (fuel load), and aerosol and trace gas emissions also from satellite sensor data (Kaufman *et al.* 1998b). These figures are not currently included in estimates of fossil fuel consumption.

In addition to the CDIAC data used earlier, the International Energy Agency provides data on national levels of fossil fuel consumption. These data are also broken down into estimates of fossil fuel types used for electrical power generation, industrial, commercial, transportation and residential use. This is a very useful starting point but only addresses part of the problem. These data still have to be disaggregated. This can be done by incorporating further data sets such as power station locations for the electrical power generation component and road networks for the transportation component. In particular, night-time lights can be used to weight this distribution around bright urban centres. The spatial distribution from international air travel may be distributed according to the sectors with the densest traffic (Penner *et al.* 1999). Other emission sources may be distributed according to the location and brightness of night-time light imagery. This results in four emission submaps that may be easily combined in a GIS to produce a global, high spatial resolution emission map.

15.4   Other applications for night-time satellite imagery

Night-time light imagery has been used for a number of other applications besides mapping socio-economic parameters. The examples below represent three of the most recent developments using DMSP OLS data.

15.4.1   DMSP OLS data for urban mapping

Any regional scale image of night-time light data will reveal lit areas extending along transportation networks and along coastlines, which even the untrained observer can associate with human settlement. Using these data as a measure of urban land cover is therefore one of its most obvious applications. Doll and Muller (1999b) have compared it with the Digital Chart of the World's urban polygons, while Elvidge *et al.* (1997) and Imhoff *et al.* (1997) describe how DMSP OLS data relate to urban boundaries from US Census data. Urban land cover is notoriously difficult to map from conventional optical remotely sensed data at the appropriate scale for global applications. There are numerous issues where an accurate map of urban areas would be hugely beneficial. Night-time imagery acquired from the DMSP OLS sensor has gone a long way to addressing this requirement. Owen *et al.* (1998) used DMSP OLS data for urban categorization in order to correct meteorological records for urban heat island bias. In the same way urban land cover as described by DMSP OLS data have been used by Imhoff *et al.* (2000) to assess the impact of urbanization on primary productivity.

Frequency composite DMSP OLS data from 1994–95 were used to identify urban and peri-urban land cover. Using a technique the authors developed in a previous study (Imhoff *et al.* 1997), wholly urban areas were classified as those being detected as lit 89 per cent of the time from cloud-free observations. Peri-urban areas were classified as those lit 5–88 per cent of the time. This map of urbanized lands was then geo-spatially combined with a US Geological Survey land cover map and a year's worth of Normalized Difference Vegetation Index data (NDVI) within a GIS. The NDVI is defined as the sum of the visible and near-infrared channels divided by its difference and has been shown to provide a reasonable estimate of absorbed photosynthetically active radiation (Asrar *et al.* 1984). Monthly maximum NDVI data sets over a year were averaged within the three classes of urban land cover. Each month's data were multiplied by the number of days in that month to provide a comparable unit (NDVI * days) which when summed over the whole year represents annual primary production. The effect of urbanization at the continental scale is generally found to cause a decrease in photosynthetic activity when compared with surrounding areas. Analysing individual cities, however, revealed that urbanization could in fact have a positive effect on primary productivity especially in cities with a strongly seasonal climate. The urban heat island effect was found to maintain a higher level of photosynthetic production for places like Chicago. Increased primary productivity in urban areas was also attributed to new species introduced and supported by irrigation in cities located in an otherwise resource-limited environment.

The effect of urbanization on primary productivity was found to be related to the climatic and physiographic environment of the city. Peri-urban areas had the highest degree of primary productivity and are also those that lie in the path of urban sprawl. Imhoff *et al.* (2000) estimates a loss of 10 NDVI * days of photosynthetic production should these lands become fully urbanized.

15.4.2 *DMSP OLS data for creation of a night-time atlas of light pollution*

Probably the most intuitive use of night-time data is to use it to map and quantify the amount of light pollution. Our limited view of the Universe has had a profound influence on human development. It serves as a constant reminder to consider the nature of our existence and begs questions, which we may never be able to answer. However, a clear view of the night sky for those who live in the developed world is a rapidly disappearing occurrence. Cinzano *et al.* (2001) set about using the DMSP OLS radiance-calibrated night-time light data set to quantify artificial night sky brightness. Light propagation from the top of the atmosphere radiances present in the DMSP OLS product was modelled through Rayleigh scattering by molecules, Mie scattering by aerosols, atmospheric extinction along light paths and Earth curvature. Hence, many areas that appear dark in the night-time light product due the absence of a ground level light source are in fact affected by light pollution from adjacent bright areas. In some cases the source of the pollution is from a neighbouring country. Cinzano *et al.* (2001) intersected their atlas of light pollution with the LandScan2000 global population density database (Dobson *et al.* 2000) to assess the number of people affected. The extent of this pollution is so widespread across the developed world that more than 99 per cent of the population of the EU and USA, and 66 per cent of the entire world population suffer from some degree of light pollution. In particular about half the population of the developed world do not have the possibility of viewing the Milky Way with the naked eye.

15.4.3   *Modelling the development of night-time lights*

The modelling of the development of night-time light is of use to all the applications discussed in this and previous chapters. The ability to predict future scenarios of the spatial distribution of night-time lights and, by association, economic activity could greatly aid transport and energy infrastructure planning, as well as environmental impact assessments. This would be particularly beneficial for developing countries. Night-time satellite imagery is well suited for cell-based modelling, not only because of its raster format, but also because it is flexible within a range of rules and assumptions that can be incorporated into a model.

Plutzar *et al.* (2000) used a cellular automata model to simulate the growth of night-time lights in China. A cellular automata model consists of a regular, discrete lattice of cells. Each cell is characterized by a state taken from a finite set of states. Evolution takes place at discrete time steps and each cell evolves according to the same rule, which depends only on the state of the cell and a finite number of neighbouring cells (Weimar 1998). The model incorporated not only night-time lights but also terrain, population and transport infrastructure data sets. Probability surfaces were generated and the weighted mean was computed to combine these surfaces. A probability surface was also generated from the night-time light data. This was based on an annular region around a cell. The higher the focal sum of this region, the higher the probability the target cell will change. The focal sum of the region can be weighted such that cells at the edge of the region have less influence in the computation. The model was run to calculate pixels that have both intensified in value, and changed state from being unlit to lit. The generated layers of changed cells are summed with non-changing cells and compared with a predefined limit to determine whether the layer will serve as an input for the next iteration of the model or be output as the result. The model shows an intensification of night-time light between Beijing and Shanghai, and expansion along the coast and around Hong Kong.

15.5 Prospects

The discussion up to now has exclusively dealt with night-time satellite imagery from the low-resolution DMSP OLS sensor as it is currently the only spaceborne sensor that is sufficiently sensitive to detect anthropogenic lighting activity from space. However, the Airborne Visible/Infrared Imaging Spectrometer (AVIRIS) can also be used to acquire high spatial resolution data over individual cities. The AVIRIS sensor is a hyperspectral imaging system which senses in 224 very narrow bands (~ 10 nm) from 0.41–2.45 mm. It is designed to fly on board NASA's U2 aircraft where, at an altitude of 20 km, it can image 20-m pixels over a 10-km swathewidth (Porter and Enmark 1987). This additional data source offers not only the advantage of an enhanced spatial resolution, but also of enhanced spectral resolution too. The exclusivity of DMSP OLS data as a source of night-time imagery has forced researchers to use it at the continental–global level without fully understanding the contributions of different light sources at the subpixel level. AVIRIS data could address this issue. A test flight over Las Vegas in 1998 suggests that there are distinctive spectral signatures over the city (Elvidge and Jansen 1999). Combining these two data sources would be of use to help understand what the DMSP OLS data is really showing at the small scale, and therefore aid the assumptions one makes in macroscale models using night-time imagery. There are various types of lighting used in cities. Each has distinct spectral characteristics depending on the element used. Commonly used types of high-intensity discharge light are high-pressure sodium used for street lights, mercury vapour and metal halide used for illuminating car parks and sports stadiums. Figure 15.5 shows a sample spectral plot from a low-altitude (around 4,000 m) acquisition over Las Vegas. Of note is a very strong signal in Band 18 (536 nm). Plots such as these could aid better estimation of energy usage. Mapping spectral patterns over cities could help to identify patterns of residential, commercial and industrial land use (Elvidge and Jansen 1999). This could be one way of filtering out the population component if concerned with assessing areas of high economic activity. In addition to this, the vastly superior spectral resolution of AVIRIS may be of use to correct night-time data atmospherically.

Temporal data sets of night-time lights will permit the analysis of relationships and help to identify urban growth patterns, which will serve as a record on urbanization patterns across the Earth. This will in turn allow

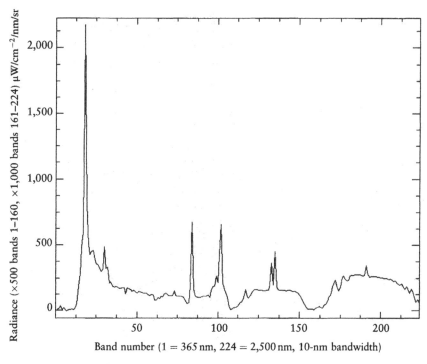

Figure 15.5 Spectral plot for a lit pixel over 224 10-nm bands from AVRIS taken over Las Vegas on the night of 5 October 1998.

more accurate assumptions to be made when setting up the models. Cellular automata modelling techniques provide a useful means to examine the future impacts of urban growth. A study has been carried out to model the urban growth of San Francisco (Clarke *et al.* 1997). In this model, the control parameters (transportation, topography, etc.) were allowed to develop with urban area, thereby providing an extra feedback into the model. It also used a time series of historical boundary data to calibrate the model. Night-time light data could also be used for this application. While the estimation of socio-economic parameters mentioned in this chapter is an important application of night-time imagery, it must also be realized that this can be used in a predictive capacity when coupled to a growth model. Taken together, these activities represent the best chance to advance research and provide better results from DMSP OLS night-time data.

15.6   Conclusions

The research presented here has extended the current body of results showing correlations between night-time lights and various socio-economic and environmental parameters at the country level to include finer scale relationships. Generally, it was found that the simple relationships built up at the country level are not as strong when compared with data at the subnational level. This has been attributed to the oversimplicity of the model to describe the disparities in the subregional economy. The use of a weighted radiance model is regarded as a possible solution.

If night-time imagery is to be used for mapping, then the conceptual issue of what the lights are actually showing must be considered. High spatial resolution night-time data could help to solve this problem. Once this has been set in a framework one can begin to deal with the range of assumptions for which there is often no data publicly available. Night-time light imagery can be used to estimate a range of parameters, but there is a requirement for much ancillary data, to distribute and calibrate relationships. GIS plays a fundamental role throughout the whole process. Its value is not limited to mapping, but has been shown to be of great help in the development of more complex models where a range of external data sources are required. With these considerations in mind, this unique remotely sensed data source provides the potential to advance mapping beyond what we currently assume to be the remit of environmental remote sensing.

15.7   Acknowledgements

The author wishes to thank Chris Elvidge for supplying the DMSP OLS and AVIRIS data, and Prof. Jan-Peter Muller and Jeremy Morley for their comments and advice in preparing this manuscript. Acknowledgement is also due to the Office of National Statistics for supplying data tables on economic statistics for the UK and European power consumption.

15.8   References

Andres, R. J., Marland, G., Fung, I., Matthews, E. and Brenkert, A. L., 1996, Geographic patterns of carbon dioxide emissions from fossil-fuel burning, hydraulic cement production, and gas flaring on a one degree by one degree grid cell basis: 1950 to 1990. *Global Biogeochemical Cycles*, 10, 419–29.

Asrar, G. S., Fuchs, M., Kanemasu, E. T. and Hatfield, J. L., 1984, Estimating absorbed photosynthetically active radiation and leaf area index from spectral reflectance in wheat. *Agronomy Journal*, 76, 300–6.

Cinzano, P., Falchi, F. and Elvidge, C. D., 2001, The first world atlas of the artificial night sky brightness. *Monthly Notices of the Royal Astronomical Society*, 328(3), 689–707.

Clarke, K. C., Gaydos, L. and Hoppen, S., 1997, A self-modifying cellular automaton model of historical urbanization in the San Francisco bay area. *Environment and Planning B*, 24, 247–61.

Croft, T. A., 1978, Night-time images of the Earth from space. *Scientific American*, 239, 68–79.

Dobson, J. E., Bright, E. A., Coleman, P. R., Durfee, R. C. and Worley, B. A., 2000, LANDSCAN: A global population database for estimating population at risk. *Photogrammetric Engineering and Remote Sensing*, 66, 849–57.

Doll, C. N. H. and Muller, J-P., 1999a, The use of radiance calibrated data to improve remotely sensed population estimation. Paper given at Conference of the Remote Sensing Society, Cardiff, pp. 127–33.

Doll, C. N. H. and Muller, J-P, 1999b, An evaluation of global urban growth via comparison of DCW and DMSP-OLS satellite data. *Proceedings of the IEEE International Geoscience and Remote Sensing Symposium, IGARSS '99*, Hamburg, Germany (Piscataway, NJ: IEEE), 1134–6.

Doll, C. N. H., Muller, J-P. and Elvidge, C. D., 2000, Night-time imagery as a tool for mapping socio-economic parameters and greenhouse gas emissions. *Ambio*, 29, 159–64.

Elvidge, C. D. and Jansen, W. T., 1999, AVIRIS observations of nocturnal lighting. *AVIRIS Airborne Geosciences Workshop Proceedings* (http://popo.jpl.nasa.gov/docs/workshops/99docs/16.pdf).

Elvidge, C. D. and Wlarave, W. T., 1999, AVIRIS observations of nocturnal lighting. *AVIRIS Airborne Geosciences Workshop Proceedings* (http://popo.jpl.nasa.gov/docs/workshops/99_docs/16.pdf).

Elvidge, C. D., Baugh, K. E., Kihn, E. A., Kroehl, H. W., Davis, E. R. and Davis, C. W., 1997, Relation between satellite observed visible–near infrared emissions, population, economic activity and power consumption. *International Journal of Remote Sensing*, 18, 1373–9.

Elvidge, C. D., Imhoff, M. L. and Sutton, P. C., 2000, Relation between fossil fuel trace gas emissions and satellite observations of nocturnal lighting. Paper given at Conference of IAPRS, Vol. XXXIII, Part B7 (Amsterdam: GITC), pp. 397–401.

Elvidge, C. D., Hobson, V. R., Baugh, K. E., Dietz, J. B., Shimabukuro, Y. E., Krug, T., Novo, E. M. L. M. and Echavaria, F. R., 2001, DMSP-OLS estimation of tropical forest area impacted by surface fires in Roraima, Brazil: 1995 versus 1998, *International Journal of Remote Sensing*, 22, 2661–73.

Imhoff, M. L., Lawrence, W. T., Stutzer, D. C. and Elvidge, C. D., 1997, A technique for using composite DMSP/OLS "city lights" satellite data to accurately map urban areas. *Remote Sensing of Environment*, 61, 361–70.

Imhoff, M. L., Tucker, C. J., Lawrence, W. T. and Stutzer, D. C., 2000, The use of multisource satellite and geospatial data to study the effect of urbanization on primary productivity in the United States. *IEEE Trans. Geoscience and Remote Sensing*, 38, 2549–56.

Kaufman, Y. J., Herring, D. D., Ranson, K. J. and Collatz, G. J., 1998a, Earth observing system AM1 mission to Earth. *IEEE Trans. Geoscience and Remote Sensing*, 36, 1045–55.

Kaufman, Y. J., Justice, C. O., Flynn, L. P., Kendall, J. D., Prins, E. M., Giglio, L., Ward, D. E., Menzel, W. P. and Setzer, A. W., 1998b, Potential global fire monitoring from EOS MODIS. *Journal of Geophysical Research-Atmospheres*, 103, 32215–38.

Li, Y-F., 1996, *Global Population Distribution (1990), Terrestrial Area and Country Name Information on a One by One Degree Grid Cell Basis*, ORNL/CDIAC-96, DB1016 (Oak Ridge, TN: Carbon Dioxide Analysis Center) (http://cdiac.esd.ornl.gov/ndps/db1016.html).

Li, Y-F., McMillan, A. and Scholtz, M. T., 1996, Global HCH with 1° × 1° longitude/latitude resolution. *Environment Science Technology*, 30, 3525–33.

Matthews, E., 1983, Global vegetation and land use – new high-resolution databases for climate studies. *Journal of Climate and Applied Meteorology*, 22, 474–87.

ONS, 2002, The Office of National Statistics (http://www.statistics.gov.uk).

ORNL/CDIAC, NDP-058a. *Carbon Dioxide Emission Estimates from Fossil-fuel Burning, Hydraulic Cement Production, and Gas Flaring for 1995 on a One-Degree Grid Cell Basis.* (Oak Ridge, TN: Carbon Dioxide Analysis Center) (http://cdiac.esd.ornl.gov/ftp/ndp058a/).

Openshaw, S., 1984, *The modifiable areal unit problem. Concepts and Techniques in Modern Geography 38* (Norwich, UK: GeoBooks).

Owen, T. W., Gallo, K. P., Elvidge, C. D. and Baugh, K. E., 1998, Using DMSP-OLS light frequency data to categorize urban environments associated with US climate observing stations, *International Journal of Remote Sensing*, 19, 3451–6.

Penner, J., Lister, D., Griggs, D. J., Dokken, D. and McFarland, M. (eds), 1999, *Aviation and the Global Atmosphere: Special Report on the Intergovernmental Panel on Climate Change* (Cambridge: Cambridge University Press).

Porter, M. and Enmark, H. T., 1987, A system overview of the Airborne Visible/Infrared Imaging Spectrometer (AVIRIS) (http://popo.jpl.nasa.gov/docs/aviris87/A-PORTER.PDF).

Plutzar, C., Grubler, A., Stojanovic, V., Reidl, L. and Pospishil, W., 2000, A GIS approach for modelling the spatial and temporal development of night-time lights. In J. Stobl, T. Blasshke and G. Griesebner (eds) *Angewandte Geographische Informationsverarbeitung XII* (Heidelberg, Germany: Wichman Verlag), pp. 389–94.

Southwell, K., 1997, Remote sensing; night lights. *Nature*, 390, 21.

Sullivan III, W. T., 1989, A 10 km image of the entire night-time Earth based on cloud-free satellite photographs in the 400–1100 nm band. *International Journal of Remote Sensing*, 10, 1–5.

Sutton, P. C., Roberts, D., Elvidge, C. and Baugh, K., 2001, Census from heaven: An estimate of the global human population using night-time light satellite imagery. *International Journal of Remote Sensing*, 22, 3061–76.

Tobler, W. R., 1969, Satellite confirmation of settlement size coefficients. *Area*, 1, 31–4.

UNEP/UNDP, 1994, *Report of the International Symposium on Core Data Needs for Environmental Assessment and Sustainable Development Strategies* (Bangkok, Thailand: UN).

Weimar, J., 1998, Simulations with cellular automata (http://www.tu-bs.de/institute/WiR/weimar/Zascriptnew/intro.html).

Welch, R., 1980, Monitoring urban population and energy utilization patterns from satellite data. *Remote Sensing of Environment*, 9, 1–9.

World Resources Institute, Earthtrends Database (http://earthtrends.wri.org).

16 Does night-time lighting deter crime?

An analysis of remotely sensed imagery and crime data

John R. Weeks

16.1 Introduction

It is perhaps intuitive that a well-lit area will reduce crime at night. Police departments in many communities in the USA, the UK and other countries have initiated programmes to encourage a reduction in night-time crime by increasing the amount of light in neighbourhoods. Several studies over the past two decades have sought to document the role that improved lighting of urban places can play a part in reducing crime. The theory is that since criminals want to avoid arrest they prefer to avoid detection by working in a poorly lighted area (Page 1977; Painter 1996). The theory is tested in the literature largely by before-and-after field studies, by comparing crime rates in neighbourhoods before an increase in public lighting with the crime rates after lighting was increased, and at least some of the available evidence is reasonably consistent with the theory that improved lighting lowers the crime rate (Ditton and Nair 1993; Herbert and Davidson 1994; Painter 1996). The literature also suggests that people feel more secure at night in better lighted areas than they do in less well-lit areas (Ditton and Nair 1992; Painter 1996).

Despite the obvious importance of the relationship between night-time lighting and crimes, I am not aware of any study that has attempted to measure the actual amount and location of night-time lighting and relate that lighting to the incidence of crime. It could be that the perceived depressing effect on crime of improved lighting in an area may have less to do with crime rates *per se*, than with the perception (whether correct or not) that lighting reduces crime (a self-fulfilling prophecy). Furthermore, when an area is provided with additional lighting, it is likely that other changes are occurring simultaneously that could be equally or more effective crime deterrents than the light itself. For example, the literature suggests that increased surveillance of an area is an effective method for reducing crime (Painter and Tilley 1999), and increased illumination may improve surveillance and thus indirectly contribute to a lowering of the crime rate. Unless the amount and location of light is measured, we can never know what the actual contribution of improved lighting might be to the reduction in crime rates.

Measurements of light at night are obtainable from remote sensors (see, for example, Elvidge *et al.* 1997; Elvidge *et al.*, Chapter 13 in this book; Doll and Muller 2000), but thus far little has been done with these data in the social sciences beyond attempts to measure population density (Sutton 1997; Sutton, Chapter 14 in this book) and to evaluate the use of night-time imagery as a proxy for levels of economic development in urban areas (Doll and Muller 2000; Doll, Chapter 15 in this book). My study seeks to quantify the relationship between the amount and location of lighting at night and the incidence of crime in an urban community in the USA. The analysis will focus on a comparison of crime in well-lighted areas compared with levels in crime in less well-lighted areas. From a conceptual perspective, this research contributes to the literature on the role of environmental context as a factor in human behaviour, using the amount of lighting in an area as the index of the environmental context, and reported criminal activity as the type of human behaviour under investigation.

16.2　Background

The literature evaluating the relationship between light and crime is actually rather sparse. Pope (1977) analysed burglary incidents over the course of a year to identify potentially recurring patterns. It was found that residential burglaries tend to occur in daylight hours during the weekdays, since homes are more likely to be unoccupied during those times. Rengert and Wasilchick (1986) drew similar conclusions about the timing of suburban residential burglaries. Consistent with the opportunistic nature of crime, non-residential burglaries were more likely to occur at night and on weekends, when those premises were least apt to be occupied. However, Pope (1977) found that deterrent features including artificial lighting did not, in fact, seem to influence the success of the burglary.

If we accept the idea that residential burglaries are most likely to occur during the day when surveillance is low and the risk of a resident being home is low, we can anticipate that night-time lighting will be less of an issue as a crime deterrent in residential areas than it will be in commercial or other non-residential areas. In Portland, Oregon, USA, a project was initiated in the late 1970s to reduce crime in a particularly derelict commercial area through a variety of environmental changes including improved lighting. An evaluation of the programme suggested that commercial burglaries had decreased over time after these changes were implemented, but that the street lighting programme had not produced a reduction in the fear of crime, nor in the incidence of street crime (Wallis and Ford 1980; Kashmuk 1981).

A clearer positive connection between lighting and crime was found after a lighting programme was initiated in two areas of Glasgow, Scotland. Analysis of data from before and after the improvement of street lighting in

1990 showed an across-the-board reduction in crime, and an increase in the perception of safety in being outside after dark (Ditton and Nair 1992, 1993, 1994). An evaluation of three locations in suburban London also revealed that improved street lighting in otherwise relatively dilapidated and high-crime areas had a noticeable effect in reducing street crime and lowering pedestrians' fear at night (Painter 1996). The methodology was similar to the study in Glasgow, in that local residents were interviewed 6 weeks prior to the street lighting programme, and then again 6 weeks after the lighting had been improved in the area. Results of another set of street lighting experiments in Cardiff and Hull in the UK suggested that the effect of improved street lighting on crime rates was difficult to assess, but that improved lighting clearly enhanced the perception of increased safety and heightened residents' enjoyment of their neighbourhood (Herbert and Davidson 1994). Painter and Farrington evaluated two other neighbourhood relighting projects (where "relighting" refers to an area that has had its night-time illumination increased) in the UK, one in Dudley, in the Birmingham metropolitan area (Painter and Farrington 1997), and one in Stoke-on-Trent, to the north of Birmingham (Painter and Farrington 1999). Both of these studies employed 12-month periods of before-and-after investigation, as well as in the inclusion of control areas against which to compare the changes occurring in the relit areas. In both Dudley and Stoke-on-Trent, the relit areas had a considerable drop in crime as reported in victimization surveys, compared with little change in crime victimization in the control areas. Both these studies, which represent the most comprehensive assessments of the impact of lighting on crime, seem clearly to suggest that lighting reduces crime rates, lowers the fear of crime and improves the general attitude that residents have about the quality of their neighbourhood environment.

However, in reviewing the literature on the effectiveness of street lighting improvements in deterring crime, Pease (1999) has concluded that while there is a general relationship between increased street lighting and crime prevention, the relationship is by no means universal. In particular, the effect seems to be stronger in the UK than in the USA, although this difference could be due simply to the small number of studies available for comparison. Furthermore, since improved street lighting is rarely targeted as a single strategy for crime prevention, it is not always easy to disentangle the influence of lighting from other measures that may have been taken, such as increased police presence, or heightened surveillance of the street scene by local residents and business operators. Of the studies in the literature, the Stoke-on-Trent project (Painter and Farrington 1999) seemed most clearly to have night relighting as the only change in the environment and, under that scenario, the effect on crime was clearly apparent. Yet, even in that study, the crimes reported to the police did not show an impact from relighting and the crimes occurring during the day also decreased in the experimental relit area. Both of these findings suggest a more complex relationship than might first be apparent.

If the effect of night-time lighting on crime is to be adequately assessed, we must have some way of quantifying the amount of light. None of the studies reviewed in the literature have, in fact, gone beyond inferring that at Time 2 (the "after" period) there was more light present in an area than there had been at Time 1 (the "before" period). Within a GIS (Geographical Information System) environment it would be possible to infer the pattern of light by using a city's digitized map of street light locations. For example, in my study, the City of Carlsbad's GIS Department was the source for GIS map layers in Arc/Info coverage and ArcView shapefile formats, and the map layers included one for city-owned street lights. However, there are immediately several problems that would be associated with the use of a simple geographic point file of light poles, including:

- Although the location of each light pole may be known, and the luminance of each lamp may be determined, there is still no sure way of knowing how much area was actually lit by each lamp (e.g., how large the buffer around the light pole should be).
- It is impossible to know the exact shape of any given lighted area without knowing the placement of the pole in relation to buildings, vegetation and other environmental features that might influence where the light would be directed.
- More importantly, the city's street lights represent only one source of night-time illumination. Businesses have neon signs and advertising, as well as privately owned security lights, and in residential areas many homes will have porch lights and other security lights available to illuminate the night. The only way to remedy these deficiencies in light coverage is to obtain and analyse night-time images for the study area, from which locations could be extracted that were geographically covered by light.

The study outlined in this chapter has the advantage that it is possible to quantify the existence of light in specific places and then to analyse the relationship of crime only to night-time lighting, thereby eliminating any residual or collateral study design effects, given that it is not evaluating a programme designed to influence crime. On the other hand, the study has the disadvantage of not providing for a before-and after examination of the effect that a change in lighting may have on the pattern of crime. Therefore, cause and effect can only be inferred indirectly.

16.3 Study site

The study area selected is located within the City of Carlsbad, CA, a suburban community in San Diego County, north of the City of San

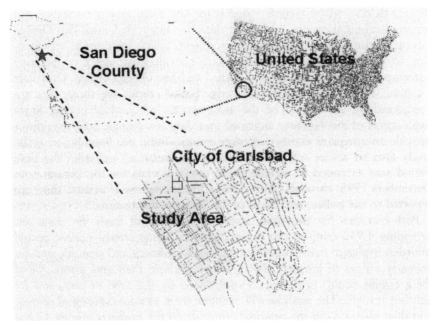

Figure 16.1 Study area within the City of Carlsbad, California (within San Diego County).

Diego (Figure 16.1). The entire City of Carlsbad covers an areal extent much greater than needed for the purposes of this project, and so a subset of the city was selected for study. The study area represents the originally settled, older portion of the city. The area was settled in the late nineteenth century and was incorporated in 1952. Extensive areas of the city outside the study area represent recent development and are dominated by single land use categories and were therefore excluded from the study area. The western portion of the city, which was selected for study, provided the desired diversity of land use categories as well as having adequate GIS map layers. Because the City of Carlsbad has one of the lowest crime rates in the County of San Diego, 4 years' worth of crime data were combined to generate a sufficient number of crimes for the purposes of the study.

16.4 Data and methods

16.4.1 Crime data

The methodology required two types of data for the analysis: (i) geo-referenced crime data and (ii) geo-referenced night-time light data. With respect to the crime data, the City of Carlsbad's Police Department has a well-integrated Computer Aided Dispatch (CAD) and Report Management

System (RMS), which records requests for law enforcement assistance and reported criminal events, and these data are integrated into The Omega Group's Crime View® software environment, which operates as an extension to ESRI's ArcView GIS. Data were provided through a cooperative agreement with both the City of Carlsbad and The Omega Group. Only Part I crimes were considered in the study. Part I crimes are those that are considered by the FBI to be the most serious of criminal events. Early exploration of the database indicated that data for a single year would not provide an adequate number of crime events within the boundaries of the study area to assure sufficient numbers for statistical analysis. The time period was extended to include crime event records for the period from December 1995 through June 1998. These represent crimes that are reported to the police, regardless of the ultimate disposition.

Part I crimes for the study area were extracted from the data set, providing 4,950 crime events for analysis, including crimes against people (murder, negligent manslaughter, rape, armed robbery and assault), and the property crimes of burglary, larceny–theft, vehicle theft and arson. All of these crimes could, in theory, be influenced by the time of day, and by lighting at night. The analysis will examine them as a total group of crimes, and then separate out the personal crimes from the property crimes. Crime events were geo-coded at the address or nearest intersection. The ArcView algorithm for geo-referencing addresses is to interpolate addresses along a street segment, and then to locate the coordinates at a point that is 40 m from the centreline of the street. More precise GPS (Global Positioning System) coordinates were not available for this study. Thus, it is evident that a perpetrator may have forcibly entered a building at the rear where lighting is inadequate, but the address may geo-reference the crime closer to the front of the building, where lighting is abundant. Similarly, a street crime may have occurred in a dark area of pavement, but the geo-coding may inadvertently place the crime within the path of lighted area. In addition, not all crimes are light sensitive, of course. A murder may occur within a building regardless of the amount of lighting outside that building. There is obviously no way of controlling for these kinds of issue.

The crime data used represent a period covering almost 4 years, whereas the night-time image was generated for a single night and, due to delays in obtaining the image because of a consistent night-time marine layer of clouds that limited flying over the area for purposes of capturing images, the night represented by the image is not contained within the dates of the crime data. These limitations may potentially affect the interpretation of our results, although it is impossible to know what bias that may introduce. Furthermore, since the crime data represent all days of the year, but crime at night is the focus, the definition of a night-time crime has to be somewhat arbitrary. Because of its location in the mid-latitudes, the variability in the time of sunrise and sunset in San Diego County is not only slight throughout the year, especially when daylight savings time is taken into consideration.

Based on data from the US Naval Observatory, sunset during the year varies in San Diego County from 4 : 42 p.m. (PST, Pacific Standard Time) in early December to 8 : 01 p.m. Pacific Daylight Time (PDT) in late June. Sunrise varies from 5 : 30 a.m. (PST) during the first week of April to 7 : 00 a.m. (PDT) during the last week of October. However, it is not just time, but human daily activities that will affect the opportunities for crime. For these reasons, night-time crimes are defined as those occurring during dark hours that were between the time that most businesses are closed (9 : 00 p.m.) to the time that people begin to awaken in the morning (5 : 00 a.m.). Since crimes at businesses during the night are often not discovered until employees arrive at work in the morning, this binary categorization of night-time permits reasonable inferences even of those crimes for which the exact time is unknown.

16.4.2 Night-time image data

The night-time light data were derived from remotely sensed images. However, night-time satellite sensor data from the Defense Meteorological Satellite Program (DMSP) at NOAA (the National Oceanic and Atmospheric Administration) are available only at a resolution of 1 km (Elvidge *et al.* 1997; Elvidge *et al.*, Chapter 13 in this book), which is unusable for a local application. Instead, the decision was made to obtain tailored data. Twenty-two frames of scanned night-time aerial photographs at a spatial resolution of 1 m were acquired with a Nikon F3 camera mounted on a light aircraft flown at an altitude of 4,000 feet between the hours of 8 : 45 p.m. and 10 : 30 p.m. on 20 September 1999. Three flight lines were used to cover the study area. Each frame covers an area of 840 m in width by 560 m in length. The overlap of adjacent images is 97.5 m along the flight line and 30 m between flight lines. The film was developed and each aerial photograph was then scanned to an 840 by 560 pixel digital raw image. In order to use these night-time images, a registration process had to be applied to match the image coordinates to those used in the crime data. A 3-m DOQQ (Digital Orthophoto Quarter Quadrangle) image with the State Plane projection was used as the reference image in conjunction with the city's point coverage of street light poles.

Given that there were no common spectral values in the night-time images and the DOQQ, the identification of Ground Control Points (GCPs) was indirect. The location of most GCPs such as the intersections of streets, highways, corners of buildings was dependent upon visual interpretation of light patterns and intensity in the night-time image and the street and building patterns in the DOQQ. It was a relatively simple task to locate a control point in the commercial area with high light intensity and density. However, in some of the residential areas, it was somewhat difficult to expand the GCP set due to the lower light intensity and density. The street light GIS point layer provided by the City of Carlsbad was used to assist in

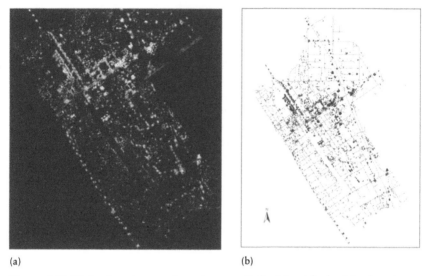

(a) (b)

Figure 16.2 Night-time image of study area. Left: Mosaicked night-time image. Right: Final binary polygon coverage of night-time light with street coverage.

the estimation of the location of some of the GCPs. The average number of GCPs used in each registration procedure was between 10 and 15, and the general total r.m.s. error for each registration was less than 5 m. After all night-time images had been registered, a mosaic was constructed by joining the night-time images together. The resulting image had a certain amount of background "noise" – transient light that was not associated with obvious locations of street lights and/or commercial areas. This was reduced through a thresholding procedure. The threshold used for each flight line was determined by visual examination of the image. A range of brightness values for three bands was tested, using ERDAS's Imagine software. The brightness value that provided the best visual discrimination between lighted and non-lighted areas was chosen. Figure 16.2(a) illustrates the final mosaicked image after thresholding.

In order to demonstrate the impact of street lighting on criminal activities, the thresholded night-time image was reduced to three levels of lighting intensity (high, medium and low) by running a simple ISODATA (Iterative Self-Organizing Data Analysis) unsupervised classification. In the final stage, a binary image was created by assigning the high- and medium-lighting-intensity areas to a value of 1 and all others to a value of 0. This binary image was vectorized to a polygon coverage, and an area threshold was used to get rid of the remaining background noise. Figure 16.2(b) shows the final lighting coverage applied to the coverage of the study area.

By overlaying the street light coverage with the night light data, it was possible to determine that there was, in fact, considerable variability in the spatial buffer within which light is cast from the city-owned street lights. In general, all light appearing to radiate from light pole locations was within a buffered zone of 60 feet in radius around the pole. However, in most instances, the shape of the street light footprint was closer to a semicircle than a full circle. Furthermore, light that appeared to be emanating from street lights accounted for only 38 per cent of the entire lighted area within the study site. Thus, any study of the effect of lighting must clearly account for all sources of light in an area. At the same time, it must be noted that the illumination was captured at a specific moment in time, and there are no assurances that all the lights observed by our images remained illuminated throughout the night. Nor are there assurances that the exact same pattern of lighting would be found on consecutive days. The city-owned streetlights would be the most consistently reliable source of lighting, but relying solely on them would yield a biased picture of the relationship between lighting and crime.

16.4.3   Other control variables

The analysis needed to account for the fact that the incidence of crime is critically influenced by crime opportunities. In particular, the locations of crimes were intersected with the parcel map coverage to determine those crimes occurring at a commercial establishment. Heads-up digitizing was conducted of all paved parking areas within the study site, and that coverage was intersected with each crime to determine which ones occurred in a parking lot.

16.4.4   Methods of analysis

The basic research question is: Does night-time lighting deter crime? Since we do not have crime data before and after an area was lighted, an impact can only be inferred on the basis of a comparison of crime occurring in places that are lighted with crime occurring in places that are not lighted. This is explored by means of three different types of analysis. First, compare the number of crimes occurring at night in lighted places with the number of crimes occurring at the same places during the day, and in relation to the pattern of day-time and night-time crimes occurring in places that are not lighted at night. In other words, the hypothesis of no difference is that the percentage of crimes taking place within areas that are lighted at night is no different than the percentage of crimes that occur at night in those places that are not lighted at night. We use the chi-square and kappa statistics to test the statistical significance and the strength of this association. The second analytical approach is to investigate whether the clustering of crimes during the day differs from the clustering at night in ways that would allow us to infer that lighting may be acting as a deterrent to crime. The *k*-function is used to test the hypotheses for clustering. The third analytical approach is to aggregate the data spatially into grid cells of uniform size and then conduct an ordinary least squares regression analysis using the number of crimes as the dependent variable, with the main predictor variable being the proportion of each grid cell that is covered by light at night, while controlling for the proportion of the cell that is comprised of commercial parcels, and the proportion that is comprised of parking areas. The ordinary least squares model, however, is misspecified because of the presence of spatial autocorrelation in the data. The Getis/Ord spatial filtering technique (Getis 1995) was used to respecify the model.

16.5   Results

16.5.1   *Tabular analysis*

In general, the results do not provide much direct support for the idea that night-time lighting is a deterrent to crime. Table 16.1 presents the findings from the tabular comparative analysis. Let us look first of all at Part I crimes. Of the 4,950 total crimes, 1,610 (33.0 per cent) occurred during the night-time hours of 9 p.m. to 5 a.m. Furthermore, 623 (12.6 per cent) of all crimes occurred in areas that were lighted at night. The hypothesis for no impact of night-time lighting was tested by comparing the percentage of all crimes occurring during the day-time in those areas lighted at night with the percentage of all night-time crimes occurring in the lighted areas. In Table 16.1 it can be seen that during the daylight hours 11.5 per cent of all crimes occurred in the lighted areas. At night, these areas would presumably be protected by the light observed from the remotely sensed images, and, so, if there was a deterrent effect, we would expect that at night the percentage of all crimes occurring in these areas would be lower than during the day. However, the data in Table 16.1 show the opposite pattern. Among all crimes occurring at night, 14.9 per cent occurred in those areas that are lighted. That difference from the day-time pattern is statistically significant at the 0.001 level, although the kappa of 0.041 shows only a weak effect.

Table 16.1 Crimes by time of day according to whether they occur within the area that is lighted at night

Did crime occur within area that is lighted at night?	Time of day crime occurred		
	Day-time	*Night-time*	*Total*
No	2,957 (88.5%)	1,370 (85.1%)	4,327 (87.4%)
Yes	383 (11.5%)	240 (14.9%)	623 (12.6%)
Total	3,340	1,610	4,950

Chi-square: 11.683 ($p = 0.001$)
Kappa: 0.041

Table 16.2 Percentage of crimes occurring in area lighted at night, according to whether they occur during the day-time or night-time

Type of crime (N in parentheses)	% occurring at night	% occurring in lighted area		Significance of chi-square	Kappa
		Day-time	Night-time		
All crimes against persons (1,219)	38.7	13.4	22.0	0.000	0.097
Murder–manslaughter (7)	28.6	0.0	0.0	n/a	n/a
Rape (24)	45.8	23.1	0.0	0.223	−0.244
Assault (1,021)	40.7	11.1	21.4	0.000	0.114
Robbery (167)	25.7	24.2	34.9	0.231	0.105
All crimes against property (3,731)	30.5	10.9	12.0	0.368	0.013
Arson (9)	44.4	0.0	0.0	n/a	n/a
Burglary (1,066)	32.8	11.5	10.6	0.756	−0.011
Larceny–theft (2,261)	28.9	10.8	12.5	0.242	0.021
Vehicle theft (395)	32.9	10.2	13.1	0.399	0.035
Crimes in parking lots (717)	18.7	3.1	23.9	0.000	0.274
Crimes on commercial property (1,806)	18.7	8.9	22.2	0.000	0.160
Crimes on multi-family property (1,562)	37.8	9.7	7.8	0.235	−0.022
Crimes on single-family property (790)	32.2	2.6	2.8	1.000	0.004
Crimes on main commercial street (786)	22.4	16.2	44.9	0.000	0.285
Robbery on main commercial street (52)	15.4	29.5	50.0	0.413	0.140
Assault on main commercial street (121)	51.2	32.2	66.1	0.000	0.339

The data thus show that there is a slight, but statistically significant increase in the likelihood of crime occurring at night in well-lit areas; a pattern that is the opposite of what we expected to find.

This pattern is quite consistent throughout the data, regardless of the type of crime under consideration, and even when taking into account whether the crime occurred in a parking lot or on commercial property. These details are shown in Table 16.2. The data show that crimes against persons are more highly related to the lighted areas than are property crimes. Of all crimes against persons, for example, 38.7 per cent occurred at night. Among those taking place during the day, 13.4 per cent occurred in the lighted area, whereas 22.0 per cent of those taking place at night were in the lighted areas. These differences are statistically significant beyond the 0.001 level. Robbery is especially noteworthy in taking place in the lighted areas, although the increase at night is not large enough to represent a statistically significant increase. The results seem to suggest that crimes occur in the lighted areas because the areas that are lighted are those with high crime rates. The analysis thus suggests that lighting follows crime, rather than being a deterrent. Note in Table 16.2 that 22.6 per cent of all crimes on the main commercial street in Carlsbad occurred in the lighted area, rising to 44.9 per cent at night. Indeed, 50 per cent of the robberies on the main commercial street occurred in lighted areas. Even more dramatically, more than half of all assaults along the main commercial street occurred at night, and, of those, two-thirds took place in the lighted areas. The lighted areas are lighted precisely because they are the locations where crimes such as these – assaults against persons – are taking place.

Among property crimes, 10.9 per cent of those occurring during the daylight hours took place in the lighted areas, and that increased slightly to 12 per cent at night. However, that difference was not statistically significant. In general, property crimes were less likely to occur in the lighted areas than were crimes against persons. Furthermore, among the 790 crimes occurring in single-family property, for example, outside night-time lighting was essentially an insignificant factor. In Carlsbad, at least, night-time lighting is mainly an issue in the commercial areas. Commercial areas make up 11 per cent of the geographic area in the study site. Although only 13 per cent of the commercial is lighted at night, that accounts for 24 per cent of the total land area that is lighted at night. Parking lots account for an additional 5 per cent of the area in the study site, but they account for 12 per cent of the total area that is lighted at night. Furthermore, nearly one-third (31 per cent) of the parking area falls within the region that is lighted at night. Overall, then, more than one-third (36 per cent) of the total lighted area is covered either by commercial property or by parking lots, although those land uses account for only 16 per cent of the area within the study site. Of some interest is the idea that crimes occurring in parking lots may well represent a situation in which lighting enhances rather than deters crime. Almost all such crimes (86 per cent) are property crimes, mostly theft from parked cars. It is probable that light in a parking lot at night helps a potential thief to decide which cars have something worthy of stealing, even though the lighting may deter crimes against persons such as assault, rape and robbery.

16.5.2 *k-function analysis*

In order to demonstrate the different patterns of crime clustering in the lighted and non-lighted areas, the *k*-function statistic was calculated. This compares the observed pattern of point data with that which would be expected under a homogeneous Poisson process. The expected number of cases within a given distance (d) of any given case is compared with the observed number of cases, and then the resulting calculations are summed over all cases in the distribution. A summary measure, $L(d)$ is calculated as the square root of the *k*-function divided by π, and $L(d)$ is then plotted at each distance (d). The null hypothesis is that there is no difference between observed and expected values [$y = x$; or $L(d) = d$]. Values of $L(d)$ that lie

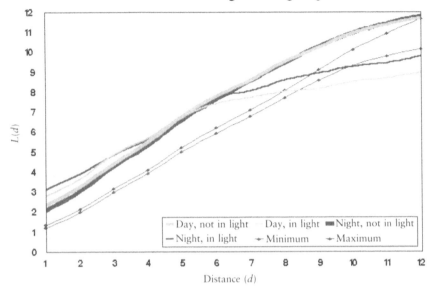

Figure 16.3 Results of *k*-function analysis.

above the line created by $y = x$ indicate a clustering of events (more observed events than expected), whereas values lying below the line indicate a dispersal of events (fewer observed events than expected) (Bailey and Gatrell 1995). Around the $y = x$ line is drawn an envelope that defines a 95 per cent confidence interval. If the *k*-function distribution falls within that envelope, then the null hypothesis of no difference is not rejected.

The clustering of crimes can be compared under four different scenarios: (i) those crimes occurring during the day in areas that are not lighted at night, (ii) those crimes occurring during the day in areas that are lighted at night, (iii) those occurring at night in areas that are not lighted and (iv) those occurring at night in areas that are lighted. If lighting makes a difference, then the only difference in clustering ought to be observed under scenario (iv). The results are shown in Figure 16.3.

It can be seen in Figure 16.3 that the crimes occurring in non-lighted areas are clustered, and the pattern of clustering is nearly identical for those crimes occurring during the day and those occurring at night. The *k*-function distributions are all above the maximum line of the envelope and so we would reject the null hypothesis and conclude that these crimes are, indeed, clustered. The crimes occurring within those areas that are lighted are also significantly clustered, although the clustering peaks at a shorter distance, probably because the size of the area that is lighted is smaller than the area that is not lighted. Crimes occurring at night in the lighted areas are slightly more clustered than crimes occurring in those areas during the day. We interpret this finding to suggest that there are fewer targets of crime at night, thus increasing the clustering. It could also be that the light focuses attention on some areas and leads to clustering of crime events for that reason.

16.5.3   Regression analysis

Finally, an analysis of data aggregated into 500-foot (152-m) square cells or quadrats. Organizing data in this way permits modelling with regression techniques to test hypotheses about the environmental influence of light in and on the density of crime events. Given the potential inaccuracies in the geo-coding of crime events relative to the polygons of light, such an aggregation perhaps reflects a reasonable approximation of the environmental influence of lighting on criminal activity. Using ArcView, a 500-foot grid was laid over the study area, the number of crimes in each cell were added up and then the percentage of each cell that was comprised of light, of parking and of commercial property was calculated. If lighting deters crime, then we should find a negative association between the number of crime events and the percentage of the cell that is lighted. On the other hand, since parking lots and commercial areas are crime attractors, the higher the percentage of those land uses, the greater should be the number of crimes. These expectations were tested using stepwise regression.

The initial model (not shown) showed that the combination of the percentage of the cell that is devoted to parking and the percentage of the cell that is lighted explain 34 per cent of the variability in the number of crime events per cell. The parking variable, as expected, was positively related to crime events, but the percentage of an area that was lighted was also positively associated with crime events. This is contrary to theoretical expectations, but is consistent with the previous analysis, in which the results suggest that lighting follows crime, rather than necessarily deterring it. The original model is, however, misspecified because there was a statistically significant level of spatial autocorrelation in the residuals. The analysis controlled for spatial autocorrelation through the technique of spatial filtering using the Gi* statistic, as described in Getis and Ord (1996) and Getis (1995). This method allows us to decompose each predictor variable into that portion that is spatial in origin, and that portion that is non-spatial. These are then reintroduced into the regression equation as separate predictor variables. The results of the revised model are shown in Table 16.3. The top panel of Table 16.3 shows that four variables emerged as being statistically significant predictors of the number of crimes per cell. In order of importance, they were the spatial component of the per cent of the cell that was commercial, followed by the spatial component of the percentage of the cell that was lighted at night, the non-spatial component of the percentage commercial, and the non-spatial component of the percentage lighted at night. The adjusted R^2 for this model was 0.311. Note, however, that all the coefficients have a positive sign. Crimes are positively associated with commercial areas, as we would expect, but they are also –

Table 16.3 Results of spatially filtered regression analysis

Predictor variable	Standardized beta coefficient	Statistical significance	
		t-score	p-value
Model A: all cells			
Percent commercial – spatial component	0.334	4.361	0.000
Percent lighted at night – spatial component	0.286	4.777	0.000
Percent commercial – non-spatial component	0.242	3.728	0.000
Percent lighted at night – non-spatial component	0.170	3.949	0.000
Dependent variable = number of crimes in grid cell			
Adjusted $R^2 = 0.311$			
Model B: after removing seven outliers			
Percent lighted at night – spatial component	0.637	18.211	0.000
Percent lighted at night – non-spatial component	0.336	9.599	0.000
Dependent variable = number of crimes in grid cell			
Adjusted $R^2 = 0.511$			

and independently – positively associated with the amount of light at night in the area. This latter finding reinforces the idea that light follows crime.

Although the model in the top panel of Table 16.3 was a properly specified model because it accounted for the spatial autocorrelation, there were several significant outliers, and they turned out to be quite revealing. Two of the grid cells near the main commercial area with high levels of lighting and large fractions of space devoted to parking had substantially fewer crimes than the model predicted. In other words, in those areas it seems as though the lighting really might have deterred crime. On the other hand, there were five other cells that had significantly more crime events than would have been predicted, and four of those five were located near a freeway entrance – an environmental context variable that we did not otherwise account for in our analysis. The spatially filtered regression model was rerun without those seven outliers to observe their impact on the results. The bottom panel of Table 16.3 shows that, without the outliers, the commercial variable drops out of the equation. In other words, the relationship between the percentage commercial in an area and the amount of crime, when controlling for the amount of light, seems largely to be due to the fact that commercial areas that are near freeway exits have high rates of crime, regardless of lighting. That leaves the amount of lighting as the statistically significant predictor of crime, albeit again in a positive direction. The beta coefficients suggest that about two-thirds of the influence is derived from the spatial component of lighting (the clustering of lighting in particular places), and about one-third of the effect is due to the non-spatial component (the amount of light in one area, regardless of the amount of light in surrounding areas). These two variables combine to explain 51 per cent of the variability in the amount of crime from one cell to another.

16.6 Discussion and conclusion

For the community under study, the evidence indicates that night-time lighting is brightest in those places where crime events are most numerous. The supposition is that lighting follows crime, rather than the other way around. This is actually consistent with the scant literature that was described earlier in which evaluations had been conducted of areas before and after lighting was improved in an area. Virtually all the study sites where such lighting was implemented were higher crime areas in which lighting was viewed as one of several strategies by which crime might be reduced. It is well known that crimes tend to cluster geographically, and so it makes sense to apply local environmental solutions to reduce crime in those "hot spots". The flip side of that, as Painter (1996: 200) has noted, is that "a badly lit environment does, of itself, cause crime. It would be foolish for policy-makers to believe that all that is required to reduce crime and fear is to find a badly lit site and relight it."

Thus, although lighting might have reduced crime in those areas, the areas themselves were chosen for lighting because of the known high rates of crime. In Carlsbad, it is probable that crime rates in well-lighted areas are lower than they would otherwise be if they were not well lighted. Nonetheless, it should be kept in mind that Painter and Farrington (1999) found in Stoke-on-Trent that the crimes reported to police did not change in the period before and after the relighting experiment, whereas their crime victimization survey did indicate a reduction. They offer the suggestion that the police reports are not strictly comparable in the before and after period, but that may not be a wholly satisfactory explanation (Pease 1999). Painter and Farrington (1999) also note that in Dudley and Stoke-on-Trent the relighting projects led to a decline in crime during the day, as well as at night. To the extent that community revitalization accompanies a relighting, then we might expect that criminals find an area to be less attractive for criminal activity than it previously was. This could help to account for the finding of relatively few differences between night-time and day-time probabilities of crime in well-lit areas.

The analysis is also consistent with the literature in the sense that lighting seems to be most intense in those areas where crimes against persons are highest. The main commercial areas in Carlsbad are those places where crime events such as assault and robbery are apt to occur, and so it is certainly not a coincidence that lighting is most intense in these areas in order to lower the perception of fear of crime victimization among people utilizing those areas.

In conclusion, this study has demonstrated the feasibility of quantifying the amount of night-time lighting using remotely sensed images, and then relating that to the location of criminal activity. Future research on the relationship between crime and lighting would be well served by adding quantification of light, as demonstrated here, to the before and after profiles of an area that is scheduled to have lighting infrastructure added to its environment.

16.7 Acknowledgements

This research was supported in part by a grant from the NASA Earth Science Enterprise Applications Division to the Affiliated Research Center in the Department of Geography at San Diego State University (Douglas Stow, Principal Investigator). The author wishes to thank John Kaiser, Dongmei Chen, Tim Dolan, Milan Mueller, Chris Langevin, Lloyd Coulter, Jeffrey Vandersip, Xiaoling Yang and especially Arthur Getis for their valuable assistance. An earlier version of this research was presented at the US Department of Justice Crime Mapping Conference, San Diego, December 2000.

16.8 References

Bailey, T. C. and Gatrell, A. C., 1995, *Interactive Spatial Data Analysis* (Harlow, UK: Longman Scientific and Technical).

Ditton, J. and Nair, G., 1992, *Street Lighting and Crime: The Strathclyde Twin Site Study* (Glasgow: Criminology Research Unit, Glasgow University).

Ditton, J. Nair, G., 1993, Crime in the dark: A case study of the relationship between street lighting and crime. In H. Jones (ed.) *Crime and the Urban Environment: The Scottish Experience* (Aldershot, UK: Avebury).

Ditton, J. and Nair, G., 1994, Throwing light on crime: A case study of the relationship between street lighting and crime prevention. *Security Journal*, 5, 125–32.

Elvidge, C. D., Baugh, K. E., Kihn, E. A., Kroehl, H. W. and Davis, E. R., 1997, Mapping city lights with night-time data from the DMSP Operational Linescan System. *Photogrammetric Engineering and Remote Sensing*, 63, 727–34.

Getis, A., 1995, Spatial filtering in a regression framework: Examples using data on urban crime, regional inequality, and government expenditures. In L. Anselin and R. Florax (eds) *New Directions in Spatial Econometrics* (Berlin: Springer-Verlag).

Getis, A. and Ord, J. K., 1996, Local spatial statistics: An overview. In P. A. Longley and M. Batty (eds), *Spatial Analysis: Modelling in a GIS Environment* (Cambridge, UK: GeoInformation International).

Herbert, D. and Davidson, N., 1994, Modifying the built environment: The impact of improved street lighting. *Geoforum*, 25, 339–50.

Kashmuk, J., 1981, *A Re-evaluation of Crime Prevention through Environmental Design in Portland, Oregon: Executive Summary* (Washington DC: US Office of Justice Planning and Evaluation).

Page, C., 1977, Vandalism: It happens every night. *Nation's Cities* (Washington DC: The National League of Cities), pp. 5–10.

Painter, K., 1996, The influence of street lighting improvements on crime, fear and pedestrian street use, after dark. *Landscape and Urban Planning*, 35, 193–201.

Painter, K. and Farrington, D. P., 1997, The crime reducing effect of improved street lighting: The Dudley project. In R. V. Clarke (ed.) *Situational Crime Prevention: Successful Case Studies*, 2nd edn (Albany, NY: Harrow and Heston).

Painter, K. and Farrington, D. P., 1999, Street lighting and crime: Diffusion of benefits in the Stoke-on-Trent project. In K. Painter and N. Tilley (eds), *Surveillance of Public Space: CCTV, Street Lighting and Crime Prevention* (Monsey, NY: Criminal Justice Press).

Painter, K. and Tilley, N. (eds), 1999, Editors' introduction: Seeing and being seen to prevent crime. *Surveillance of Public Space: CCTV, Street Lighting and Crime Prevention* (Monsey, NY: Criminal Justice Press).

Pease, K., 1999, A review of street lighting evaluations: Crime reduction effects. In K. Painter and N. Tilley (eds), *Surveillance of Public Space: CCTV, Street Lighting and Crime Prevention* (Monsey, NY: Criminal Justice Press).

Pope, C. E., 1977, *Crime-specific Analysis: The Characteristics of Burglary Incidents* (Washington, DC: US National Criminal Justice Information and Statistics Service).

Rengert, G. and Wasilchick, J., 1986, *Suburban Burglary: A Time and a Place for Everything* (Springfield, IL: C. C. Thomas).

Sutton, P., 1997, Modeling population density with night-time satellite imagery and GIS. *Computing, Environment and Urban Systems*, 21, 227–44.

Wallis, A. and Ford, D., 1980, *Crime Prevention through Environmental Design: The Commercial Demonstration in Portland, Oregon* (Washington DC: US National Institute of Justice).

Essential reading

Introduction to Remote Sensing, 3rd edn
James B. Campbell

Taylor & Francis   Pb: 0-415-28294-2

Introduction to Remote Sensing:
Digital Image Processing and Applications
Paul Gibson, University of Ireland, Maynooth and Claire Power,
University of Greenwich, UK

Hb: 0-415-18961-6
Taylor & Francis   Pb: 0-415-18962-4

Introduction to Remote Sensing:
Principles and Concepts
Paul Gibson, University of Ireland, Maynooth with contributions from
Claire Power, University of Greenwich, UK

Hb: 0-415-17024-9
Taylor & Francis   Pb: 0-415-19646-9

Remote Sensing and Urban Analysis
Edited by J.-P. Donnay, University of Liege, Belgium, M. J. Barnsley,
University of Wales, Swansea, UK and Paul Longley,
University College London

Taylor & Francis   Hb: 0-7484-0860-6

Classification Methods for Remotely Sensed Data
Brandt Tso and Paul Mather, University of Nottingham, UK

Hb: 0-415-25908-8
Taylor & Francis   Pb: 0-415-25909-6

Information and ordering details
For price availability and ordering visit our website **www.gis.tandf.co.uk**
Alternatively, our books are available from all good bookshops

Index